Intelligent Systems Reference Library   47

Editors-in-Chief

Prof. Janusz Kacprzyk
Systems Research Institute
Polish Academy of Sciences
ul. Newelska 6
01-447 Warsaw
Poland
E-mail: kacprzyk@ibspan.waw.pl

Prof. Lakhmi C. Jain
School of Electrical and Information
Engineering
University of South Australia
Adelaide
South Australia SA 5095
Australia
E-mail: Lakhmi.jain@unisa.edu.au

T0138173

For further volumes:
http://www.springer.com/series/8578

Witold Pedrycz and Shyi-Ming Chen (Eds.)

Time Series Analysis, Modeling and Applications

A Computational Intelligence Perspective

 Springer

Editors
Prof. Witold Pedrycz
Electrical & Computer Engineering
University of Alberta
Edmonton AL
Canada

Prof. Shyi-Ming Chen
Department of Computer Science and
Information Engineering
National Taiwan University of Science
and Technology
Taipei
Taiwan

ISSN 1868-4394
ISBN 978-3-642-43700-7
DOI 10.1007/978-3-642-33439-9
Springer Heidelberg New York Dordrecht London

e-ISSN 1868-4408
ISBN 978-3-642-33439-9 (eBook)

Preface

Temporal and spatiotemporal data are commonly encountered in a variety of areas of applications. We are faced with data coming from numerous sensors, data feeds, recordings coming numerous domains of application. A thorough analysis and interpretation of time series associated with sound mechanisms of their interpretation is highly demanded. While there has been a continuous progress in the area with a truly remarkable wealth of highly diversified models of time series, some fundamental challenges are still vividly present today. The ongoing quest for accurate and highly interpretable models of time series has never been that timely as in the recent years calling for the use of advanced technologies of system modeling, especially those emerging in Computational Intelligence.

The principles, methodology, and practical evidence of Computational Intelligence have been poised to play a vital role in the analysis, design, and interpretation of time series. As a matter of fact, it has already exhibited a visible position in this realm. In particular, in this area we may capitalize on the important facets of learning, structural design and interpretability along with human-centricity, where all of these facet are vigorously supported by the leading technologies of Computational Intelligence. A quick bird's eye view at the area done by running Google Scholar reveals interesting figures that speak for themselves. As of the middle of July 2012, the search returned 49,900 entries for the query "time series and Computational Intelligence". The significant raise of the interest in the area is reflected in the numbers reported over time: for the same query as used above in the period of 2007–2012 the number of hits is 18,600 while just for the five years, 2000–2005 we see the increase with the number of 10,500 hits. This speaks loudly to the rapid pace of progress visible in the area.

Considering the vital synergy among neurocomputing, fuzzy sets and evolutionary optimization where this synergy plays a pivotal role in the realization of CI constructs, the same synergistic linkages become crucial when working with time series. The nonlinear character of time series is captured in the models originating in the setting of neural networks and fuzzy models. Fuzzy sets and Granular Computing, in general, bring a highly desirable facet of transparency of models of time series; along with "standard" rule based-systems, we also encounter recurrent models of time series, which help capture the facet of dynamics of time series. Chaotic time series help gain a deeper insight into the dynamics and complexities of time series and quantify these phenomena. Evolutionary optimization and population-based optimization are of relevance in the context of the design of models of time series, especially when dealing with their structural optimization. We witness a broad spectrum of applications to diversified areas of physical and human-generated phenomena such as those dealing e.g., with financial markets and Internet activities.

The contributions to this volume are highly reflective of the wealth of the technologies of CI by bringing together ideas, algorithms, and numeric studies, which convincingly demonstrate their relevance, maturity and visible usefulness.

This volume is aimed at a broad audience of researchers and practitioners. Owing to the nature of the material being covered and a way it has been arranged, we are convinced that it helps establish a comprehensive and timely picture of the ongoing pursuits in the area and stimulate further progress.

We hope that this book will appeal to a broad spectrum of readers engaged in various branches of operations research, management, social sciences, engineering, and economics.

We would like to take this opportunity to express our sincere thanks to the authors for reporting on their innovative research and sharing their insights into the area. The reviewers deserve our thanks for their constructive input. We highly appreciate a continuous support and encouragement from the Editor-in-Chief, Professor Janusz Kacprzyk whose leadership and vision makes this book series a unique vehicle to disseminate the most recent, highly relevant and far-fetching publications in the domain of CI.

We hope that the readers will find this edited volume of genuine interest and the research reported here will trigger further progress in research, education, and numerous practical endeavors.

Contents

Chapter 1
The Links between Statistical and Fuzzy Models for Time Series Analysis and Forecasting

José Luis Aznarte and José Manuel Benítez

Abstract. Traditionally, time series have been a study object for Statistics. A number of models and techniques have been developed within the field to cope with time series of increasing difficulty. On the other hand, fuzzy systems have been proved quite effective in a vast area of applications. Researchers and practitioners quickly realized that time series could also be approached with fuzzy and other soft computing techniques. Unfortunately, for a long time both communities have somehow ignored each other, disregarding interesting results and procedures developed in the other area. We addressed the problem of digging in the links between Statistical and fuzzy models for time series analysis and forecasting. In this chapter we present some of the most relevant results we have found in this area. In particular we introduce a new procedure based on statistical inference to build fuzzy systems devoted to time series modelling.

Keywords: Time series, autoregression, regime-switching, fuzzy rule-based models, functional equivalence.

1 Introduction

Time series analysis is a prominent area within mathematical statistics, data analysis, stochastic finance and econometrics. During the last years, it has been a prolific field of study in terms of research and applications.

Traditionally, time series have been studied within statistics, a field where most advances have been obtained. A milestone in the formalization of the idea of forecasting future values of a time series as a combination of past values, was due to Box and Jenkins and materialized into their AR, MA, ARMA and ARIMA family of models. While it has become a standard reference, a key limitation of ARIMA is its linear nature, which makes it not very effective when approaching nonlinear time series. Statisticians have

W. Pedrycz & S.-M. Chen (Eds.): Time Series Analysis, Model. & Applications, ISRL 47, pp. 1–30.
DOI: 10.1007/978-3-642-33439-9_1

developed more advanced nonlinear models. A widely known family of models are Threshold AR (TAR). This is a rather extensive family of models with increasing complexity: STAR, LSTAR, NCSTAR, ... These models are distinctive representatives of the statistical approach to time series analysis.

On the other hand, time series analysis is a problem which has always attracted the attention of Computational Intelligence (CI) researchers and practitioners. Forecasting future values of a series is usually a very complex task, and many CI methods and models have been used to tackle it, including Artificial Neural Networks and Fuzzy-Rule Based Systems in their various formulations. Notwithstanding, a common characteristic of those approaches is that they usually consider time series as just another data set which requires some small adaptions to be cast into the regression or classification form of which most CI models were created. They represent a data-drive approach towards time series analysis.

Statistics and Computational Intelligence methods represent different approaches toward a common goal. In this sense, it is natural to think about how these two approaches can interact with each other. A first step necessary in this line is to deepen into the links connecting these two fields. In particular, we have researched into the connection of TAR models and fuzzy rule-based systems. A number of equivalence results have been found. Each of these results is a straight link which allows us to exchange properties and procedures between the two areas. This opens an important line of research and applications with the cooperative combination of methods and models from both fields much in the spirit that led to the birth of Soft Computing from its constituent techniques.

As an example of the usefulness of the equivalence results, we show how one can use the hypothesis testing framework to determine the number of fuzzy rules required to model a given series.

The structure of this chapter is as follows: first we will define the notation of the fuzzy rule-based models (FRBM) that we consider, in Section 2. In Section 3, the family of the regime-switching models is briefly described whereas in Section 4 the relations that link those models with FRBM is established. Section 5 covers the linearity tests developed for FRBM, which are applied to determine the number of fuzzy rules required to model a given problem in Section 6. Finally, the results of some experiments are shown in Section 7.1 and the chapter ends with Section 8 where the main conclusions are drawn.

2   Fuzzy Rule-Based Models for Time Series Analysis

For the sake of clarity, let us first note the expression of the fuzzy rule-based model considered here. When dealing with time series problems (and, in general, when dealing with any problem for which precision is more important than interpretability), the Takagi-Sugeno-Kang paradigm is preferred over other variants of FRBM. A fuzzy rule of type TSK has the following shape:

IF x_1 IS A_1 AND x_2 IS A_2 AND ... AND x_p IS A_p

$$\text{THEN } y = \mathbf{b}\mathbf{x}_t = b_0 + b_1 x_1 + b_2 x_2 + \ldots + b_p x_p \quad (1)$$

where x_i are input variables, A_j are fuzzy sets for input variables and y is a linear output function.

Concerning the fuzzy reasoning mechanism for TSK rules, the *firing strength* of the ith rule is obtained as the t-norm (usually, multiplication operator) of the membership values of the premise part terms of the linguistic variables:

$$\mu_i(\mathbf{x}) = \prod_{j=1}^{d} \mu_{A_j^i}(x_j), \quad (2)$$

where the shape of the membership function of the linguistic terms $\mu_{A_j^i}$ can be chosen from a wide range of functions. One of the most common is the Gaussian bell,

$$\mu_A(x) = \exp \frac{-(x-c)^2}{2\sigma^2}, \quad (3)$$

but it can also be a logistic function,

$$\mu_A(x) = \frac{1}{1 + \exp\left(-\gamma(x-c)\right)}, \quad (4)$$

and also non-derivable functions as a triangular or trapezoidal function.

The overall output is computed as a weighted average or weighted sum of the rules output. In the case of the weighted sum, the output expression is:

$$y_t = G\left(\mathbf{x}_t; \psi\right) = \sum_{i=1}^{R} \mu_i(\mathbf{x}_t) \cdot \mathbf{b}_i \mathbf{x}_t, \quad (5)$$

where G is the general nonlinear function with parameters ψ, and R denotes the number of fuzzy rules included in the system. While many TSK FRBS perform a weighted average to compute the output, additive FRBS are also a common choice. They have been used in a large number of applications, for example [11, 24, 26, 15].

When applied to model or forecast a univariate time series $\{y_t\}$, the rules of a TSK FRBM are expressed as:

IF y_{t-1} IS A_1 AND y_{t-2} IS A_2 AND ... AND y_{t-p} IS A_p

$$\text{THEN } y_t = b_0 + b_1 y_{t-1} + b_2 y_{t-2} + \ldots + b_p y_{t-p}. \quad (6)$$

In this rule, all the variables y_{t-i} are lagged values of the time series, $\{y_t\}$.

3   Regime Switching Autoregressive Models

As stated above, in statistical time series modeling, one of the oldest and most successful concepts is to forecast future values of a time series as a combination of its past values. This is a quite natural idea that we apply on every day's life, and it was popularized in 1970 after [10]. In that work, Box and Jenkins formalized the use of the *autoregressive* (AR) model, which assumes that future values of a time series can be expressed as a linear combination of its past values[1].

An AR model of order $p \geq 1$ is defined as

$$y_t = a_0 + a_1 y_{t-1} + \ldots + a_p y_{t-p} + \varepsilon_t \tag{7}$$

where $\{\varepsilon_t\} \sim N(0, \sigma^2)$, usually known as Gaussian *white noise* (equivalent to a random signal with a flat power spectral density). For this model we write $\{y_t\} \sim AR(p)$, and the time series $\{y_t\}$ generated from this model is called the $AR(p)$ process.

Such a simple model proved to be extremely useful and suited to series which, at first sight, seemed to be too complex as to be linear. Applications of the Box and Jenkins methodology spread in the following decades, covering various scientific areas such as Biology, Astronomy or Econometrics.

However, there were still many problems which could not be addressed using linear models. In 1978, taking a step towards nonlinearity, Tong [39] proposed a *piece-wise linear* model: the threshold autoregressive (TAR) model. The success of this model in Econometrics gave birth to a new family of models, the autoregressive regime switching models, which are based on the idea of partitioning the state-space into several sub-spaces, each of which is to be modeled by an AR model.

A general autoregressive regime switching model with k ($k \geq 2$) regimes can be defined as

$$y_t = \sum_{i=1}^{k} \mathbf{a}'_i \mathbf{x}_t \cdot \Phi_i(\mathbf{z}_t; \boldsymbol{\psi}_i) + \varepsilon_t, \tag{8}$$

where $\mathbf{x}_t = (1, y_{t-1}, y_{t-2}, \ldots, y_{t-p})$ is an input vector containing p lagged values of the series and $\mathbf{a}_i$ defines the local autoregressive model i (note that $\mathbf{a}'_i \mathbf{x}_t$ encodes the skeleton of the autoregressive model defined by (7)). The variable controlling the transition is $\mathbf{z}_t$ and normally is composed of a subset of the elements of $\mathbf{x}_t$ (hence $\mathbf{z}_t \in \mathbb{R}^q$ with $q \leq p$). The vector of parameters $\boldsymbol{\psi}_i$ defines the location and shape of the transition functions Φ_i, whose functional form is one of the main differences among the models of the family.

---

[1] This section is taken from [1].

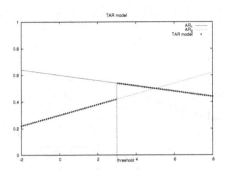

Fig. 1 An example of TAR model

3.1 Threshold Autoregressive model (TAR)

As mentioned above, the TAR is the seminal regime switching model, and is characterized by using the indicator function as transition function, i.e. $\Phi_i = \mathbf{I}_{A_i}$. This function, described below in detail, marks the sharp changes from one linear model to another through a set of thresholds defined on one of the variables involved. This variable can be an exogenous variable associated to the process being modeled or one of the lagged values of the series, in which case the model is called self-exciting.

A *self exciting threshold autoregressive* (SETAR) model is defined as

$$y_t = \sum_{i=1}^{k} \mathbf{a}_i' \mathbf{x}_t \cdot \mathbf{I}_{A_i}(y_{t-d}) + \varepsilon_t, \qquad (9)$$

where y_{t-d} is the value of the series at time $t - d$ and is usually known as the threshold variable, $\mathbf{I}_{A_i}$ is an indicator (or *step*) function (which takes the value zero below the threshold and one above it) and $\{A_i\}$ forms a partition of $(-\infty, \infty)$, with $\cup_{i=1}^{k} A_i = (-\infty, \infty)$ and $A_i \cap A_j = \emptyset, \forall i \neq j$.

We define the interval $A_i = (c_{i-1}, c_i]$, with $-\infty = c_0 < c_1 < \ldots < c_k = \infty$, where the c_i's are called thresholds. The ordering of the thresholds is required in order to guarantee the identifiability of the model. Figure 1 shows the graphical representation of a two regimes SETAR model.

3.2 Smooth Transition Autoregressive Model (STAR)

A key feature of TAR models is the discontinuous nature of the AR relationship as the threshold is passed. Taking into account that nature is generally continuous, in 1994 an alternative model called *smooth transition autoregressive* (STAR) was proposed by Teräsvirta [37]. In STAR models there is a smooth continuous transition from one linear AR to another, rather than a sudden jump.

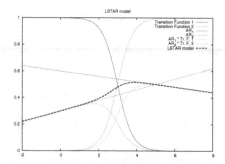

Fig. 2 An example of 2 regime STAR model using logistic transition function

In this model and variants, the indicator function is substituted by a smooth function with sigmoid characteristics. The STAR model is defined as

$$y_t = \sum_{i=1}^{k} \mathbf{a}_i' \mathbf{x}_t \cdot \Phi_i(y_{t-d}; \boldsymbol{\psi}_i) + \varepsilon_t. \tag{10}$$

The transition functions $\Phi_i(y_{t-d}; \boldsymbol{\psi}_i)$ are continuous functions, bounded between 0 and 1, with parameters $\boldsymbol{\psi}_i$. The regime that occurs at time t is determined by the observable lagged variable y_{t-d} and the associated value of $\Phi_i(y_{t-d}; \boldsymbol{\psi}_i)$. Different choices for the transition functions give rise to different types of regime-switching behaviour. A popular choice is when $\Phi_1 = 1$ (the function constantly equal to 1) and $\Phi_2 = \ldots = \Phi_k = f$, where f is the first-order logistic function with parameters $\boldsymbol{\psi}_i = (\gamma_i, c_i)'$ for regime i:

$$f(y_{t-d}; \boldsymbol{\psi}_i) = (1 + \exp(-\gamma_i(y_{t-d} - c_i)))^{-1}. \tag{11}$$

The resultant model is called the Logistic STAR (LSTAR). Figure 2 shows a STAR model with two regimes for which $\Phi_1 = 1 - f$ and $\Phi_2 = f$.

The parameters c_i in (11) can be interpreted as the threshold between two regimes, in the sense that the logistic function changes monotonically from 0 to 1 as y_{t-d} increases and $f(c_i; \gamma_i, c_i) = 0.5$.

The parameter γ_i determines the smoothness of the transition from one regime to another. As γ_i becomes very large, the logistic function approaches an indicator function and hence the change of $f(y_{t-1}; \gamma_i, c_i)$ from 0 to 1 becomes instantaneous at $y_{t-d} = c_i$. Consequently, the LSTAR nests threshold autoregressive (TAR) models as a special case. Furthermore, when $\gamma \to 0$ the LSTAR model reduces to a linear AR model.

In the LSTAR model, the regime switches are associated with small and large values of the transition variable y_{t-d} relative to c_i. In certain applications it may be more appropriate to specify a transition function such that the regimes are associated with small and large absolute values of y_{t-d} (again

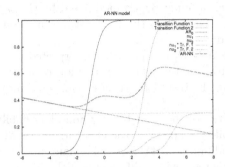

Fig. 3 An example of an AR-NN with 2 hidden units

relative to c_i). This can be achieved by using, for example, the exponential function, in which case the model may be named ESTAR.

As it is the case for the TAR model, symmetries of the parameter space cause unidentifiability of the STAR —that is, it cannot be uniquely identified. Enforcing the ordering of the regimes ($c_1 < c_2 < \ldots < c_k$) partially solves this problem. Notwithstanding, the logistic activation function (which verifies that $f(x) = 1 - f(-x)$), is another source for unidentifiability, so the restriction $\gamma_i > 0$ must also be respected for every i.

3.3 Autoregressive Neural Network Model (AR-NN)

After the success of Artificial Neural Networks in so many fields including Time Series Analysis, some researchers [32, 38] considered them as statistical nonlinear models and applied statistical inference to the problem of their specification. They devised a "bottom-up" strategy which allowed for proper statistical inference, as well as an in-sample evaluation of the estimated model.

The autoregressive single hidden layer neural network (AR-NN) model [32] is defined as

$$y_t = \mathbf{a}_0' \mathbf{x}_t + \sum_{i=1}^{k} \alpha_i \Phi_i(\mathbf{b}_i' \mathbf{z}_t; \boldsymbol{\psi}_i) + \varepsilon_t \qquad (12)$$

being α_i the connection weights and $\mathbf{b}_i$ a vector of real valued parameters defining a linear transformation on $\mathbf{z}_t$. For this autoregressive regime-switching model, the functions Φ_i are assumed to be logistic in this paper, $\Phi_1 = \ldots = \Phi_k = f$, as defined in equation (11). Although in the Soft Computing field it is frequent to take $\mathbf{a}_0 = 0$, the original formulation of the AR-NN included this "linear unit."

The geometric interpretation of this model considers that the AR-NN divides the p-dimensional Euclidean space with hyper-planes (defined by $\mathbf{b}_i' \mathbf{z}_t$) resulting in several polyhedral regions. It computes the output as the sum of the contribution of each hyper-region modulated by the smoothing function f. Figure 3 shows an example of the shape of the function generated by an AR-NN with two hidden units.

Following [32], an AR-NN can be either interpreted as a semi-parametric approximation to any Borel-measurable function or as an extension of the LSTAR model where the transition variable can be a linear combination of stochastic variables.

Three characteristics of the model imply non-identifiability. The first one is the interchangeability property of the elements of the AR-NN model. The value in the likelihood function of the model remains unchanged if we permute the hidden units. This results in $k!$ different models that are indistinguishable from one another and in $k!$ equal local maxima of the log-likelihood function. The second characteristic is that, for the transition function, $f(x) = 1 - f(-x)$. This yields two observationally equivalent parameterizations for each hidden unit. Finally the presence of irrelevant hidden units is also a problem. If model (12) has hidden units such that $\alpha_i = 0$ for at least one i, the parameters $\mathbf{b}_i$ remain unidentifiable. Conversely, if $\mathbf{b}_i = 0$ then α_i can take any value without the likelihood function being affected.

The approach devised by [32] overcomes these limitations by imposing some restrictions on the parameters. The first problem is solved by enforcing $\alpha_1 > \cdots > \alpha_k$ or $b_{10} < \cdots < b_{k0}$. The second problem is solved by enforcing $b_{i1} > 0$ for every i. Finally, the third problem is dealt with by applying statistical inference in the model specification.

3.4 Linear Local Global Neural Network (L^2GNN)

Another member of the regime switching family, and a recent statistical approach to artificial neural networks, is the Local Global Neural Network (LGNN) model [36]. The central idea of LGNN is to express the input-output mapping by a piece-wise structure. The model output is constituted by a combination of several pairs, each of those composed by an approximation function and by an activation-level function. The activation-level functions are equivalent to the transition function of the general autoregressive regime switching model, and define the role of an associated approximation function for each subset of the domain. Partial superposition of activation-level functions is allowed. In this way, the problem of approximation functions is faced through the specialization of neurons in each of the sectors of the domain. In other words, the neurons are formed by pairs of activation-level and approximation functions that emulate the generator function in different parts of the domain.

The LGNN is defined as

$$y_t = \sum_{i=1}^{k} L(\mathbf{x}_t; \boldsymbol{\chi}_i)\Phi_i(\mathbf{z}_t; \boldsymbol{\psi}_i) + \varepsilon_t \qquad (13)$$

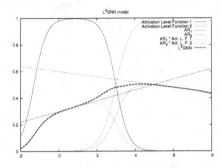

Fig. 4 An example of a 2 regime L^2GNN

where the functions L and Φ_i are the approximation and transition functions, respectively.

In the original formulation [36], $\Phi_i(\mathbf{z}_t; \boldsymbol{\psi}_i)$ is noted as $B(\mathbf{z}_t; \boldsymbol{\psi}_i)$ and is defined as the difference between two opposed logistic functions:

$$B(\mathbf{z}_t; \boldsymbol{\psi}_i) = -\left(f(\mathbf{b}_i'\mathbf{z}_t; \gamma_i, c_i^{(1)}) - f(\mathbf{b}_i'\mathbf{z}_t; \gamma_i, c_i^{(2)}) \right) \qquad (14)$$

where $\boldsymbol{\psi}_i = \left(\mathbf{b}_i, \gamma_i, c_i^{(1)}, c_i^{(2)} \right)$, representing a linear transformation of $\mathbf{z}_t$ encoded by $\mathbf{b}_i$, a steepness parameter γ_i and two location parameters $(c_i^{(1)}, c_i^{(2)})$.

This model is closely related to the *mixture-of-experts* approach [21] and offers a great deal of flexibility in the functional form of the approximation function $L(\mathbf{x}_t; \boldsymbol{\chi}_i)$. This flexibility has not been fully explored so far, but there have been attempts to combine in the same model linear approximators with nonlinear ones [34], for example.

A special case of the LGNN model is the Linear-Local Global Neural Network (L^2GNN) [36]. In this case, the approximation functions are linear, that is, $\boldsymbol{\chi}_i = \mathbf{a}_i$ is a vector of linear parameters and $L(\mathbf{x}_t; \boldsymbol{\chi}_i) = \mathbf{a}_i'\mathbf{x}_t$. Hence, the L^2GNN is closely related with the general autoregressive regime switching model of equation (8), and is defined as

$$y_t = \sum_{i=1}^{k} \mathbf{a}_i'\mathbf{x}_t B(\mathbf{z}_t; \boldsymbol{\psi}_i) + \varepsilon_t. \qquad (15)$$

It is worth noting that, as the previous models, this model is neither locally nor globally identifiable. In [36] the restrictions which ensure identifiability are stated: for $i = 1, \ldots, k$ the ordering of the thresholds is given by $c_i^{(1)} < c_{i+1}^{(1)}$ and $c_i^{(2)} < c_{i+1}^{(2)}$ together with $c_i^{(1)} < c_i^{(2)}$, whereas the identifiability problems posed by the symmetry of the transition function are solved by enforcing $\gamma_i > 0$ and $b_{i1} > 0$. Figure 4 shows a simplified L^2GNN model with two regimes.

3.5 Neuro-Coefficient Smooth Transition Autoregressive model (NCSTAR)

One of the latest developments in threshold-based models is the Neuro-Coefficient STAR [30]. This model is a generalization of some of the previously described models and can handle multiple regimes and multiple transition variables. This model can be seen as a linear model whose parameters change through time and are determined dynamically by a single hidden layer feed-forward neural network.

Consider a linear model with time-varying coefficients expressed as in equation (7) and let the coefficients vary through time: $a_0(t), a_1(t), \cdots, a_p(t)$. The time evolution of such coefficients is given by the output of a single hidden layer neural network with k hidden units:

$$a_j(t) = \sum_{i=1}^{k} \alpha_{ij} f\left(\mathbf{b}_i' \mathbf{z}_t; \gamma_i, c_i\right) - \alpha_{0j}, \qquad (16)$$

where $j = 0, \cdots, p$, α_{ji} and α_{j0} are real coefficients (connection weights) and f is a logistic function as defined in expression (11).

Substituting the p realizations of (16) in the linear model, we obtain the general form of the NCSTAR model:

$$y_t = \boldsymbol{\alpha}_0' \mathbf{x}_t + \sum_{i=1}^{k} \boldsymbol{\alpha}_i' \mathbf{x}_t f(\mathbf{b}_i' \mathbf{z}_t) + \varepsilon_t. \qquad (17)$$

Similarly to a_i in the previous models, $\boldsymbol{\alpha}_i$ represents a vector of real coefficients, called *linear parameters*. In this model, the value of the slope parameter γ_i is taken to be the norm of $\mathbf{b}_i$. In the limit, when the slope parameter approaches infinity, the logistic function becomes a step function.

As happened with previous models, this model is neither locally nor globally identifiable, and this is due to the special characteristics of its functional form that cause non-identifiability. In order to guarantee identifiability, we need to impose some restrictions, namely $c_i < c_{i+1}$ and $b_{i1} > 0$. Also, it is important to ensure that no irrelevant units are included, which can be achieved by using the incremental building procedure proposed in [31].

The choice of the elements of $\mathbf{z}_t$, which determines the dynamics of the process allows a number of special cases. An important one is when $\mathbf{z}_t = y_{t-d}$. In this case, model (17) becomes a LSTAR model with k regimes. It should be noticed as well that this model also nests the SETAR model. When $\gamma_i \to \infty$ $\forall i$, the LSTAR model becomes a SETAR model with k regimes.

Another interesting case is when $\boldsymbol{\alpha}_i' = (\alpha_{i0}, 0, \ldots, 0), \forall i > 0$. Then the model becomes an AR-NN model with k hidden units. Finally, this model is related to the Functional Coefficient Autoregressive (FAR) model [14], to the Single-Index Coefficient Regression model [41], and to Fuzzy Rule-based Models, as we shall see below.

4 Relations with Fuzzy Rule-Based Models

As stated above, establishing the equivalence of different models has been important in the neural networks field since its establishment, see for example [27, 28, 23, 35, 25, 9, 13]. These results imply some useful consequences as the possibility of interpreting one family of models in terms of the others or the transfer of properties and algorithms. Concerning neural networks, these results allowed to overcome the "black-box" characteristic as they led to knowledge extraction methods.

In [5] we explored the links existing between an AR model and a fuzzy rule used in the time series framework and that STAR models can be seen as a particular case of a fuzzy rule-based system. In [1] we extended those results to the neural autoregressive models listed in Section 3. Let us recall those results.

4.1 The AR Model and the TSK Fuzzy Rules

Fuzzy rules are the core element of fuzzy systems. When applied to time series, as seen in equation (6), fuzzy rules can describe the relationship between the lagged variables in some parts of the state-space. A close look into this equation suggested the following

Proposition 1. *When used for time series modelling, a TSK fuzzy rule can be seen as a local AR model, applied on the state-space subset defined by the rule antecedent.*

This connection between the two models opened the possibility of an exchange of knowledge from one field to another, enabling us to apply what we know about AR models to fuzzy rules and vice versa. From the point of view of Box-Jenkins models, each of these rules represents a local AR model which is applied only when some conditions hold. These conditions are defined by the terms in the rule antecedent. The output of the autoregressive system is modulated by the membership degree of the lagged variables to some fuzzy sets describing parts of the state-space domain. This scheme is closely related to the structure of the Threshold Autoregressive family of models, as shown below.

4.2 STAR Model and Fuzzy Rule-Based Models

After the previous result, we were able to go further in the exploration of the relationships between threshold models and fuzzy logic-based models. On the one hand, we have seen that AR models are good linear models applicable to prediction problems. As well, we know that a TAR model is basically a set of local AR models, and that it allows for some nonlinearity in its computations. On the other hand, we have seen how a fuzzy rule relates to an AR model, in Proposition 1. Knowing that fuzzy rule-based models contain sets of fuzzy rules, we were interested in considering the relationship existing between threshold models and fuzzy rule-based models.

It is rather clear that there is some parallelism between the two afore-mentioned families of models. At a high level, models from both sides are composed of a set of elements (AR – fuzzy rules) which happen to be closely related, as stated above. On a lower level, both families of models rely on building a hyper-surface on the state-space which tries to model the relation-ship between the lagged variables of a time series. Moreover, both define this hyper-surface as the composition of hyper-planes which apply only in certain parts of the state-space.

Indeed, the following

Proposition 2. *The STAR model is functionally equivalent to an Additive TSK FRBS with only one term in the rule antecedents.*

was proved. For a deeper discussion on these basic facts, refer to [5].

4.3   Neural Autoregressive Models and Fuzzy Rule-Based Systems

As described in Section 3, some of the most recent developments of the thresh-old autoregressive family of models include the AR-NN, the LGNN and the NCSTAR models. We will now explore the consequences of Proposition 1 regarding those models.

4.3.1   Autoregressive Neural Network (AR-NN)

Recalling equation (12), it is clear that the AR-NN is composed of an AR linear term plus a neural network. The neural network term is a regular multilayered perception, and, as such, is interpretable as a fuzzy additive system, in the way shown in [9]. This work states as well that, by using the interactive-or operator, it is possible to view artificial neural networks as Mamdani-type fuzzy rule-based models.

Furthermore, under the FRBS paradigm, the AR term of the AR-NN can be considered as a *generic* rule, that is, a rule which applies on the whole do-main of the problem. Such generic rules, which fire unconditionally, produce a default answer which is added to the values of the fired rules on those areas covered by them. This type of rules has been used previously by researchers and practitioners to encode knowledge which is domain-wide applicable.

Thus, we can prove the following

Proposition 3. *The Autoregressive Neural Network (AR-NN) model is func-tionally equivalent to a TSK FRBS with a default rule.*

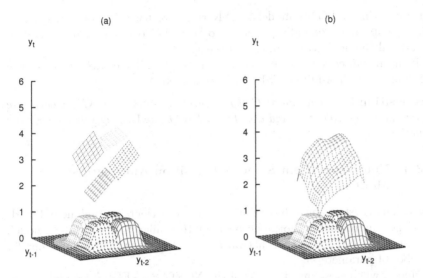

Fig. 5 (a) Four local AR models (or fuzzy rules) (b) The L^2GNN model (or the fuzzy inference system) derived from them

Proof. Using the result in [9], which states that a neural network is functionally equivalent to an FRBS, and considering the AR term as a rule of type

$$\text{IF true THEN } y_t = \mathbf{a}_0'\mathbf{x}_t, \tag{18}$$

the proof is trivial.

Viewing the AR-NN as a combination of an AR model and a fuzzy inference system allows for linguistic interpretation of the system. In addition, this let us include *a priori* expert knowledge into the model.

4.3.2 Local Global Neural Network

The more general approach of LGNN models, closely related to mixtures of experts model, satisfies the following

Proposition 4. *Local Global Neural Networks are a generalization of Additive TSK FRBS.*

Proof. It is straightforward after considering the expression of TSK rules (6), and the expression for the LGNN (13). Since $L(\mathbf{x}_t; \chi_i)$ can take any form, it can also be a linear function of the inputs, which is exactly a TSK rule. As the aggregation rule for LGNN is additive, we can conclude that the LGNN model is a generalization of Additive TSK FRBS.

For the same reason that after setting the general original formulation of the LGNN, researchers straightforwardly focused on linear approximation functions (linear LGNN models), TSK rules are basically used with linear consequents. It is generally preferred to keep the consequents linear and to encode all the nonlinearity in the antecedents.

If linear consequents were used (i.e. in the L^2GNN model), though, the relationship with Additive TSK FRBS is immediate:

Proposition 5. *Linear Local Global Neural Network (L^2 GNN) models are functionally equivalent to Additive TSK FRBS using* $\mathrm{B}(\mathbf{z}_t; \boldsymbol{\psi}_i)$ *as membership function.*

4.3.3   Neuro-coefficient Smooth Transition Autoregressive Models

This kind of systems introduces time varying coefficients to combine AR models. Their mathematical formulation is quite similar to the L^2GNN model, varying only the form of their activation level functions (which has a smaller number of parameters).

Hence, when studying the links of the NCSTAR to FRBS, we find similar results to those obtained for the previous statistical models. They can be expressed in terms of the following

Proposition 6. *Neuro-Coefficient Smooth Transition Autoregressive (NC-STAR) models are functionally equivalent to Additive TSK FRBS with logistic membership function.*

Proof. We must recall equations (5) (Additive TSK FRBS) and (17) (NC-STAR model). Considering that the multidimensional logistic function is obtained as the product of uni-dimensional logistic functions, it is easy to see that the firing degree of a rule is equivalent to the transition function of a hidden unit of the NCSTAR, and hence, that both models are functionally equivalent.

Finally, the following Theorem condenses the results drawn above:

Theorem 1. *The TSK FRBS is a generalization of the regime switching autoregressive models TAR, STAR, AR-NN, L^2 GNN and NCSTAR.*

Proof. Trivial in the light of propositions 1, 2, 3, 5 and 6.

5   Hypothesis Testing in the Framework of FRBM

As stated before, a fundamental objection argued by scientists with a classical statistical background against Soft Computing models in general and neural networks and FRBM in particular was the lack of a sound statistical theory behind them. Not being able to prove *a priori* if such models had good statistical properties (related to their much widespread 'black-box' condition) prevented them to be accepted by wide parts of the scientific community despite its good performance in practical situations. Fuzzy-related researchers' and practitioners' attitude towards this has usually been to work from an engineering point of view and to further extend the practical applications of the models and methods in hope that their empirical benefits were at some point good enough as to finally convince the scientific community.

The above equivalence results have an immediate impact on this question, as they permit the derivation of a statistical approach to a family of Soft Computing models, namely the FRBM family, considering them as nonlinear time series models.

This includes *a priori* proofs of their statistical properties, such as stationarity or identifiability [6] which throw some light into their inner behavior and *a posteriori* diagnostic checks which closely examine the residuals as shown in [3]. Also, the use of log-likelihood based estimation methods allow us to guarantee existence, convergence, consistence and asymptotic normality of the estimators [1]. These properties are important for a statistical model to be accepted. Furthermore, the development of linearity tests grant the ability to decide, based on the data, if a series can be modeled with a single linear autoregressive model or if an FRBM seems appropriate instead [2].

Before deriving these tests, a word on notation must be said. In the standard FRBM framework, the residuals are considered as an information source about the 'goodness of fit' of the model. They are looked at once the model is built, as they are the basis for computing the so-called *error measures*: mean squared error, mean average error and so on.

In the statistical field, on the other hand, the time series formed by the residuals, $\{\varepsilon_t\}$, is a fundamental piece of the modeling process, and as such it is always included in the definition of the models. Hence, we will redefine the Additive TSK FRBM, Equation (5), in the time series framework as

$$y_t = G\left(\mathbf{x}_t; \boldsymbol{\psi}\right) + \varepsilon_t = \sum_{i=1}^{R} \boldsymbol{\mu}_i(\mathbf{x}_t) \cdot \mathbf{b}_i \mathbf{x}_t, + \varepsilon_t, \qquad (19)$$

where $\boldsymbol{\psi}$ is the parameter vector, including the consequent (linear) parameters, $\boldsymbol{\psi}_p = (\mathbf{b}_1, ..., \mathbf{b}_r)$ and the antecedent (nonlinear) parameters, $\boldsymbol{\psi}_\omega$, whose number depends on the type of membership function, μ, used. The residuals, ε_t, are hence included in the definition of the FRBM.

In this section, we will consider membership functions of Gaussian type, being the most common derivable membership functions used in this context. It is usually expressed as

$$\boldsymbol{\mu}(\mathbf{x}_t; \boldsymbol{\psi})) = \prod_i \exp\left(-\frac{(x_i - c_i)^2}{2\sigma^2}\right) \qquad (20)$$

but we will rewrite it as

$$\boldsymbol{\mu}(\mathbf{x}_t; \boldsymbol{\psi}) = \prod_i \exp\left(-\gamma(x_i - c_i)^2\right), \qquad (21)$$

where $\boldsymbol{\psi} = (\gamma, \mathbf{c})$.

Since FRBM can be seen as nonlinear regression models, the standard procedures for testing parameter significance, like LM-tests, should be applicable, in principle. To perform these tests, however, the asymptotic distribution of the model parameters must be known [1].

In the fuzzy literature, however, no attention has been paid to hypothesis testing up to now. While it is obvious that a linear time series should be modeled with a linear model, i.e. a single (default) rule, to our knowledge there is no testing procedure to avoid the mistake of using highly complex structures to model simple problems.

The problem of deciding upon the linearity of a given problem was already treated in [2], where an LM-based linearity test against an FRBM was proposed. Next we extend the application of such test to introduce a way to decide if a given problem can be solved using a linear model or if we need a combination of rules to model it. Let us suppose that we have an FRBM composed of a single linear model or default rule which applies to the whole input space:

$$y_t = \mathbf{b}_0 \mathbf{x}_t + \varepsilon_t. \tag{22}$$

Now we want to know if the use of an extra rule with Gaussian membership function would increase the performance of the model. We would add that rule as follows:

$$y_t = G(\mathbf{x}_t; \boldsymbol{\phi}_p, \boldsymbol{\phi}_\omega) = \mathbf{b}_0 \mathbf{x}_t + \mathbf{b}_1 \mathbf{x}_t \boldsymbol{\mu}(\mathbf{x}_t; \boldsymbol{\psi}) + \varepsilon_t. \tag{23}$$

Our goal is to test for the significance of the extra rule, so in this case, recalling (21), an appropriate null hypothesis could be

$$H^0 : \gamma = 0, \tag{24}$$

being the alternative $H^1 : \gamma > 0$. Hypothesis (24) opens up the possibility of studying linearity in the Lagrange Multiplier (LM) testing framework. Under this null hypothesis, the contribution of the extra rule is identically equal to a constant and merges with the intercept b_{00} of the default rule, that is, the rule is not necessary.

We assume that, under (24), the maximum likelihood estimators of the parameters of (21) are asymptotically normal and hence can be estimated consistently.

As it was thoroughly discussed in [6], model (23) is only identifiable under the alternative hypothesis, i.e., if the null is true, the parameters are not locally unique and thus the estimator does not follow an asymptotic normal distribution. This issue is known as the problem of 'hypothesis testing when a nuisance parameter is present only under the alternative,' and was first studied by [16]. In this situation the test statistic of the LM-test does not follow a known distribution and thus the standard asymptotic distribution theory for the likelihood ratio is not available.

However, we can avoid this difficulty and obtain a χ^2-statistic by following the method first suggested in [29] and then widely applied to neural network-based models by [32, 31, 36] amongst others. This method proposes the expansion of the expression of the firing strength of a fuzzy rule into a Taylor series around the null hypothesis $\gamma = 0$:

$$\mu(\mathbf{x}_t; \gamma, \mathbf{c}) \approx \mu(\mathbf{x}_t; 0, \mathbf{c}) + \left. \frac{\partial \mu}{\partial \gamma} \right|_{\gamma=0} \gamma + R(\mathbf{x}_t; \gamma, \mathbf{c})$$

$$= \gamma \sum (x_i - c_i)^2 + R(\mathbf{x}_t; \gamma, \mathbf{c}) \quad (25)$$

which for the expression of the contribution of the extra rule yields

$$C \approx \mathbf{b}_1 \mathbf{x}_t \left[\gamma \sum (x_i - c_i)^2 \right] = \sum_{i=1}^{q} \theta_i x_i +$$

$$\sum_{i=1}^{q} \sum_{j=i}^{q} \theta_{ij} x_i x_j + \sum_{i=1}^{q} \sum_{j=i}^{q} \sum_{k=j}^{q} \theta_{ijk} x_i x_j x_k.$$

In this case, contrary to what happens when using the sigmoid membership function (as in the STAR model, [29]), the first order Taylor approximation is enough for our needs, as all the $\theta_i, \theta_{ij}, \theta_{ijk}$ depend on the intercept, b_{10}, of (23). The first linear term merges with the system's default rule, while the remainder of the Taylor expansion adds up to the error term, becoming $\varepsilon^\star = \varepsilon + \mathbf{b}_1 \mathbf{x}_t R(\mathbf{x}_t; \gamma, \mathbf{c})$, which means that $\varepsilon^\star = \varepsilon$ under the null. Thus the expansion results in the following model:

$$y_t = \boldsymbol{\pi}' \mathbf{x}_t + \sum_{i=1}^{q} \sum_{j=1}^{q} \theta_{ij} x_i x_j + \sum_{i=1}^{q} \sum_{j=i}^{q} \sum_{k=j}^{q} \theta_{ijk} x_i x_j x_k + \varepsilon_t^\star. \quad (26)$$

The null hypothesis can hence be defined as

$$H_0 : \theta_{ij} = 0 \wedge \theta_{ijk} = 0 \ \forall \ i, j, k \in 1, \ldots, q. \quad (27)$$

This null hypothesis circumvents the identification problem, and allows us to obtain a statistical test concerning the use of the extra rule. This test is based on the local approximation to the log-likelihood for observation t, which takes the form (ς is the variance of ε):

$$l_t = -\frac{1}{2} \ln (2\pi) - \frac{1}{2} \ln \varsigma^2 - \frac{1}{2\varsigma^2} \times$$

$$\left\{ y_t - \boldsymbol{\pi}' \mathbf{x}_t - \sum_{i=1}^{q} \sum_{j=1}^{q} \theta_{ij} x_i x_j - \sum_{i=1}^{q} \sum_{j=1}^{q} \sum_{k=1}^{q} \theta_{ijk} x_i x_j x_k \right\}^2. \quad (28)$$

Following [31], we must rely on the following assumptions:

Assumption 2. *The $((r + 1) \times 1)$ parameter vector defined by $[\boldsymbol{\psi}', \varsigma^2]'$ is an interior point of the compact parameter space $\boldsymbol{\Psi}$ which is a subspace of $\mathbb{R}^r \times \mathbb{R}^+$, the r dimensional Euclidean space.*

Assumption 3. *Under the null hypothesis, the data generating process (DGP) for the sequence of scalar real valued observations $y_{t\,t=1}^{T}$ is an ergodic stochastic process, with true parameter vector $\psi \in \boldsymbol{\Psi}$.*

Assumption 4. $E|z_{t,i}|^{\delta} < \infty, \forall i \in \{1, \ldots, p\}$ *for some* $\delta > 8$.

Under H$_0$ and Assumptions 2, 3 and 4 we can compute the standard Lagrange Multiplier or score-type test statistic given by

$$\text{LM} = \frac{1}{\hat{\sigma}^2} \sum_{t=1}^{T} \hat{\varepsilon} \hat{\boldsymbol{\tau}}_t' \times$$

$$\left\{ \sum_{t=1}^{T} \hat{\boldsymbol{\tau}}_t \hat{\boldsymbol{\tau}}_t' - \sum_{t=1}^{T} \hat{\boldsymbol{\tau}}_t \hat{\mathbf{h}}_t' \times \left(\sum_{t=1}^{T} \hat{\mathbf{h}}_t' \hat{\mathbf{h}}_t \right)^{-1} \times \sum_{t=1}^{T} \hat{\mathbf{h}}_t \hat{\boldsymbol{\tau}}_t' \right\} \times$$

$$\sum_{t=1}^{T} \hat{\boldsymbol{\tau}}_t' \hat{\varepsilon} \quad (29)$$

where $\hat{\varepsilon} = y_t - \hat{\pi}' \mathbf{x}_t$ are the residuals estimated under the null hypothesis,

$$\hat{\mathbf{h}}_t = \frac{\partial G(\mathbf{x}_t; \boldsymbol{\psi}_p, \boldsymbol{\psi}_\omega)}{\partial \hat{\boldsymbol{\psi}}_p \partial \hat{\boldsymbol{\psi}}_\omega} \bigg|_{\boldsymbol{\psi}_p = \hat{\boldsymbol{\psi}}_p \wedge \boldsymbol{\psi}_\omega = \hat{\boldsymbol{\psi}}_\omega} \quad (30)$$

is the gradient of the model (in this case $\hat{\mathbf{h}}_t = \mathbf{x}_t$) and $\hat{\boldsymbol{\tau}}_t$ contains all the nonlinear regressors in (26). This statistic has an asymptotic χ^2 distribution with m degrees of freedom, where $m = \|\hat{\boldsymbol{\tau}}_t\|$.

Although it might seem complicated at first sight, this test can be easily carried out in three stages:

1. Regress y_t on $\mathbf{x}_t$ and compute the residual sum of squares $SSR_0 = \sum_{t=1}^{T} \hat{\varsigma}_t^2$
2. Regress $\hat{\varsigma}_t$ on $\mathbf{x}_t$ and on the m nonlinear regressors of (26). Compute the residual sum of squares $SSR_1 = \sum_{t=1}^{T} \hat{\tau}_t^2$.
3. Compute the χ^2 statistic

$$\text{LM}_{\chi^2}^l = T \frac{SSR_0 - SSR_1}{SSR_0}$$

or the F version of the test (recall that $p = \|\mathbf{x}_t\|$)

$$\text{LM}_F^l = \frac{(SSR_0 - SSR_1)}{m} \left(\frac{SSR_1}{(T - p - 1 - m)} \right)^{-1}.$$

If the value of the test statistic exceeds the appropriate value of the χ^2 or F distribution, the null hypothesis is rejected.

6 Determining the Number of Rules of an FRBM

Once we have developed the statistical theory for the FRBM, including the linearity tests, we are closer to establishing a sound statistical procedure to specify the structure of an FRBM. This specification includes the determination of the number of fuzzy rules that are sufficient to model a given time series.

Knowledge included in an FRBM is represented by fuzzy rules. Obtaining these rules is a fundamental problem in the design process of an FRBM. When an expert on the system or domain under study is available, he or she can deliver the rules. Its elicitation is a knowledge acquisition process which is affected by many well-known problems described in the literature [20] (chap. 5), [18] (chap. 3).

This is the reason why, opposed to traditional interview-based techniques, some alternatives are proposed, based in automatic learning methods. The idea is to use one of these techniques to capture or learn a set of examples that describe the behaviour of the system. Later, when this set is fixed, we translate this knowledge into fuzzy rules. This approach has given birth to a myriad of procedures to extract fuzzy rules, based on diverse algorithms or machine learning models, including classification trees, evolutionary algorithms, clustering techniques, and neural networks. For a review on developments on this issue, see [19].

Notwithstanding, one of the most common procedure for automatic rule base determination remains the one proposed by Wang and Mendel [40] in 1992. This procedure is based in a combinatorial approach, and divides the universe of discourse into fuzzy regions, assigning rules to those regions which cover the available data.

Some disadvantages of this procedure are the high number of rules it produces and, again, its lack of a mathematical foundation that justifies it. In this section, we will propose an alternative method, based on the formal developments that we have carried out throughout this Chapter, that overcomes these limitations. The method relies on hypothesis testing and produces parsimonious models because it proceeds in a bottom-up manner.

Suppose that we have an FRBS with a default rule and $r + 1$ fuzzy rules. It can be written as follows:

$$y_t = G\left(\mathbf{x}_t; \boldsymbol{\psi}\right) + \varepsilon_t = \mathbf{b}_0\mathbf{x}_t + \sum_{i=1}^{r} \mathbf{b}_i\mathbf{x}_t \cdot \boldsymbol{\mu}_i\left(\mathbf{x}_t; \boldsymbol{\psi}_{\mu_i}\right) +$$

$$\mathbf{b}_{r+1}\mathbf{x}_t \cdot \boldsymbol{\mu}_{r+1}\left(\mathbf{x}_t; \boldsymbol{\psi}_{\mu_{r+1}}\right) + \varepsilon_t, \quad (31)$$

where $\boldsymbol{\mu}(\mathbf{x}_t; \boldsymbol{\psi}_\mu)$ is defined as in (21). Let us assume that we have accepted the hypothesis of model (31) containing r rules and we want to test whether the $(r + 1)$-th rule is necessary. An appropriate null hypothesis could be

$$H_0 : \gamma_{r+1} = 0, \tag{32}$$

being the alternative $H_1 : \gamma_{r+1} > 0$.

As it was the case for hypothesis (24) of the linearity test, hypothesis (32) allows us to work in the Lagrange Multiplier (LM) testing framework. Under this null hypothesis, the contribution of the $(r+1)$-th rule, C_{r+1}, is identically equal to a constant and merges with the intercept in the default rule, that is, the rule is not necessary.

At this point, the test is similar to the linearity one, so we proceed in the same way: to avoid unidentified parameters we expand the contribution of the extra rule into a Taylor series around the null hypothesis $\gamma = 0$ as in (25), which allows us to operate on the expression of the contribution of $(r+1)$-th rule:

$$C_{r+1} \approx \mathbf{b}_{r+1}\mathbf{x}_t \left[\frac{\gamma_{r+1}}{\sigma_{r+1}^2} (\boldsymbol{\omega}_{r+1}\tilde{\mathbf{x}}_t - c_{r+1})^2 \right] = \ldots$$

$$= \theta_0 + \sum_{i=1}^{q} \theta_i x_i + \sum_{i=1}^{q}\sum_{j=1}^{q} \theta_{ij} x_i x_j + \sum_{i=1}^{q}\sum_{j=1}^{q}\sum_{k=1}^{q} \theta_{ijk} x_i x_j x_k. \tag{33}$$

The intercept θ_0 and the first linear term merge with the system's default rule, while the remainder of the Taylor expansion adds up to the error term, becoming $\varepsilon^{\star} = \varepsilon + \mathbf{b}_{r+1}\mathbf{x}_t R(\tilde{\mathbf{x}}; \gamma, \boldsymbol{\omega}, c, \sigma)$, which means that $\varepsilon^{\star} = \varepsilon$ under the null. Thus the expansion results in the following model:

$$y_t = \boldsymbol{\pi}_0 \mathbf{x}_t + \sum_{i=1}^{r} \mathbf{b}_i \mathbf{x}_t \mu_i \left(\mathbf{x}_t; \boldsymbol{\psi}_{\mu_i}\right)$$

$$+ \sum_{i=1}^{q}\sum_{j=1}^{q} \theta_{ij} x_i x_j + \sum_{i=1}^{q}\sum_{j=1}^{q}\sum_{k=1}^{q} \theta_{ijk} x_i x_j x_k + \varepsilon_t^{\star}. \tag{34}$$

The null hypothesis can hence be defined as

$$H_0 : \theta_{ij} = 0 \wedge \theta_{ijk} = 0 \ \forall \ i, j, k \in 1, \ldots, q. \tag{35}$$

As before, this new null hypothesis circumvents the identification problem, and allows us to obtain a statistical test concerning the use of the $r + 1$-th rule.

Under H_0 and Assumptions 2, 3 and 4 we can compute the standard Lagrange Multiplier or score-type test statistic given by (29) where the residuals estimated under the null hypothesis are

$$\hat{\varepsilon}_t = y_t - \sum_{i=1}^{r} \mathbf{b}_i \mathbf{x}_t \cdot \boldsymbol{\mu}_i \left(\mathbf{x}_t; \boldsymbol{\psi}_{\mu_i}\right), \tag{36}$$

while the gradient matrix $\hat{\mathbf{h}}_t$ (defined as in (30)) contains the gradient for each linear and nonlinear parameter of the model under the null (that is, the model with r rules) and $\hat{\boldsymbol{\tau}}_t$ now contains all the nonlinear regressors corresponding to the Taylor series expansion of the new rule added in (31), i.e. those multiplying the θ_{ij} and θ_{ijk}.

This statistic has an asymptotic χ^2 distribution with m degrees of freedom, being m the number of nonlinear regressors of the model under the alternative hypothesis, with the Taylor series expansion which leads to (33).

This test can again be carried out in stages:

1. Estimate the FRBM with a default rule and r fuzzy rules. If the data set is small and the model is difficult to estimate, instead of regressing y_t on all the regressors of the model under the null (as we did in the linearity test), we will use the procedure suggested by [31, 17] and regress the residuals $\hat{\varepsilon}_t$ on $\hat{\mathbf{h}}_t$ and compute the residual sum of squares $SSR_0 = \sum_{t=1}^{T} \tilde{\varepsilon}_t^2$.

   This approach is known to be more robust against numerical problems that can arise in applying the nonlinear least squares procedure. In that case, the residual vector is not precisely orthogonal to the gradient matrix and this has an adverse effect on the empirical size of the test.
2. Regress $\hat{\varepsilon}_t$ on $\hat{\mathbf{h}}_t$ and $\hat{\boldsymbol{\tau}}_t$. Compute the residual sum of squares $SSR_1 = \sum_{t=1}^{T} \hat{\nu}_t^2$.
3. Compute the χ^2 statistic

$$\text{LM}_{\chi^2}^l = T \frac{SSR_0 - SSR_1}{SSR_0}$$

or the F version of the test $(p = \|\mathbf{x}_t\|)$

$$\text{LM}_F^l = \frac{(SSR_0 - SSR_1)}{m} \left(\frac{SSR_1}{(T - p - 1 - m)} \right)^{-1}.$$

If the value of the test statistic exceeds the appropriate value of the χ^2 or F distribution, then the null hypothesis is rejected.

7 Experiments

7.1 Montecarlo Experiments

The use of synthetic data sets has been recently studied in the framework of Soft Computing. For a detailed state-of-the-art, see [8]. Nonetheless, in the statistical field, it is a common practice to use this type of experiments to check the modelling capabilities of the proposals.

The basic assumption is that any series is considered to be generated by a usually unknown data generating process (DGP) to which a noise component is added. As a reverse result of this, to generate an artificial time series, we need to define a DGP and a noise distribution, whose sum in iterative application will produce the data. This artificial series could then be studied under the chosen modelling scheme, identifying and estimating a model for it. If the parameters of this model are (or tend to be) equal to the parameters of the original DGP, we obtain a clear evidence that the modelling scheme is correct.

In order to simulate a series according to the aforementioned basic assumption, we must go back again to the expression of the general model, equation (8). The first part of the right hand side of that expression is called in this context the *model skeleton*, and of course is the part which is to be modelled. Having defined a model skeleton or DGP, we generate the series by seeding a random starting point $\mathbf{x}_{t_0}$ and successively obtaining the y_t, $t = 1...T$, by applying the skeleton function and adding a n.i.d. value given by the random series ε_t. It is usually a good idea to discard the first N observations to avoid initialization effects.

In this study, we generated six synthetic time series. The first five of them are the ones used by [31], that we reuse in order to test them in the FRBM framework. The sixth one is similar, but it only has a higher number of rules. The experiments were run on 500 replications of each model and we applied the iterative testing procedure to determine the number of rules needed for each model.

7.2 *Experiment 1*

We start by simulating a stationary linear autoregressive model:

$$y_t = 0.8 - 0.5y_{t-1} + 0.3y_{t-2} + \varepsilon_t, \quad \varepsilon_t \sim \mathrm{NID}(0, 1^2). \quad (37)$$

Knowing that this series is linear, we first wanted to check if the null hypothesis of the linearity test would be accepted or not. By using the skeleton and the random noise series, we simulated 500 replications of the model and applied the test to them. The results were conclusive: over the 500 series, the null hypothesis was accepted in 95.2% of the cases. There were just 24 series where the test failed.

Then, to compare this result with standard FRBM modelling, suppose that, when faced to this series, we decide to apply a standard fuzzy model like ANFIS [22], in its basic grid partitioning style. If 3 labels were assigned to each of the two input variables, y_{t-1} and y_{t-2}, the model would have 9 rules and a total of 39 parameters (9×3 linear parameters and 6×2 nonlinear parameters).

To make the comparison more fair, we also tried to model the series with an ANFIS using, instead of grid partition, a substractive clustering method with default parameters to determine the number of rules. In this case, the model ended up with just 3 fuzzy rules (also fixing 3 linguistic labels per input), counting a total of 21 parameters (3×3 linear and 6×2 nonlinear).

Comparing the complexity of this ANFIS model with the DGP, which has a total of 3 parameters, the importance of the linearity tests becomes evident.

7.3 Experiment 2

The second simulated model is similar to the specification studied by [37], and is a two regime STAR model with two extreme regimes:

$$y_t = 1.8y_{t-1} - 1.06y_{t-2} +$$
$$(0.02 - 0.9y_{t-1} + 0.795y_{t-2}) \times \mu_S(\mathbf{x}_t; \psi) + \varepsilon_t,$$
$$\varepsilon_t \sim NID(0, 0.02^2) \quad (38)$$

where the nonlinear parameters are $\psi = [\gamma, \omega, c] = [20, (1,0), 0.02]$.

The first regime of this model, corresponding to $\mu_S(\mathbf{x}_t; \psi) = 0$, is explosive, while the other regime, determined by $\mu_S(\mathbf{x}_t; \psi) = 1$, is not. For the long term behaviour, the model has a unique stable stationary point, $y_\infty = 0.036$.

We applied the linearity test to this series, obtaining a 98.3% of correct rejections of the null hypothesis. Then, assuming the alternative hypothesis to be true, we applied the number of rules determination procedure, and obtained the correct number of rules in the 97.7% of the cases. Over the 500 replications, only 6 were determined to have more than 2 rules.

7.4 Experiment 3

The third simulated model corresponds to a three regime STAR model:

$$y_t = -0.1 + 0.3y_{t-1} + 0.2y_{t-2} +$$
$$(-1.2y_{t-1} + 0.5y_{t-2}) \times \mu_S(\mathbf{x}_t; \psi_1) +$$
$$(1.8y_{t-1} - 1.2y_{t-2}) \times \mu_S(\mathbf{x}_t; \psi_2) + \varepsilon_t,$$
$$\varepsilon_t \sim NID(0, 0.5^2) \quad (39)$$

where the nonlinear parameters are $\psi_1 = [\gamma_1, \omega_1, c_1] = [20, (1,0), -0.6]$ and $\psi_2 = [\gamma_2, \omega_2, c_2] = [20, (1,0), 0.6]$.

This model has three limiting regimes, of which the "lower" one corresponds to $\mu_S(\mathbf{x}_t; \psi_1) = \mu_S(\mathbf{x}_t; \psi_2) = 0$ and is stationary, the "middle" regime has $\mu_S(\mathbf{x}_t; \psi_1) = 1$ and $\mu_S(\mathbf{x}_t; \psi_2) = 0$ and is explosive, while the "upper" regime is determined by $\mu_S(\mathbf{x}_t; \psi_1) = \mu_S(\mathbf{x}_t; \psi_2) = 1$ and is also explosive. Concerning its long term behaviour, the model has a limit cycle with a period of 8 time units.

In this case, the linearity test determined the nonlinearity of the series in the 100% of the cases. On the other hand, the incremental building procedure fixed the correct number of regimes in 90% of the cases, adding extra rules in 49 of the 500 models.

7.5 Experiment 4

The model simulated in this fourth experiment is a two regime NCSTAR:

$$y_t = 0.5 + 0.8y_{t-1} - 0.2y_{t-2} +$$
$$(-0.5 - 1.2y_{t-1} + 0.8y_{t-2}) \times \mu_S(\mathbf{x}_t; \psi) + \varepsilon_t$$
$$\varepsilon_t \sim \text{NID}(0, 0.5^2) \quad (40)$$

with $\psi = [\gamma, \omega, c] = [11.31, (0.7071, -0.7071), 0.1414]$.

This model has two extreme regimes, given by $\mu_S(\mathbf{x}_t; \psi) = 0$ which is a stationary regime and $\mu_S(\mathbf{x}_t; \psi) = 1$ which has a unit root and hence is non-stationary. Still, for its long term behaviour, the process has two stable stationary points, 0.38 and -0.05.

In this case, the linearity test worked in the totality of the 500 runs of the experiment and the iterative building strategy found the proper number of rules in the 98.2% of the cases. Only 9 series were set to be modelled with more than 2 regimes, proving again the effectiveness of the testing procedure.

7.6 Experiment 5

The fifth simulated model is a full three regime NCSTAR, given by

$$y_t = 0.5 + 0.8y_{t-1} - 0.2y_{t-2} +$$
$$(1.5 - 0.6y_{t-1} - 0.3y_{t-2}) \times \mu_S(\mathbf{x}_t; \psi_1) +$$
$$(-0.5 - 1.2y_{t-1} + 0.7y_{t-2}) \times \mu_S(\mathbf{x}_t; \psi_2) + \varepsilon_t,$$
$$\varepsilon_t \sim NID(0, 1^2) \quad (41)$$

where $\psi_1 = [\gamma_1, \omega_1, c_1] = [8.49, (0.7071, -0.7071), -1.0607]$ and $\psi_2 = [\gamma_2, \omega_2, c_2] = [8.49, (0.7071, -0.7071), 1.0607]$.

This model also has three limiting regimes: in the "lower" one, $\mu_S(\mathbf{x}_t; \psi_1) = \mu_S(\mathbf{x}_t; \psi_2) = 0$ and it is stationary. The "middle" regime, given by $\mu_S(\mathbf{x}_t; \psi_1) = 1$ and $\mu_S(\mathbf{x}_t; \psi_2) = 0$, is also stable, as well as the "upper" regime, characterized by $\mu_S(\mathbf{x}_t; \psi_1) = \mu_S(\mathbf{x}_t; \psi_2) = 1$. This process has an unique stable stationary point at $y_\infty = 0.99$.

We applied the linearity test to the 500 series of this data set, and linearity was rejected in 100% of the cases. The results were not so precise when the test was applied iteratively to determine the appropriate number of regimes of the model: of the 500 series, an 83.2% of the models were set to have 3 regimes, while up to 84 models were built with only 2 regimes. This conservative behaviour contrasts with what happened in the other 3 regime model, in Experiment 3, where the mistaken models had more rules instead of less, and could be related to the higher number of parameters which make estimation harder.

7.7 Experiment 6

The sixth model was built in order to test the behaviour of the modelling cycle when dealing with more complicated models. It contains five regimes, and is given by

$$
\begin{aligned}
y_t = 0.5 + 0.8y_{t-1} - 0.2y_{t-2} + \\
(1.5 - 0.6y_{t-1} - 0.3y_{t-2}) \times \mu_S(\mathbf{x}_t; \psi_1) + \\
(0.2 + 0.3y_{t-1} - 0.9y_{t-2}) \times \mu_S(\mathbf{x}_t; \psi_2) + \\
(-1.2 + 0.6y_{t-1} + 0.8y_{t-2}) \times \mu_S(\mathbf{x}_t; \psi_3) + \\
(-0.5 - 1.2y_{t-1} + 0.7y_{t-2}) \times \mu_S(\mathbf{x}_t; \psi_4) + \varepsilon_t, \\
\varepsilon_t \sim NID(0, 0.2^2) \quad (42)
\end{aligned}
$$

where

$$
\begin{aligned}
\psi_1 &= [\gamma_1, \omega_1, c_1] = [8.49, (0.7071, -0.7071), -1.0607] \\
\psi_2 &= [\gamma_1, \omega_1, c_1] = [8.49, (0.7071, -0.7071), -0.59] \\
\psi_1 &= [\gamma_1, \omega_1, c_1] = [14.23, (0.7071, -0.7071), 0.59] \\
\psi_4 &= [\gamma_2, \omega_2, c_2] = [14.23, (0.7071, -0.7071), 1.0607].
\end{aligned}
$$

Testing for linearity, not surprisingly 100% of the series were determined to be nonlinear. The problem arose when determining the number of regimes, as only in 30% of the 500 series the procedure found the correct number of regimes.

Nevertheless, for a model with such a big number of parameters (15 linear plus 16 nonlinear), it is clear that the length of the series (500 points) is insufficient. In order to remove the influence of the length of the series, we created a new set of 500 series with 5000 points each. We repeated the experiment with these longer series, and the results were much better: of the 500 series, 96% were found to have 5 regimes. There were only 21 series that yielded models with an incorrect number of regimes, always higher than 5 except for one, which was fixed to have 4 regimes.

7.8 Real-World Examples

In order to show the applicability of the proposal in a real situation, we have applied it to three real-world examples: the lynx captures in a period of time in a region of Canada, the calls received in an emergency call centre in the region of Castilla y León and the airborne pollen concentration in the atmosphere of the city of Granada, Spain. All these series were described in detail in [4].

For the lynx series, [33] proposed an AR(2) model considering the sample correlogram, and second order autoregression was also chosen by [12] in a harmonic-autoregressive combined model and by [31] for the NCSTAR model. We fix the order of our model also to 2, for these reasons.

The linearity test against a NCSTAR threw a p-value of 0.000259. This indicates that the series is nonlinear and suggest the use of advanced models. The modelling cycle ended when the second regime was added, and the estimated model has two regimes given by

$$y_t = 0.9599 + 1.2514y_{t-1} - 0.3398y_{t-2} +$$
$$(2.5466 + 0.3764y_{t-1} - 0.7973y_{t-2})\mu_S(\mathbf{x}_t; \psi_S) + \varepsilon_t \quad (43)$$

with $\psi_S = (\gamma, \omega, c) = (103.1266, [0.4630, 0.8863], 9.4274)$.

Regarding the emergency call centre problem, figure 6 shows the histogram of the transformed series, which shows a long right tail, being the computed kurtosis -0.5452, while the skewness had a value of 0.1285. The figure also shows the ACF and PACF functions, of which the first one shows a clear cyclic behaviour with period 7 and the second one shows significant partial autocorrelation in the first 7 lags. Attending to these diagrams, and the weekly nature of the series, we decided to fix to 7 the order of our models.

The linearity test against the FRBM threw a fairly low value: $1.920692e-08$. The test indicates that the series is nonlinear and that it could be explained with a complex nonlinear model.

As it was the case for the lynx problem, the modelling cycle ended up by assigning two regimes to the model, which were fixed to:

$$y_t = -0.0112 - 0.1827y_{t-1} - 0.1404y_{t-2} - 0.1663y_{t-3} - 0.1643y_{t-4}$$
$$- 0.0666y_{t-5} - 0.0315y_{t-6} + 0.1110y_{t-7} +$$
$$(0.03271 - 0.1952y_{t-1} - 0.3390y_{t-2} - 0.2442y_{t-3} - 0.0654y_{t-4}$$
$$- 0.2683y_{t-5} - 0.09218y_{t-6} + 0.2282y_{t-7})$$
$$\times \mu_S(\mathbf{x}_t; \psi_S) + \varepsilon_t \quad (44)$$

in the sigmoid case, with $\psi_S = (\gamma, \omega, c) = (53.5264, [0.5529, 0.4822, 0.3892, -0.1493, -0.3677, 0.2943, -0.2248], -0.0114)$.

For the airborne polen series, we applied the same transformation as in [7].

Once the preprocessing was done, we turned our attention to variable selection. We considered the autocorrelation function (acf) and the partial autocorrelation function (pacf) for the transformed dataset (Figure 7). These diagrams indicate that present values are influenced by previous days' values, decreasing its influence as the time lag increases. Concretely, the strongest partial autocorrelation is found in the previous six days, while the most recent three days are those showing a stronger ascendancy over the actual value. For this reason, and taking into account computational efficiency considerations, only three autocorrelation steps were considered here as inputs for the models.

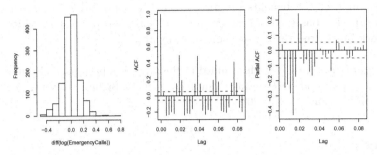

Fig. 6 Histogram, autocorrelation and partial autocorrelation functions for the transformed call centre series

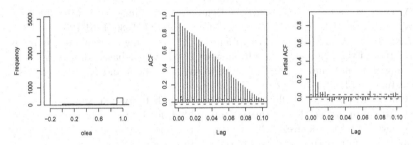

Fig. 7 Histogram, autocorrelation and partial autocorrelation functions for the transformed airborne pollen series

Not surprisingly, the linearity test threw a really low p-value ($3.2391e-94$), so the iterative procedure was applied to fix the number of required rules. The number of rules was found to be quite high: 11 rules were estimated in both the NCSTAR and the NCGSTAR. This is coherent with the expected complexity of the series.

8 Conclusions

The same spirit that inspired the philosophy behind Soft Computing is used to to motivate the cooperation between Soft Computing and Statistics, particularly, with respect to time series analysis and forecasting. We have studied some of the most remarkable models for time series analysis, the Threshold Autoregressive family and researched their links with fuzzy systems. A number of functional equivalence results were discovered. These are gates which allow the easy exchange of other results and procedures between Statistics and Fuzzy sytems, inducing a mutual profit. In particular, fuzzy system can get a sound foundation in the same way that statistical models are stated. Hence the effective practical performance of fuzzy system would be backed up by solid theory.

As an example of the benefits of this connection, a procedure for fuzzy system construction based on statistical inference has been proposed and its effective work illustrated through a number of examples.

Acknowledgements. The research included in this work has been partially supported by the Ministerio de Ciencia e Innovación of the Spanish Government through Research Grant Ref. TIN2009-14575.

References

1. Aznarte M., J.L., Benítez, J.M.: Equivalences between neural-autoregressive time series models and fuzzy systems. IEEE Transactions on Neural Networks 21(9), 1434–1444 (2010), doi:10.1109/TNN.2010.2060209
2. Aznarte M., J.L., Medeiros, M.C., Benítez, J.M.: Linearity testing for fuzzy rule-based models. Fuzzy Sets and Systems 161(13), 1836–1851 (2010)
3. Aznarte M., J.L., Medeiros, M.C., Benítez, J.M.: Testing for remaining autocorrelation of the residuals in the framework of fuzzy rule-based time series modelling. International Journal of Uncertainty, Fuzziness and Knowledge-Based Systems 18(4), 371–387 (2010)
4. Aznarte M., J.L.: Modelling time series through fuzzy rule based models: a statistical approach. Ph.D. thesis, Universidad de Granada (2008)
5. Aznarte M., J.L., Benítez, J.M., Castro, J.L.: Smooth transition autoregressive models and fuzzy rule-based systems: Functional equivalence and consequences. Fuzzy Sets Syst. 158(24), 2734–2745 (2007),
doi:http://dx.doi.org/10.1016/j.fss.2007.03.021
6. Aznarte M., J.L., Benítez Sánchez, J.M.: On the identifiability of TSK additive fuzzy rule-based models. In: Soft Methods for Integrated Uncertainty Modelling. Advances in Soft Computing, pp. 79–86. Springer (2006)
7. Aznarte M., J.L., Benítez Sánchez, J.M., Nieto Lugilde, D., Linares Fernández, C.D., Díaz de la Guardia, C., Alba Sánchez, F.: Forecasting airborne pollen concentration time series with neural and neuro-fuzzy models. Expert Syst. Appl. 32(4), 1218–1225 (2007),
doi:http://dx.doi.org/10.1016/j.eswa.2006.02.011
8. Basu, M., Ho, T.K.(eds.): Data Complexity in Pattern Recognition. Springer (2006)
9. Benítez, J., Castro, J., Requena, I.: Are artificial neural networks black boxes? IEEE Transactions on Neural Networks 8, 1156–1164 (1997)
10. Box, G.E.P., Jenkins, G.M.: Time Series Analysis: Forecasting and Control. Holden–Day, San Francisco (1970)
11. Byun, H., Lee, K.: A decision support system for the selection of a rapid prototyping process using teh modified topsis method. Intern. Journal of Advanced Manufacturing Technology 26(11-12), 1338–1347 (2005)
12. Campbell, M., Walker, A.: A survey of statistical work on the McKenzie river series of annual Canadian lynx trappings for the years 1821 - 1934, and a new analysis. J. Roy. Statist. Soc. A 140, 411–431 (1977)
13. Castro, J., Mantas, C., Benitez, J.: Interpretation of artificial neural networks by means of fuzzy rules. IEEE Transactions on Neural Networks 13(1), 101–116 (2002), doi:10.1109/72.977279

14. Chen, R., Tsay, R.: Functional-coefficient autoregressive models. Journal of the American Statistical Association 88, 298–308 (1993)
15. Coito, F.V., Palma, L.B., da Silva, R.N.: Robust fault diagnosis approach using analytical and knowledge based techniques applied to a water tank system. International Journal of Engineering Intelligent Systems for Electrical Engineering and Communications 13(4), 237–244 (2005)
16. Davies, R.B.: Hypothesis testing when a nuisance parameter is present only under the alternatives. Biometrika 74(1), 33–43 (1987), doi:http://www.jstor.org/stable/2336019
17. Eitrheim, O., Teräsvirta, T.: Testing the adequacy of smooth transition autoregressive models. Journal of Econometrics 74, 59–76 (1996)
18. Grzymala-Busse, J.: Managing uncertainty in expert systems. Kluwer Academic Publishers, Dordrecht (1991)
19. Hellendoorn, H., Driankov, D. (eds.): Fuzzy model identification: selected approaches. Springer, London (1997)
20. Ignizio, J.: Introduction to Expert Systems. McGraw-Hill, Inc., New York (1991)
21. Jacobs, R., Jordan, M., Nowlan, S., Hinton, G.: Adaptive mixtures of local experts. Neural Computation 3, 79–87 (1991)
22. Jang, J.-S.R.: ANFIS: Adaptive-network-based fuzzy inference system. IEEE Transactions on Systems, Man and Cibernetics 23(3) (1993)
23. Jang, J.-S.R., Sun, C.-T.: Functional equivalence between radial basis function networks and fuzzy inference systems. IEEE Transactions on Neural Networks 4(1), 156–159 (1993), doi:10.1109/72.182710
24. John, R., Innocent, P.: Modeling uncertainty in clinical diagnosis using fuzzy logic. IEEE Transactions on Systems, Man, and Cybernetics, Part B 35(6), 1340–1350 (2005), doi:10.1109/TSMCB.2005.855588
25. Kolman, E., Margaliot, M.: Are artificial neural networks white boxes? IEEE Transactions on Neural Networks 16(4), 844–852 (2005), doi:10.1109/TNN.2005.849843
26. Lee, I., Kosko, B., Anderson, W.F.: Modeling gunshot bruises in soft body armor with an adaptive fuzzy system. IEEE Transactions on Systems, Man, and Cybernetics, Part B 35(6), 1374–1390 (2005)
27. Li, H.X., Chen, C.: The equivalence between fuzzy logic systems and feedforward neural networks. IEEE Transactions on Neural Networks 11(2), 356–365 (2000), doi:10.1109/72.839006
28. Liang, X.B.: Equivalence between local exponential stability of the unique equilibrium point and global stability for hopfield-type neural networks with two neurons. IEEE Transactions on Neural Networks 11(5), 1194–1196 (2000), doi:10.1109/72.870051
29. Luukkonen, R., Saikkonen, P., Teräsvirta, T.: Testing linearity against smooth transition autoregressive models. Biometrika 75, 491–499 (1988)
30. Medeiros, M., Veiga, A.: A hybrid linear-neural model for time series forecasting. IEEE Transactions on Neural Networks 11(6), 1402–1412 (2000)
31. Medeiros, M., Veiga, A.: A flexible coefficient smooth transition time series model. IEEE Transactions on Neural Networks 16(1), 97–113 (2005)
32. Medeiros, M.C., Teräsvirta, T., Rech, G.: Building neural network models for time series: a statistical approach. Journal of Forecasting 25(1), 49–75 (2006), http://ideas.repec.org/a/jof/jforec/v25y2006i1p49-75.html

33. Moran, P.: The statistical analysis of the Canadian lynx cycle. i: structure and prediction. Aust. J. Zool. 1, 163–173 (1953)
34. Fariñas, M., Pedreira, C.E.: Mixture of experts and Local-Global Neural Networks. In: ESANN 2003 Proceedings - European Symposium on Artificial Neural Networks, pp. 331–336 (2003)
35. de Souto, M., Ludermir, T., de Oliveira, W.: Equivalence between ram-based neural networks and probabilistic automata. IEEE Transactions on Neural Networks 16(4), 996–999 (2005), doi:10.1109/TNN.2005.849838
36. Suarez-Farinas, M., Pedreira, C.E., Medeiros, M.C.: Local global neural networks: A new approach for nonlinear time series modeling. Journal of the American Statistical Association 99, 1092–1107 (2004), http://ideas.repec.org/a/bes/jnlasa/v99y2004p1092-1107.html
37. Teräsvirta, T.: Specification, estimation and evaluation of smooth transition autoregresive models. J. Am. Stat. Assoc. 89, 208–218 (1994)
38. Teräsvirta, T., van Dijk, D., Medeiros, M.C.: Linear models, smooth transition autoregressions, and neural networks for forecasting macroeconomic time series: A re-examination. International Journal of Forecasting 21(4), 755–774 (2005), http://ideas.repec.org/a/eee/intfor/v21y2005i4p755-774.html
39. Tong, H.: On a threshold model. Pattern Recognition and Signal Processing (1978)
40. Wang, L., Mendel, J.M.: Generating fuzzy rules by learning from examples. IEEE Transactions on Systems, Man and Cybernetics 22(6), 1414–1427 (1992)
41. Xia, Y., Li, W.: On single-index coefficient regression models. Journal of the American Statistical Association 94, 1275–1285 (1999)

Chapter 2
Incomplete Time Series: Imputation through Genetic Algorithms

Juan Carlos Figueroa-García, Dusko Kalenatic, and César Amilcar López

Abstract. Uncertainty in time series can appear in many ways, and its analysis can be performed based on different theories. An important problem appears when time series is incomplete since the analyst should impute those observations before any other analysis.

This chapter focuses on designing an imputation method for multiple missing observations in time series through the use of a genetic algorithm (GA), which is designed for replacing these missed observations in the original series. The flexibility of a GA is used for finding an adequate solution to a multi-criteria objective, defined as the error between some key properties of the original series and the imputed one. A comparative study between a classical estimation method and our proposal is presented through an example.

1 Introduction and Motivation

The analysis of time series includes the handling of nonlinear behavior, heteroscedasticity and incomplete series. Data loss is an important problem for univariate time series analysis since most of the available estimation methods require either complete information or covariates to estimate missing observations. Moreover, when the series has a large number of missing observations or there is a subset of missing observations in a row, then classical estimation methods cannot produce a reasonable solution, so the use of GAs arises as an alternative for problems involving multiple missing data.

Juan Carlos Figueroa-García · Cesar Amilcar López-Bello
Universidad Distrital Francisco José de Caldas,
Bogotá - Colombia
e-mail: {jcfigueroag,clopezb}@udistrital.edu.co

Dusko Kalenatic
Universidad de La Sabana, Chía - Colombia
e-mail: duskokalenatic@yahoo.com

W. Pedrycz & S.-M. Chen (Eds.): Time Series Analysis, Model. & Applications, ISRL 47, pp. 31–52.
DOI: 10.1007/978-3-642-33439-9_2   © Springer-Verlag Berlin Heidelberg 2013

Thus, the scope of this chapter is to present an evolutionary algorithm for imputing all missing observations of an incomplete time series. The main focus is to preserve some key properties of available data after imputation.

Nowadays, evolutionary algorithms are efficient computational intelligence tools which provide fast and efficient exploration of the search space of complex problems. To do so, a multi-criteria fitness function derived from the autocorrelation function, mean and variance of the series, is minimized.

This chapter is divided into six sections, Section 1 presents the Introduction and Motivation; in Section 2 some useful statistical measures for time series analysis are introduced; in Section 3 the proposed genetic algorithm is described and its methodological issues are presented; in Section 4, we apply the genetic algorithm to a weather prediction case; Section 5 presents a statistical analysis to verify the obtained results; and finally in Section 6, some concluding remarks of the proposal are presented.

1.1 A Review

The missing data problem is mainly presented in financial and biological time series. In fact, it is an uncontrollable phenomenon which conduces to get biased results on posterior analysis such as identification and prediction.

There exist some methods to impute missing data, some of them based on optimal estimators, as the *EM Algorithm* proposed by Dempster [16] and Gaetana & Yao [22], and its modifications. Other approaches are based on averages, expected values or simple prediction structures, and some advanced methods are based on both covariates and additional information of the series, which leads to new directions to estimate those missed observations. For further information see González, M. Rueda & A. Arcos [24], Qin, Zhang, Zhu, Zhang & Zhang [41] Ibrahim & Molenberghs [30], Tsiatis [46], Chambers & Skinner [15] and Hair, Black, Babin & Anderson [26].

The mathematical treatment of time series is different to multivariate or longitudinal data since it has some special properties such as autocorrelated structures, trend and/or seasonal components and ergodic behavior. Basically, a time series is analyzed for forecasting, so an incomplete series does not allow to obtain the best predictors. Most of classical estimation methods do not provide good results when there are no covariates, complementary information, multiple missing observations in multiple locations, or even when the time series is volatile.

A univariate time series has no covariates for prediction, and in most cases there is no additional information available. If the time series has multiple missing data, then it is impossible to obtain its decomposition into Autoregressive (AR) and Moving Average (MA) processes.

Some applications of GAs to missing data problems were reported by Figueroa-García, Kalenatic & Lopez [19, 20], who used GAs to weather time series; Mussa Abdella & Tshilidzi Marwala [2], who used neural networks trained by genetic algorithms to impute missing observations in databases; Siripitayananon, Hui-Chuan & Jin Kang-Ren [44] treated the missing data problem by using Neural Networks, Parveen & Green [39] solved a similar problem using Recurrent networks and Broersen, de Waele & Bos [12] found an optimal method to estimate missing data by means of autoregressive models and its spectral behavior.

Other interesting works are proposed by Nelwamondo, Golding and Marwala [37] who use dynamic programming to train neural networks; Kalra and Deo [32] who applied genetic algorithms for imputing missing data in biological systems; Zhong, Lingras and Sharma [48] who compared different imputation techniques for traffic problems; Londhe [35] who design a real-time framework for impute missed observations of wave measures; Ssali and Marwala [45] proposed a theoretical approach based on computational intelligence tools and decision trees to missing data imputation; JiaWei, TaoYang and YanWang [31] used fuzzy clustering for array problems; Abdella and Marwala [1] provide basic key features for implementing neural networks and genetic algorithms in missing data problems; and Eklund [18] computed the confidence interval of missed observations for spatial data problems.

Given this background, we present an evolutive algorithm for imputing multiple missing observations applied to a study case with multiple missing observations, where classical algorithms cannot solve the problem properly. Now, some basic definitions about time series are provided in next section.

2 Statistical Definitions for Time Series Analysis

The main purpose of a statistical analyst when analyzing a time series is to extract information about its behavior in order to make a decision based on the available information so far.

Now, a classical scenario starts from the definition of some basic statistical measures which represent the properties of the series before using any forecasting method. This reasoning is based on the concept of a *stochastic time series process*, which is defined as follows.

Definition 2.1. *Consider a set of observations of the variable x, where $x \in S$ and S is a metric space in which x is measured. This set x is said to be a Stochastic Process $\{X_t\}$ if it is a random sequence of observations recorded at a specific time t, $t \in T$ where T is the time space described by its probability density function (pdf). The pdf is a function in the form $f(X; \theta | S, \omega)$ where ω is the probability space of $f(X; \theta)$ and θ is a vector of parameters that characterizes its behavior.*

Remark 2.1. *Indeed, a stochastic process $\{X_t\}$ has the following property: Its pdf can vary at different times t_1 and t_2, although the metric space S is the same at all instants $t \in T$, then the probability that a specific value x occurs at different times x_{t_1} and x_{t_2}, is different.*

As usual, the most important order statistics for obtaining optimal models as ARIMA (Autoregressive, Co-integrated and Moving Average), ARCH (Autoregressive Conditional Heteroscedastic) and GARCH (Generalized Autoregressive Conditional Heteroscedastic) are the mean, the variance and the autocorrelation function, defined as follows

Definition 2.2. *The expected value of a random variable $E(X_t)$ is a measure of concentration of $\{X_t\}$ in ω defined as:*

$$E(X) = \int_{-\infty}^{\infty} x \, f(x; \theta) \, d(x) \tag{1}$$

Let $\{x_1, x_2, \cdots, x_n\}$ be observations of a time series. An unbiased estimator of $E(X_t)$ assuming large samples is the sample mean:

$$\bar{x} = \sum_{t=1}^{n} \frac{x_t}{n} \tag{2}$$

where n is the sample size.

Definition 2.3. *The variance of a random variable $Var(X)$ is a measure of form of $\{X_t\}$ in ω and is defined as:*

$$Var(X_t) = \int_{-\infty}^{\infty} (x - E(X))^2 \, f(x; \theta) \, d(x) \tag{3}$$

An unbiased estimator of $Var(X_t)$ assuming large samples is:

$$Var(X) = \sum_{t=1}^{n} \frac{(x_t - \bar{x})^2}{n - 1} \tag{4}$$

On the other hand, a *time series model* is a model that tries to infer some key properties of the series. According to Brockwell and Davis [10, 11], Hamilton [27] and Anderson [3], a time series model is

Definition 2.4. *A time series is a set of observations x_t, each one being recorded at a specific time t. A time series model for the observed data $\{x_t\}$ is a specification of joint distribution (or possibly the means and covariances) of a sequence of random variables $\{X_t\}$ for which $\{x_t\}$ is postulated to be a realization.*

The *sample autocovariance* and *sample autocorrelation* of the series is a linear relation between the variable at a specific time $\{x_t\}$ to itself at a lag h, $\{x_{t+h}\}$. Graybill & Mood [36], Wilks [47], Huber [29], Grimmet [25], Ross [42], Brockwell and Davis [10, 11], and Harville [28] defined them as follows

Definition 2.5. *The sample autocovariance function* $\hat{\gamma}(h)$ *is:*

$$\hat{\gamma}(h) = \sum_{t=1}^{n-|h|} \frac{(x_{t+|h|} - \bar{x})(x_t - \bar{x})}{n}, \quad -n < h < n \tag{5}$$

Definition 2.6. *The sample autocorrelation function* $\hat{\rho}(h)$ *is:*

$$\hat{\rho}(h) = \frac{\hat{\gamma}(h)}{\hat{\gamma}(0)}, \quad -n < h < n \tag{6}$$

When the series is incomplete, we cannot obtain *sufficient statistics*, which leads to misspecification problems of posterior models. In fact, the autocorrelation function defined in (6) is one of the most important measures of the behavior of the series, so the imputation of all missed observations is a key step before computing $\hat{\rho}(h)$.

In the next sections, we show all methodological aspects for imputing those missing observations through a GA. The main reason to use GAs is its flexibility and speed for finding solutions to nonlinear and complex problems.

3 The Proposed Genetic Algorithm (GA)

GAs are simple structures (For further information about GAs see Kim-Fung Man, Kit-Sang Tang & Sam Kwong [41]). An individual in a population can be seen as a set of missed data, so it should be imputed in the incomplete series. Figueroa-García, Kalenatic & Lopez [19, 20] use this principle to find an adequate solution to missing data problems, and this approach improves its fitness function based on some key properties of available data.

Our approach is based on six steps which guarantee an adequate solution: 1) a statistical preprocessing of the original series to obtain a stationary process 2) define a fitness function for comparing all individuals 3) generate of a population of individuals where each one is a solution of all missed observations 4) apply evolutionary operators for exploring the search space 6) evaluate the fitness function to select the best solution.

3.1 Why GA for Imputing Missing Data in Time Series?

An incomplete time series is a complex problem in the sense that classical imputation algorithms depend on covariates or additional information, and in many cases we have no this information. On the other hand, this is a multicriteria problem, which has no easy solutions. In this way, GAs are an interesting option for imputing multiple missing data in time series by the following reasons:

- Multicriteria capability.
- Nonlinear capability.
- Flexibility and computational simplicity.
- Efficiency of the solutions.

As shown in the Introduction, some learning-based techniques like neural networks and hill-climbing methods are commonly used to solve this problem. We propose an alternative method which learns from statistical properties of available data to cases where the series has *multiple missed observations* and/or *no covariates*. Those cases are particularly complex because optimal estimation techniques do not produce results when either multiple observations are lost or no covariates exist. Thus, evolutionary optimization arises as a flexible tool for finding solutions to complex cases, as the proposed one.

The following sections describe some general aspects of genetic algorithms applied to imputation in time series.

3.2 Preprocessing of Available Data

Some computational aspects should be kept in mind before applying any genetic operator, among them we have: Linear transformations, lag operators, seasonal and trend decompositions.

Firstly, we standardize data by using a linear transformation, removing the effect of units, and then we apply a lag operator to remove the effect of the mean of the process, obtaining a stationary series. These transformations reduce the complexity of the mean, the variance and the autocorrelation function of the series by removing its units, so its interpretation is easier and its search space is reduced, improving the performance of the algorithm. In this way, the following transformation is applied to available data $\{x_i^a\}$ in order to obtain a new standardized series $\{z_i^a\}$:

$$z_i^a = \frac{x_i^a - \bar{x}^a}{\sqrt{Var(x^a)}} \qquad (7)$$

where $\{x_i^a\}$ is a vector of size $(n - m)$ of available data of the series.

Here, the mean $\bar{x}^a$ and variance $Var(x^a)$ of available data $\{x_i^a\}$ are obtained by removing the missing observations from its original one, as follows:

$$\bar{x}^a = \sum_{i=1}^{n-m} \frac{x_i^a}{n - m} \qquad (8)$$

$$Var(x^a) = \sum_{i=1}^{n-m} \frac{(x_i^a - \bar{x}^a)^2}{n - m - 1} \qquad (9)$$

Now, we compute a lag operator of order d, ∇_d, defined as the difference between z_t and itself at period d,

$$\nabla_d(z_t) = z_t - z_{t-d} \qquad (10)$$

The main idea of this transformation is to obtain a stationary series with no effect of the units of the series, so $\nabla_d(z_t)$ should be used as the target of the genetic algorithm. To do so, the following definition is given,

Definition 3.1 (Target series). *Hereinafter, we will refer to $\{z_t\}$ as a standardized series after applying (7) and (10) until reach a stationary series with zero mean and if possible, constant variance.*

The autocorrelation function of Z_t, $\hat{\rho}(h)$ cannot be computed when the series is incomplete, so we use a subset of Z_t, $\{z_t^l\}$ defined as the largest and most recent subset from available data, that is:

$$\hat{\gamma}(h)^l = \sum_{t=n_1}^{n_2-|h|} \frac{(z_{t+|h|}^l - \bar{z}^l)(z_t^l - \bar{z}^l)}{n_2 - n_1 + 1}, (n_1 - n_2) \leqslant h \leqslant (n_2 - n_1) \qquad (11)$$

where n_1 and n_2 are lower and upper bounds of t, and $\hat{\gamma}(h)^l$ is the autocovariance of the largest and most recent subset of X_t, denoted by l.

$$\hat{\rho}(h)^l = \frac{\hat{\gamma}(h)^l}{\hat{\gamma}(0)^l} \qquad (12)$$

In this way, $\hat{\rho}(h)^l$ is an important statistical measure obtained from available data, so its use as a part of the fitness function of the genetic algorithm is essential for time series analysis.

Remark 3.1 (Index sets i and t). *The index set i is related to available data $\{x_i\}$, instead of index t which is related to the series with missing data $\{x_t\}$, so we have $i \in t \in T$.*

An elite-based strategy is combined with a multicriteria fitness function to compose the basic structure of a genetic algorithm for imputing missing data. Its methodological aspects are discussed in the following subsections.

3.3   The Fitness Function

A time series $\{z_t\}$ is said to be incomplete if there exist m missing observations located by an index vector v, where $1 \leqslant v \leqslant n$. A vector of imputations of all missed observations is called $\{y_t\}$, where $y_t = 0$ when $t \notin v$ and $y_t = z_j$ when $t \in v$, z_j is the j_{th} element of y_t, $1 \leqslant j \leqslant m$ located in the v_{th} position. A new series where all missed observations are replaced is defined as $\{\hat{z}_t\}$:

$$z_t^a + y_t = \hat{z}_t \qquad (13)$$

Now, the main goal is to find a vector y_t which does not change the properties of available data. For our purposes, the autocorrelation function, mean and variance of the available data are the goal of the genetic algorithm.

A genetic solution to the m missing observations should not change its $\hat{\gamma}(h)^l$, $\bar{z}^a$ and $Var(z^a)$ measures. To do so, we define the fitness function of the algorithm as a multicriteria function namely $\mathcal{F}$, regarding a set of H lags used for computing autocorrelations, as follows:

$$\mathcal{F} = \sum_{h \in H} \mid \hat{\rho}(h)^l - \hat{\rho}(h) \mid + \mid \bar{z}^a - \bar{\bar{z}} \mid + \mid Var(z^a) - Var(\hat{z}) \mid \qquad (14)$$

Thus, the main goal of the algorithm is to minimize $\mathcal{F}$ and if possible, reach zero as optimal solution. Note that our proposal is based on the design of a fitness function that minimize the differences among the statistical measures of the available data and the series after imputation $\hat{z}_t$ in three ways:

- Significative autocorrelations, $h \in H$.
- Sample mean.
- Sample variance.

As shown in Definition 2.4, a time series can be described through its mean, variance and covariances, so (14) tries to characterize the time series after imputation of missing data. In this proposal, we aggregated different units in a single function without problems, since $\{z_t\}$ is a standardized variable.

Remark 3.2 (Magnitude of $\mathcal{F}$). *It is important to note that the use of* (7) *and* (10) *leads to obtain measures of* $\{z_t\}$ *with no effect of the mean and units of the original series, so* $\hat{\rho}(h)^l$, $\bar{z}^a$ *and* $Var(z^a)$ *are standardized measures that can be added in* (14) *without loss of generality.*

3.4 Individuals

An individual is defined as a vector of a population indexed on a matrix where each one is a solution itself. As always, a genetic structure contains many individuals forming a population.

Each individual represents a complete vector of missed observations, which will be located in z_t by using y_t. Thus, each individual has as much elements (genes) as missed observations exist, indexed by v. A graphical explanation of the genes and individuals of the algorithm is shown in Figure 1

In Figure 1, v is the index vector of the lost observations of z_t. Note that z_j has the same elements than v, but v only has the t_{th} position of the missed observations while z_j is a vector of solutions located by v, which is computed through GAs. Finally, $\hat{z}_t$ is defined in (13) where $y_t = z_j$ located by v.

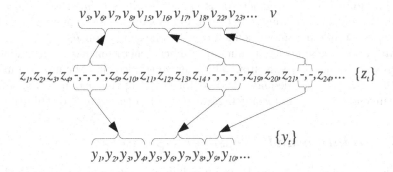

$$v_5, v_6, v_7, v_8, v_{15}, v_{16}, v_{17}, v_{18}, v_{22}, v_{23}, \ldots \quad v$$

$$z_1, z_2, z_3, z_4, -, -, -, -, z_9, z_{10}, z_{11}, z_{12}, z_{13}, z_{14}, -, -, -, -, z_{19}, z_{20}, z_{21}, -, -, z_{24}, \ldots \quad \{z_t\}$$

$$y_1, y_2, y_3, y_4, y_5, y_6, y_7, y_8, y_9, y_{10}, \ldots \quad \{y_t\}$$

$$z_1, z_2, z_3, z_4, y_1, y_2, y_3, y_4, z_9, z_{10}, z_{11}, z_{12}, z_{13}, z_{14}, y_5, y_6, y_7, y_8, z_{19}, z_{20}, z_{21}, y_9, y_{10}, z_{24}, \ldots \quad \{\hat{z}_t\}$$

Fig. 1 Individuals, genes, v, z_t and $\hat{z}_t$

3.5 Populations and Number of Generations

An important part of a GA is how to generate a population (collection of individuals), and how many populations (generations) will be used for exploring the search space. First, the population size is defined by the m missing observations and a pre-selected $k \in K$ number of individuals, creating a matrix of size $k \times m$ labelled as $P_{k,m}^g$ where g denotes the *generation index*.

Different population sizes can be selected for exploring the search space. According to Burke et. al. [13], Goldberg [23], Bäck [8], Bagchi [9], and Fonseca and Fleming [21], the selection of a higher population sizes together with fitness-based operators may reduce the performance of the algorithm, even when the search space would be better covered.

In this way, Figueroa-García, Kalenatic & Lopez [19, 20] used three population sizes: $k \in [100, 500, 1000]$, so based in their experimental evidence, we recommend to use a size of $k = 100$ in order to increase the speed of the algorithm, with no loss of ability of exploration of the solution space.

Another important parameter of the algorithm is the number of generations G, which is commonly used as stopping criterion. In this approach, this parameter operates as a controller of the iterations of the algorithm, so we set $\max_g = G$. This parameter depends on the complexity of the problem, the size and the nature of the missed observations, so the analyst should select G experimentally and using knowledge of the problem.

3.6 Population Random Generator

Definition 3.1 establishes that $\{z_t\}$ should be a standardized series, so this condition reduces the complexity of the algorithm, allowing us to use a standard uniform generator, which is computationally simpler than other generators e.g. Normal, exponential or mixed methods.

The uniform random generator is called R_j. It is defined as $R_j(a,b) = a + r_j(b-a)I_{[0,1]}(r_j)$ where a is the minimun value, b is the maximum value and r_j is a random number defined by the Index Function $I_{[0,1]}$.

An important analysis of random number generation has been made by Devroye [17] and Law & Kelton [33]. They concluded that the uniform number generator is an adequate method for covering the search space, so we recommend to use a uniform generator instead of the sample distribution.

3.7 Mutation and Crossover Operators

In Figure 1 we have explained how an individual and a gene are defined. This allows us to easily compose a population through R_j which is ranked using an elite-based method. After that, a mutation and crossover strategy can be applied to get a better exploration of the search space, as proposed below:

Mutation strategy:

1. Select a random position for each orderly individual in $P_{k,m}^g$ by its fitness function.
2. Replace the selected position with a new individual obtained by using a random generator $R_j(a,b)$.
3. Repeat (2) for the c_1 better individuals orderly for each population $P_{k,m}^g$ at the generation g.

Crossover strategy:

1. Select the c_2 first individuals in the orderly population $P_{k,m}^g$ by its fitness function.
2. Generate a new individual by replacing all even genes with their respective even gene located in the next individual.
3. Generate a new individual by replacing all odd genes with their respective odd gene located in the next individual for each one.
4. Repeat (3) for the c_2 better individuals orderly for each population $P_{k,m}^g$ at the generation g.

Remark 3.3 (Ranking of the solutions). *Figueroa-García, Kalenatic and López [19, 20] used an elite-based method for ranking the individuals of the population. Although it is a classical method which shows a good behavior, we encourage the reader to implement other ranking methods for the sake of new developments and improvements.*

3.7.1 Completing the Population

A classical strategy for exploring the space of solutions is by replacing the worst individuals by new ones, preserving the best ones at each population $P_{k,m}^g$. As usual, the number of best individuals is a free parameter, and in some cases it is involved as a random part of the algorithm.

Now, $P_{k,m}^{g+1}$ is updated by a set of random individuals, which is generated by replacing the worst individuals with new ones, in order to find better solutions. In short, the best k^1 individuals are preserved for the next generation and later it is completed by $\{k - k^1 - c_1 - c_2\}^m$ new individuals.

3.8 Stopping Strategy

There are different criteria for stopping a GA. Two of the most used methods are: A first one which uses a predefined maximum number of iterations called G, that is $g \rightarrow G$, and a second one which stops a GA when its fitness function $\mathcal{F}$ has no a significant improvement after a specific number of iterations.

Aytug and Koehler [6, 7] proposed an alternative stopping criterion for GAs based on a function of its mutation rate, the size of the population strings and the population size. Bhattacharrya and Koehler [5] generalized their results to non-binary alphabets. Pendharkar and Koehler [40] proposed a stopping criterion based on the markovian properties of a GA, and Safe et.al. [43] proposed entropy measures for constructing stopping operators.

The number of generations G is another degree of freedom of a GA. Usually, as $\mathcal{F}$ has no improvements, then G should be reduced. Finally, the best individual is selected by ranking $\mathcal{F}$ through all runs and generations, so the best individual will be imputed in the original series to complete the series.

An elite-based approach usually gets better solutions because this ensures the improvement of the solution through all generations. On the other hand, different stopping criteria can be used as long as the solutions are improved.

A brief description of the algorithm is presented in the Algorithm 1.

Algorithm 1. Genetic algorithm

Require: $v, n, n_1, n_2, m, H, c_1, c_2, k^1, \hat{\rho}(h)^l, \bar{z}^a, Var(z)^a$
    Generate an initial population of size k by using R_j
    for $g = 1 \rightarrow G$ **do**
        return $\mathcal{F}$ For each k_{th} individual
        Index $P_{k,m}^g$ by $\mathcal{F}$
        Apply the mutation operator to the c_1 better individuals
        Apply the crossover operator to the c_2 better individuals
        return $\mathcal{F}$ For each k_{th} individual
        Index $P_{k,m}^g$ by $\mathcal{F}$
        Preserve the best k^1 individuals, indexed by $\mathcal{F}$
        Complete the population with a vector of size $\{k - k^1 - c_1 - c_2\}^m$
    end for
    return $\mathcal{F}$ For each k_{th} individual
    Index $P_{k,m}^G$ by $\mathcal{F}$
    return $\min \mathcal{F}$
    Replace $P_{1,m}^G$ in the original series, indexing it by using v

---

Another common strategy for exploring the space of solutions is by running the algorithm several times, *Runs*. The number of runs should be a function of k, m, G and the computing time of each run, so we recommend to initialize with small attempts before running an adequate experiment. A graphical display of the imputation strategy is shown in Figure 2.

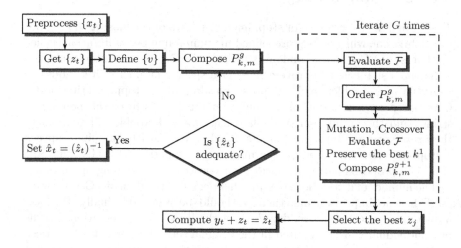

Fig. 2 Flowchart of the proposed GA

In the following section, an application of the algorithm is presented and compared to a classical imputation algorithm.

4 Application Example

The selected study case is a weather time series that has multiple missed observations produced by a failure in the measurement device. In this case, we use the minimum temperature *(MT)* recorded at the town of Chía - Colombia during 1368 days between 10 p.m. and 5 a.m. when maximum and minimum levels are registered, each one measured every half an hour throughout the night. All missed observations are displayed by discontinuities. Figure 3 shows the original series and the series after preprocessing by applying (7) and (10), where, a) is the original series and b) is the series after preprocessing.

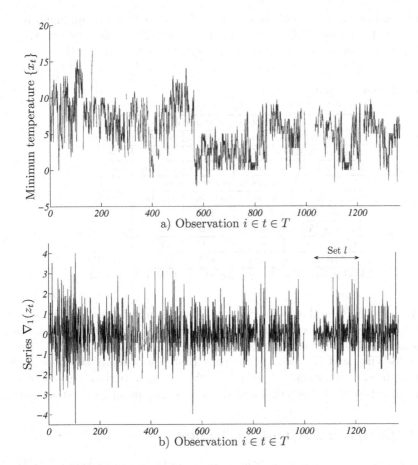

Fig. 3 Study Case measured in Chía - Colombia

4.1 Statistical Analysis

Some basic statistics obtained from available data are shown in Table 1. Its mean and variance are used as estimations of (8), (9). We obtain $\hat{\rho}(h)^l$ by using (12) for $H = \{1, \cdots, 6\}$ in order to define the fitness function $\mathcal{F}$ for each genetic structure.

Table 1 Observed statistics

Measure	$\hat{\rho}(1)$	$\hat{\rho}(2)$	$\hat{\rho}(3)$	$\hat{\rho}(4)$	$\hat{\rho}(5)$	$\hat{\rho}(6)$	$\bar{z}^a$	$Var(z^a)$	min	max
Value	-0.339	-0.052	0.004	-0.104	0.129	-0.051	-0.027	5.053	-4.436	4.015

In this case, we have 1367 observations ($N = 1367$) of the minimum temperature at the town of Chía - Colombia, where 159 are lost ($m = 159, n = 1208$). Some randomness tests done over $\{z_i^a\}$ are shown in Table 2.

Table 2 Tests of randomness

Tests on normality				
Test.	p-value	p-value	p-value	p-value
Shapiro-Wilks	0.0023	≈ 0	≈ 0	≈ 0
K-S.	≈ 0	≈ 0	0.0014	≈ 0
Tests on randomness				
Test.	p-value	p-value	p-value	p-value
Runs Test	0.001	≈ 0	≈ 0	≈ 0
Turning Points	≈ 0	≈ 0	≈ 0	≈ 0
Ljung-Box [b]	≈ 0	≈ 0	≈ 0	≈ 0
ARCH [b]	≈ 0	≈ 0	≈ 0	0.0025

[b] This test is made by using the first lag of the series

All tests conclude that the series is not a random variable. The Ljung-Box and ARCH tests reject the hypothesis that the series has no serial correlation, this means that it presents autocorrelation at least on its first lag. Both Shapiro-Wilks and Kolmogorov-Smirnov (K-S) tests reject the hypothesis that each series is normally distributed, which is an important constraint for some imputation methods that are based on strong normality assumptions.

4.2 Classical Estimation Methods

One of the most popular imputation algorithms is the expectation maximization *(EM)* algorithm, which is based on conditional expectations of a random variable, obtained from a set of auxiliary variables which give an estimate of the behavior of the missing data. Its principal objective is to maximize the *Likelihood or Log-Likelihood Function* of the *pdf* sample, obtaining an optimal estimation of the missing observations. This algorithm was proposed by Dempster [16], and Gaetana & Yao [22] proposed a variation of the EM algorithm based in a simulated annealing approach to improve its efficiency for the multivariate case. Celeux & Diebolt [14], Levine & Casella [34], Nielsen [38] have reported some modifications for a stochastic scenario, and Arnold [4] estimates the parameters of a state-dependant AR model by using the EM algorithm with no prior knowledge about state equations.

By using the EM algorithm, a maximum likelihood estimator is obtained by replacing all v positions of $\{x_i\}$ by its expected value $E(x)$, so we have that $\{x_t\} = E(x)$; and the regression method replaces all missed observations by random residuals of available data; its results are displayed in Figure 4.

Another method is based on auxiliary regressions, which consists on estimate the mean and variance of available data, and then each missed observation is replaced by the estimated mean plus a random residual obtained from a regression of available data against auxiliary variables. In the case of univariate series, this method is a simple estimation of its mean and variance.

In Figure 4, a) shows the results of the EM algorithm and b) presents the results of the regression method. According to Table 3, we can conclude that these approaches does not show desirable properties for univariate time series. Their statistical properties are presented in Table 3.

Table 3 Results of classical estimation methods

Measure	$\hat{\rho}(1)$	$\hat{\rho}(2)$	$\hat{\rho}(3)$	$\hat{\rho}(4)$	$\hat{\rho}(5)$	$\hat{\rho}(6)$	$\hat{z}$	$Var(\hat{z})$	It.	$\mathcal{F}$
EM algorithm	-0.302	-0.071	-0.046	-0.027	0.042	-0.0003	-0.0002	0.883	4	4.516
Regression	-0.266	-0.071	-0.051	-0.018	0.044	0.005	0.0128	0.997	N.A.	4.470

By evaluating the Fitness Function $\mathcal{F}$, both classical algorithms provide great differences among the obtained mean, variance, $\hat{\rho}^l(h)$ and their available values. With these evidences it is clear that the EM algorithm is not the best option to estimate missing data on a univariate time series context.

4.3 Genetic Approach

Figure 2 describes the methodology proposed in this chapter. First, we apply (7) and (10) to standardize data, then we compute $\rho(h)^l$ using (12) for the first $H = 6$ lags, and later we compute $\bar{z}^a$ and $Var(z^a)$ of available data, as shown in Table 5.

Figueroa-García, Kalenatic and Lopez [19, 20] have found better results with $k = 100$ individuals, outperforming computing time and improving the quality of solutions. The crossover, mutation and remaining parameters used in the GA for each series are shown in Table 4, where a and b are obtained from the observed series as its potential maximum and minimum values. c_1, c_2 are free parameters which modify the mutation and crossover rates of the GA, and G is selected by trial and error based on the behavior of $\mathcal{F}$. n_1 and n_2 are initial and end points of the *largest and most recent* complete dataset (See *"Set l"* in Figure 3-b), which are needed by(11) to obtain $\hat{\rho}(h)^l$ through the use of (12) for $H = \{1, \cdots, 6\}$.

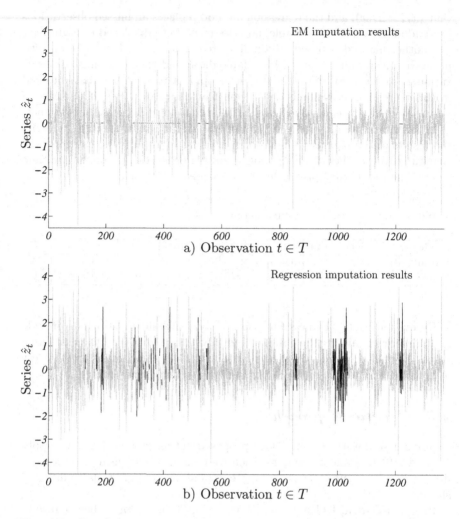

Fig. 4 Results of classical imputation methods

Table 4 Genetic algorithm parameters

Parameter	k	a	b	c_1	c_2	n_1	n_2	m	G	Av. time (sec)
Value	100	-5.5	5	4	4	1036	1211	159	5000	448.3

In this Table, the average time (in sec.) was obtained from 25 runs of the algorithm. The maximum processing time was 497.3 sec and the lower processing time was 421.1 sec. After the total 125.000 generations of the algorithm divided into 25 runs, the best solution (minimum $\mathcal{F}$), was selected. The obtained solution is displayed in black in Figure 5 and the obtained results for all imputed data are presented in Table 5.

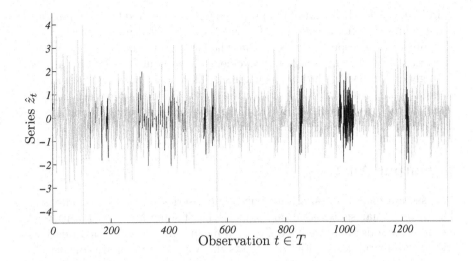

Fig. 5 Complete dataset with imputed missing data

Table 5 Results of evolutionary optimization

Measure	$\hat{\rho}(1)$	$\hat{\rho}(2)$	$\hat{\rho}(3)$	$\hat{\rho}(4)$	$\hat{\rho}(5)$	$\hat{\rho}(6)$	$\hat{z}$	$Var(\hat{z})$	$\mathcal{F}$
Value	-0.340	-0.052	0.004	-0.105	0.119	-0.052	-0.027	5.053	0.0112

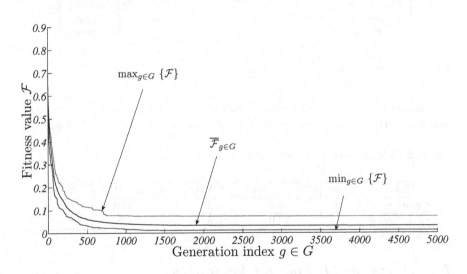

Fig. 6 Behavior of the GA for 25 runs

In Figure 6, the behavior of the proposed GA is measured by the minimum, average and maximum values of $\mathcal{F}$ over 25 runs, called $\max_{g \in G}\{\mathcal{F}\}$, $\overline{\mathcal{F}}_{g \in G}$ and $\min_{g \in G}\{\mathcal{F}\}$ respectively. Note that the GA always goes to stable values of $\mathcal{F}$, and they have no a high improvement after about $g = 2000$ iterations.

In general, the GA solution has no great differences to original data and it does not change its statistical properties. In this way, the proposed method seems to be a better method for imputing multiple missing observations in time series than other classical algorithms.

5 Output Analysis

This section focuses on analyzing the original series vs. genetic imputation. The output analysis is based on comparisons of some interesting statistical measures, among them we have: Tests on means, variances, autocorrelations and experimental design. Figure 7 shows the way all of them are connected.

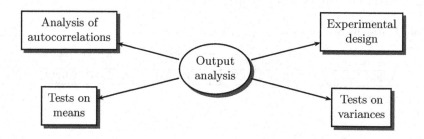

Fig. 7 Output analysis

Some descriptive statistics, randomness, differences on means, variances and autocorrelations tests were performed, as shown in Tables 6, 8 and 9.

Table 6 Tests of normality and randomness (significance)

Test	Runs	Turning-point	S-W	K-S	Ljung-Box[b]	ARCH[b]
Original series	≈ 0	≈ 0	≈ 0	≈ 0	0.0001	≈ 0
Imputed series	≈ 0	≈ 0	≈ 0	≈ 0	≈ 0	≈ 0

b This test is made by using the first lag of the series.

5.1 Tests on Means and Variances

The general hypotheses used for finding differences between original and imputed series, are as follows

Table 7 Hypothesis on means and variances

	Test on means	Test on variances
H_0 :	$\bar{\hat{z}} = \bar{z}^a$	$Var(\hat{z}) = Var(z^a)$
H_a :	$\bar{\hat{z}} \neq \bar{z}^a$	$Var(\hat{z}) \neq Var(z^a)$

The obtained results of the Tests on means are presented in Table 8.

Table 8 Tests on means (significance)

Test	ANOVA	Welch	Brown-Forsythe	K-S	Mann-Whitney
Original vs. Imputed	0.964	0.964	0.964	≈ 1	0.6404

With these statistical evidences, the hypothesis on means defined in Table 7 with a 95% confidence level, is accepted. We implement the Levene's test for contrasting their variances. Its results are shown in Table 9.

Table 9 Levene test

Test	Levene stat	Significance
Original vs. Imputed	0.00414	0.948

The ANOVA, Welch, Brown-Forsythe, K-S, Mann-Whitney and Levene tests conclude that there are no differences between $\bar{\hat{z}} \rightarrow \bar{z}^a$, and $Var(\hat{z}) \rightarrow Var(z^a)$ respectively, this means that the genetic solution has no statistical differences to available data. With these statistical evidences, the hypothesis defined in Table 7 are accepted with a 95% confidence level.

Remark 5.1 (Additional analysis). *Although we recommend the use of experimental design and autocorrelation analysis (See Figure 7), we did not perform those analysis due to the high similarity among autocorrelations and the absence of differences between means and variances. In case where means and/or variances have differences, it is recommended to perform an experimental design for finding the causes of differences among each run of the GA and the original series.*

Roughly speaking, the GA outperforms the solution provided by classical algorithms, in terms of the statistical properties of the series. The obtained results have no any statistical evidence to reject H_0, so we can accept them.

6 Concluding Remarks

The following concluding remarks can be made

1. The proposed genetic algorithm outperforms classical algorithms, providing better solutions without modifying their available properties.
2. The flexibility of evolutionary methods allows us to design efficient algorithms for finding missing observations in a time series context; its non-linear capability becomes a powerful tool for exploring the search space.
3. The use of multi-criteria fitness operators are alternative tools in front to classical imputation methods as the EM algorithm and its modifications.
4. Most of optimization techniques need additional variables to be consistent. The presented approach finds successful solutions with no additional information, which is a common issue in univariate time series.
5. Some emerging applications as multivariate data analysis, signal and image processing problems are proposed for future applications.

Finally, we encourage the reader to improve the presented results by modifying our proposal. The fitness function ($\mathcal{F}$), population size, c_1, c_2, k^1 and g can be modified, so other strategies can be used for getting better results in other missing data cases.

References

1. Abdella, M., Marwala, T.: Treatment of missing data using neural networks and genetic algorithms. In: IEEE (ed.) Proceedings of International Joint Conference on Neural Networks, pp. 598–603. IEEE (2005)
2. Abdella, M., Marwala, T.: The use of genetic algorithms and neural networks to approximate missing data in database. In: IEEE (ed.) IEEE 3rd International Conference on Computational Cybernetics, ICCC 2005, vol. 3, pp. 207–212. IEEE (April 2005)
3. Anderson, T.W.: The Statistical Analysis of Time Series. John Wiley and Sons (1994)
4. Arnold, M.: Reasoning about non-linear AR models using expectation maximization. Journal of Forecasting 22(6), 479–490 (2003)
5. Aytug, H., Bhattacharrya, S., Koehler, G.J.: A markov chain analysis of genetic algorithms with power of 2 cardinality alphabets. ORSA Journal on Computing 96(6), 195–201 (1997)
6. Aytug, H., Koehler, G.J.: Stopping criteria for finite length genetic algorithms. ORSA Journal on Computing 8(2), 183–191 (1996)
7. Aytug, H., Koehler, G.J.: New stopping criterion for genetic algorithms. European Journal of Operational Research 126(1), 662–674 (2000)
8. Bäck, T.: Evolutionary Algorithms in Theory and Practice: Evolution Strategies, Evolutionary Programming, Genetic Algorithms. Oxford University Press (1996)
9. Bagchi, T.: Multiobjective Scheduling by Genetic Algorithms. Kluwer Academic Publishers (1999)
10. Brockwell, P., Davis, R.: Time Series: Theory and Methods. Springer (1998)

11. Brockwell, P., Davis, R.: Introduction to Time Series and Forecasting. Springer (2000)
12. Broersen, P., de Waele, S., Bos, R.: Application of autoregressive spectral analysis to missing data problems. IEEE Transactions on Instrumentation and Measurement 53(4), 981–986 (2004)
13. Burke, E.K., Gustafson, S., Kendall, G.: Diversity in genetic programming: An analysis of measures and correlation with fitness. IEEE Transactions on Evolutionary Computation 8(1), 47–62 (2004)
14. Celeux, G., Diebolt, J.: The SEM algorithm: a probabilistic teacher algorithm derived from the EM algorithm for the mixture problem. Computational Statistics Quarterly 2(1), 73–82 (1993)
15. Chambers, R.L., Skinner, C.J.: Analysis of Survey Data. John Wiley and Sons (2003)
16. Dempster, A.P., Laird, N.M., Rubin, D.B.: Maximum-likelihood from incomplete data via the EM algorithm. Journal of Royal Statistical Society 39(1), 1–38 (1977)
17. Devroye, L.: Non-Uniform Random Variate Generation. Springer, New York (1986)
18. Eklund, N.: Using genetic algorithms to estimate confidence intervals for missing spatial data. IEEE Transactions on Systems, Man and Cybernetics, Part C: Applications and Reviews 36(4), 519–523 (2006)
19. Figueroa García, J.C., Kalenatic, D., Lopez Bello, C.A.: Missing Data Imputation in Time Series by Evolutionary Algorithms. In: Huang, D.-S., Wunsch II, D.C., Levine, D.S., Jo, K.-H. (eds.) ICIC 2008. LNCS (LNAI), vol. 5227, pp. 275–283. Springer, Heidelberg (2008)
20. Figueroa García, J.C., Kalenatic, D., López, C.A.: An evolutionary approach for imputing missing data in time series. Journal on Systems, Circuits and Computers 19(1), 107–121 (2010)
21. Fonseca, C.M., Fleming, P.J.: Genetic algorithms for multiobjective optimization: Formulation, discussion and generalization. Evolutionary Computation 3(1), 1–16 (2004)
22. Gaetan, C., Yao, J.F.: A multiple-imputation metropolis version of the EM algorithm. Biometrika 90(3), 643–654 (2003)
23. Goldberg, D.E.: Genetic Algorithms in Search, Optimization, and Machine Learning. Adisson-Wesley (1989)
24. González, S., Rueda, M., Arcos, A.: An improved estimator to analyse missing data. Statistical Papers 49(4), 791–796 (2008)
25. Grimmet, G., Stirzaker, D.: Probability and Random Processes. Oxford University Press (2001)
26. Hair, J.F., Black, W.C., Babin, B.J., Anderson, R.E.: Multivariate Data Analysis, 7th edn. Prentice-Hall (2009)
27. Hamilton, J.D.: Time Series Analysis. Princeton University (1994)
28. Harville, D.A.: Matrix Algebra from a Statician's Perspective. Springer-Verlag Inc. (1997)
29. Huber, P.: Robust Statistics. John Wiley and Sons, New York (1981)
30. Ibrahim, J., Molenberghs, G.: Missing data methods in longitudinal studies: a review. TEST 18(1), 1–43 (2009)
31. JiaWei, L., Yang, T., Wang, Y.: Missing value estimation for microarray data based on fuzzy c-means clustering. In: IEEE (ed.) Proceedings of High-Performance Computing in Asia-Pacific Region, 2005 Conference, pp. 616–623. IEEE (2005)

32. Kalra, R., Deo, M.: Genetic programming for retrieving missing information in wave records along the west coast of india. Applied Ocean Research 29(3), 99–111 (2007)
33. Law, A., Kelton, D.: Simulation System and Analysis. McGraw Hill International (2000)
34. Levine, L.A., Casella, G.: Implementations of the monte-carlo EM algorithm. Journal of Computational Graphic Statistics 10(1), 422–439 (2000)
35. Londhe, S.: Soft computing approach for real-time estimation of missing wave heights. Ocean Engineering 35(11), 1080–1089 (2008)
36. Mood, A.M., Graybill, F.A., Boes, D.C.: Introduction to the Theory of Statistics. Mc Graw Hill Book Company (1974)
37. Nelwamondo, F.V., Golding, D., Marwala, T.: A dynamic programming approach to missing data estimation using neural networks. Information Sciences (in press 2012)
38. Nielsen, S.F.: The stochastic EM algorithm: Estimation and asymptotic results. Bernoulli 6(1), 457–489 (2000)
39. Parveen, S., Green, P.: Speech enhancement with missing data techniques using recurrent neural networks. In: IEEE (ed.) Proceedings of the IEEE International Conference on Acoustics, Speech, and Signal Processing (ICASSP 2004), vol. 1, pp. 733–738. IEEE (2004)
40. Pendharkar, P.C., Koehler, G.J.: A general steady state distribution based stopping criteria for finite length genetic algorithms. European Journal of Operational Research 176(3), 1436–1451 (2007)
41. Qin, Y., Zhang, S., Zhu, X., Zhang, J., Zhang, C.: Semi-parametric optimization for missing data imputation. Applied Intelligence 27(1), 79–88 (2007)
42. Ross, S.M.: Stochastic Processes. John Wiley and Sons (1996)
43. Safe, M., Carballido, J., Ponzoni, I., Brignole, N.: On Stopping Criteria for Genetic Algorithms. In: Bazzan, A.L.C., Labidi, S. (eds.) SBIA 2004. LNCS (LNAI), vol. 3171, pp. 405–413. Springer, Heidelberg (2004)
44. Siripitayananon, P., Hui-Chuan, C., Kang-Ren, J.: Estimating missing data of wind speeds using neural network. In: IEEE (ed.) Proceedings of the 2002 IEEE Southeast Conference, vol. 1, pp. 343–348. IEEE (2002)
45. Ssali, G., Marwala, T.: Computational intelligence and decision trees for missing data estimation. In: IEEE (ed.) IJCNN 2008 (IEEE World Congress on Computational Intelligence), pp. 201–207. IEEE (2008)
46. Tsiatis, A.A.: Semiparametric Theory and Missing Data. Springer Series in Statistics (2006)
47. Wilks, A.: Mathematical Statistics. John Wiley and Sons, New York (1962)
48. Zhong, M., Lingras, P., Sharma, S.: Estimation of missing traffic counts using factor, genetic, neural, and regression techniques. Transportation Research Part C: Emerging Technologies 12(2), 139–166 (2004)

Chapter 3
Intelligent Aggregation and Time Series Smoothing

Ronald R. Yager

Abstract. Predicting future values of a variable from past observations is fundamental task in many modern domains. This process, often referred to as times series smoothing, involves an aggregation of the past observations to predict the future values. Our objective here is to use recent advances in computational intelligence to suggest new and better approaches for performing the necessary aggregations. We first look at some special features associated with the types of aggregations needed in times series smoothing. We show how these requirements impact on our choice of weights in the aggregations. We then note the connection between the method of aggregation used in times series smoothing and that used in the intelligent type aggregation method known as the Ordered Weighted Averaging (OWA) operator. We then take advantage of this connection to allow us to simultaneously view the problem from a times series smoothing perspective and OWA aggregation operations perspective. Using this multiple view we draw upon the large body of work on families of OWA operators to suggest families for the aggregation of time series data. A particularly notable result of this linkage is the introduction of the use of linear decaying weights for time series data smoothing.

Index Terms: Time Series Smoothing, Prediction, Aggregation, OWA Operators.

1 Introduction

An important task in time series analysis is using a sequence of past observations about some variable to predict a future value for the variable [1-3]. This process, often referred to as times series smoothing, involves an aggregation of the past observations to predict the future values. While approaches such as weighted average and exponential smoothing [1] have been used to perform this aggregation

Ronald R. Yager
Machine Intelligence Institute
Iona College
New Rochelle, NY 10801
e-mail: yager@panix.com

W. Pedrycz & S.-M. Chen (Eds.): Time Series Analysis, Model. & Applications, ISRL 47, pp. 53–75.
DOI: 10.1007/978-3-642-33439-9_3   © Springer-Verlag Berlin Heidelberg 2013

our objective here is to use recent advances in computational intelligence to suggest new and better approaches for performing the required aggregations. We first look at the time series smoothing task and indicate some key features desired in the aggregation process. Among these is a desire for an unbiased estimate that has minimal variance and one that gives preference to more recent observations to better account for a changing world. We also desire an efficient aggregation process, one that can easily use previous aggregations in the calculations as new observations come. We note the connection between the type of aggregation used in times series smoothing and that used in the intelligent type aggregation method known as the Ordered Weighted Averaging (OWA) operator [4, 5]. This connection allows us to simultaneously view this process as a time series-smoothing problem and OWA aggregation [6]. This multiple view allows us to bring tools from both perspectives to the problem of smoothing. Here we draw upon the large body of work on families of OWA operators to suggest families for the aggregation of time series data. A particularly notable result of this linkage is the introduction of the use of linear decaying weights for time series data smoothing.

2   Time Series Smoothing and Aggregation

In time series smoothing we use observations about some variable, x_t for t = 1 to n, to predict a future value for the variable, x_{n+1}, based upon some aggregation of the earlier values. A key factor that determines the form for the aggregation is our assumption about the underlying pattern generating the data. The one we shall use here is that the underlying variable is almost constant and we are observing $x_t = a + e_t$ where e_t is some error with mean zero and constant variance. Here we use the observations $x_1, ..., x_n$ to obtain an estimate of a, $\bar{a}$, and then use this estimate as our predictor of x_{n+1}. In order to obtain an unbiased estimate we use

a mean type aggregation [7]. In this case $\bar{a} = F(x_1, ..., x_n) = \sum_{j=1}^{n} u_j x_j$ where u_j

are a collection of weights $u_j \in [0, 1]$ that sum to one.

There are a number of features that are special when making these calculations in the framework of time series data. One is the repetitive nature of the task. We are constantly getting additional readings for x_t and then using these to update our estimate for $\bar{a}$. Formally to deal with this sequential updation we shall use the term a_n to indicate our smoothed value $F(x_1, ... , x_n)$. Thus a_n and $\bar{a}$ at time n are synonyms. Another special feature of the time series environment is that while we are assuming that the underlying value a is fixed it is often more realistic to allow for the possibility of slow drift, that is a is quasi-constant. We shall see these two special features of the temporal environment play an important role in the determination of the weights, the u_j.

This repetitive nature of the calculation has two immediate implications. The first is that it is beneficial to make the calculation of $F(x_1, ..., x_n)$ as simple as possible. If we could take advantage of prior calculations this would help. We should note that while beneficial, in this age of great computational power this is not as important as in the past. The second implication however is more significant. The repeated updation task requires we are going to implement many calculations of the form

$$F(x_1, ..., x_n), F(a_1, ..., a_n, x_{n+1}), F(a_1, ..., a_{n+1}, a_{n+2}), ...$$

where each of these is a mean aggregation. Since that the mean operator is not generally associative [8] this implies there is no mandated manner for performing the aggregation as we add values. However, it is important that all of these calculations be done in some kind of consistent manner. Since each of these aggregations will involve a different number of arguments we shall be using weighting vectors of different dimensions. Here then the issue of consistently calculating estimate of a forf different n involves an appropriate choice of weighting vectors of growing dimensions.

The second special feature of time series data, the allowance for possible variation in the underlying value a has as an implication that not all observation should be treated the same. In particular more weight should be assigned to the most recent observations. Thus we have a preference for weighting vectors in which $u_j \geq u_i$ for $j > i$.

In order to formalize this requirement we introduce a characterizing feature of a choice of weights called the **average age** of the data used. If n is the current time then the age of the piece of data x_t is $AGE(t) = n - t$. Using this we get as the

average of the data $\overline{AGE} = \dfrac{\sum\limits_{j=1}^{n} u_j AGE(j)}{\sum\limits_{j=1}^{n} u_j}$. Since $\sum\limits_{j=1}^{n} u_j = 1$ then

$$\overline{AGE} = \sum_{j=1}^{n} u_j AGE(j) = n - \sum_{j=1}^{n} j u_j .$$

As we previously indicated we have some preference for fresh data or *youthful* data. We have the freshest data if we select $u_n = 1$ and all other u_j equal zero. In this case $\overline{AGE} = 0$. However there is some other conflicting objective that we must consider. As our observations are of the form $x_t = a + e_t$ where e_t is assumed to be a random noise component with mean zero and variance σ^2, each piece of data has a variance of σ^2. Since our objective is to find a good estimate for a we desire to have a small variance in our estimate $\bar{a}$.

With

$$\bar{a} = \sum_{t=1}^{n} u_t x_t = \sum_{t=1}^{n} u_t(a + e_t)$$

since the sum of the $u_j = 1$ we get as our expected value, $E_x[\bar{a}] = a$, thus this is an unbiased estimate. To find the variance of our estimate we calculate

$$Var(\bar{a}) = E_x\left[\sum_{t=1}^{n}(u_t x_t - \bar{a})^2\right]$$

where E_x denotes the expected value. Under the assumption that the observations are uncorrelated we obtain

$$Var(\bar{a}) = \sum_{t=1}^{n} u_t^2 \sigma^2 = \sigma^2 \sum_{t=1}^{n} u_t^2$$

Under the objective to minimize the variance and since we have no control over the σ^2 our problem reduces to trying to make $H(u) = \sum_{t=1}^{n} u_t^2$

We now see that our objective in choosing the weights is to try to find weights, $\mu_j \in [0, 1]$ summing to one that make $H(u) = \sum_{t=1}^{n} u_t^2$ as small and while also making $\overline{AGE} = \sum_{j=1}^{n} u_j(n - j)$ small. As we shall see the goals of trying to make $H(u)$ and $\overline{AGE}$ small are essentially conflicting under the conditions that $\sum_{j=1}^{n} u_j$ and $u_t \in [0, 1]$..

We now make some relevant observations. Let $G = \langle g_1, ..., g_n\rangle$ be a collection of weights such that $\sum_{j=1}^{n} g_j = 1$ and $g_i \in [0, 1]$. Assume we are going to assign these weights to the u_j. That is we have some permutation function $\pi: [1, ..., n] \rightarrow [1, ..., n]$ so that $u_j = g_{\pi(i)}$. Each of u_j is assigned one of the g_i. The first observation we make is that the value of $H(u)$ is independent of how we assign the g_i to u_j. Thus $H(u)$ just depends only on what are the elements in G. The value of $H(u)$ is the same for any function π. On the other hand the calculation of

$\overline{AGE}(u) = \sum\limits_{j=1}^{n} u_j (n - j)$ depends on the function π we use for associating the g_i

with u_j. It can be shown that if $\tilde{\pi}$ is the assigning function such that $\tilde{u}_j \geq \tilde{u}_k$ for $j >$

k and π is any arbitrary function assigning the g_i to the u_j then $\overline{AGE}(\tilde{\pi}) \leq \overline{AGE}(\pi)$.

Thus for any collection of weights G we always obtain the smallest average age by assigning the weights such that the newer the data the more the weight, $u_j \geq u_i$ if j $> i$. Thus any weight assignment should always satisfy this condition of having u_j $\geq u_j$ if $j < j$. Thus given a collection G there is only one way to assign them to the weights. Hence the key question becomes the choice of G.

We now investigate the issue of simultaneously trying to minimize H(u) and

$\overline{AGE}(u)$. Consider the function. $\overline{AGE}(u) = \sum\limits_{j=1}^{n} u_j (n - j)$. We see the smallest

value of $\overline{AGE}(u)$ is obtained for the case where $u_n = 1$ and $u_j = 0$ for $j \neq n$. In this

case $\overline{AGE}(u) = 0$. On the hand under the restriction that the u_j are ordered so that

$u_j \geq u_k$ for $j \geq k$ we see that the largest value for $\overline{AGE}(u)$ is obtained when $u_j = \frac{1}{n}$

for all j. In this case $\overline{AGE}(u) = \frac{(n - 1)}{2}$.

Consider now the function $H(u) = \sum\limits_{j=1}^{n} u_j^2$ where the $u_j \in [0, 1]$ and $\sum\limits_{j=1}^{n} u_j = 1$.

It is well known that this attains its largest value where there is some $u_k = 1$ and

all others are zero. It attains its smallest value when $u_j = \frac{1}{n}$ for all j. More

generally the more diversely distributed the weights the smaller H(u).

Here then we see the basic conflict in choosing the weights. The function H(u) tries to attain its goal of being small by spreading the weights as uniformly as possible among the u_j while the function AGE tries to attain its minimal value by selecting the most recent weight equal to one.

3 Classical Parameterized Smoothing Techniques

Using the linear smoothing approach our objective is to develop a procedure for

obtaining the weights u_j such that $\sum\limits_{j=1}^{n} u_j x_j$ provides the best forecast for x_{n+1}.

Ideally such a procedure should be able to deal with the repeated updation required as a result of new incoming data. One commonly used approach is to use a parameterized family of weights. In this case the weights are generated using some rule characterizing the family and the values of some associated parameters. The values of the parameters are learned from the data being observed.

The use of parameterized approaches have a number of beneficial features. One major benefit of using a parameterized family of weights is the well-organized way in which they determine the generation of the weights as we obtain more observations. The importance of this feature can not be overestimated, for we note that mean operators are not generally associative which implies that the process of going from aggregating n pieces of data to aggregating to $n + 1$ pieces is not fixed. By using a parameterized family of weights we are imposing a discipline on the process of weight generation. Another important benefit is that the learning process is simplified to just the determination of the few required parameters. Furthermore often because of intuitive meaning of these parameters the process of learning their values can be completely bypassed and a user can intelligently supply the required parameter values.

However, the use of parameterized methods comes at a price. As we indicated the potential to provide good smoothing is generally effected by our ability to simultaneously minimize the variance and use recent data. As we noted these are generally conflicting objectives. The use of a parameterized method further restrict are freedom to simultaneously satisfy these two conditions.

In the following we shall compare some classic parameterized methods for building smoothing functions based upon their ability to simultaneously satisfy the two objectives, minimize the expected variance and minimize the age of the data.

To aid in this task we shall find it useful to characterize a smoothing method by what we will call its *flexibility*. We define the flexibility of a smoothing method as a curve of the Minimal Attainable Variance (MAV) vs. the Average Age (AA). In figure #1 we provide a prototypical example of a flexibility curve.

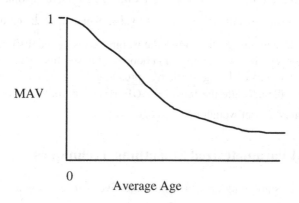

Fig. 1. Typical flexibility curve

A notable feature of these flexibility curves are their monotonically decreasing nature, the smaller the average age the larger the minimal attainable variance. We shall say that method-1 is more flexible for average age T then method-2 if the MAV for method-1 at average age T is less than the MAV for method-2 at average age T. More globally we shall say that method-1 is **more flexible** than method-2 if for all values of T the MAV of method-1 is at least as small as that of method-2. The more flexible a method the better it can provide a good smoothing function.

One classic parameterized method for obtaining the weights is the moving average [1]. Here we take the average of the last P observations This approach has as its only parameter the size of the window P. Using this method the value of $\bar{a}$ at time n, a_n, is

$$\bar{a} = \frac{x_n + x_{n-1} + x_{n-2} + \dots + x_{n-P+1}}{P} = \frac{1}{P}\sum_{j=0}^{P-1} x_{n-j}.$$

Is it assumed $P \leq n$.

In this case the associated weights are

$u_j = 0$ \qquad for $j \leq n - P$

$u_j = \frac{1}{P}$ \qquad for $n - P + 1 \leq j \leq n$

For this type of weight allocation

$$H_M(u) = \sum_{j=}^{n} u_j^2 = \frac{1}{P}$$

$$\overline{AGE}_M(u) = \sum_{j=1}^{n}(n-j)u_j = \frac{P-1}{2}$$

Here we see again the conflict, $H_M(u)$ can be made smaller by increasing P while $\overline{AGE}_M(u)$ can be made smaller by decreasing P.

Letting a_{n-1} and a_n be our smoothed value after obtaining $n-1$ and n pieces of data respectively we can express this moving average in a recursive fashion as

$$a_n = a_{n-1} + \frac{x_n - x_{n-P}}{P}$$

Here x_n and x_{n-P} are the nth and n - P observations.

Another classic approach is the exponential smoothing method [1, 9]. Using this method our estimate of a at time n is defined in a recursive manner as

$$a_n = \alpha x_n + (1-\alpha) a_{n-1}$$

where a_n and a_{n-1} are smoothed values after n and n - 1 observations respectively and x_n is the nth observation. Here $\alpha \in [0, 1]$, called the smoothing constant, is the only parameter

It can be shown [1] that

$$a_n = \alpha \sum_{k=0}^{n-2} (1-\alpha)^k x_{n-k} + (1-\alpha)^{n-1} x_1$$

From this we see that the weights, the u_j, are

$$u_1 = (1-\alpha)^{n-1}$$
$$u_j = \alpha(1-\alpha)^{n-j} \qquad \text{for } j = 2 \text{ to } n$$

It can be shown that these sum to one.

Using these weights and denoting $\beta = (1-\alpha)$ we can obtain the average age of the data used in this case of exponential smoothing as

$$\overline{AGE_E}(u) = \sum_{j=1}^{n} (n-j)u_j = 0\alpha + 1(\alpha\beta) + 2(\alpha\beta^2) + 3(\alpha\beta^3) + \ldots\ldots\ldots$$

$$** \ \overline{AGE}_E(u) = \alpha \sum_{k=0}^{n-2} k\beta^k + (n-1)\beta^n$$

Furthermore it can be shown [1] that as n gets large, $n \to \infty$, we obtain

$$\overline{AGE_E}(u) = \frac{\beta}{\alpha} = \frac{1-\alpha}{\alpha}$$

In this case of exponential smooth the expected variance is

$$H_E(u) = \sum_{j=1}^{n} u_j^2 = \sum_{k=0}^{n-2} (\alpha\beta^k)^2 + (\beta^n)^2$$

It can be shown that as n gets large

$$H_E(u) = \frac{\alpha}{1+\beta} = \frac{\alpha}{2-\alpha}$$

Again here we see the conflict, making α bigger acts to decrease $\overline{AGE_E}(u)$ while increasing $H_E(u)$. The satisfaction of the age criteria wants α big, close to one while satisfaction of the variance criteria wants α small, close to zero.

There is an interesting relationship between these two approaches. We have shown that $\overline{AGE_M}(u) = \frac{P-1}{2}$ and $\overline{AGE_E}(u) = \frac{1-\alpha}{\alpha}$. If we let α_p be the value of α that makes $\overline{AGE_E}(u) = \overline{AGE_M}(u)$ for a particular P we get $\alpha_p = \frac{2}{P+1}$. We note that in this case

$$H_E(u) = \frac{\alpha_p}{2-\alpha_p} = \frac{\frac{2}{P+1}}{2-\frac{2}{P+1}} = \frac{2}{2P} = \frac{1}{P}$$

which is exactly the same value as $H_M(u)$. Thus these two approaches have the same flexibility once we select the parameter.

The moving average and the exponential smoothing techniques as just described do not completely specify the weights that are to be used. In each of these cases we have a parameter that has to be supplied to precisely determine the weights. The parameter can be obtained in particular applications by learning it based on the performance of the system. For example we can measure the difference between a_n, the forecasted value for x_{n+1}, and the actual value of x_{n+1} and then use the square error minimization to obtain the parameter.

In introducing the moving average and exponential smoothing we have introduced parameterized families of weighting functions to be used to aggregate our time series data. The actual performance of these methods in a given application will depend on our ability to find values for the parameters that work well. While the particular performance in an application will depend upon the data as we have pointed out in a global sense our ability to obtain good forecasting will depend upon the underlying method's ability to be able to get weights that simultaneously satisfy our criteria of small variance and young data. As we have shown these two criteria are often conflicting.

Given this connection between a smoothing methods ability to simultaneously have weights that have a small variance and small average age and its possibility of obtaining weights that provide a good forecast using the observed data we may benefit from looking for more families of parameterized weight functions. In particular families that are more flexible then these two would seem to provide a more fertile terrain to look for forecasting methods in applications.

With this in mind we now turn to the OWA aggregation operators [4]. These are a class of mean like aggregation operators in which considerable work has been done in obtaining families of weighting functions and as such may provide some interesting candidates for sequential aggregations.

4 OWA Operators and Induced OWA Operators

The Ordered Weighted Averaging (OWA) operator of dimension n introduced by Yager [4] is an aggregation operator F: $R^n \to R$ defined as

$$F(y_1, y_2, ..., y_n) = \sum_{j=1}^{n} w_j b_j$$

where b_j is the jth largest of argument values and w_j is a collections of weights such that $w_j \in [0, 1]$ and $\sum_{j=1}^{n} w_j = 1$. Collectively the weights are called the OWA weighting vector and denoted as W. We can represent the ordered arguments by a vector B called the order argument vector in which b_j is the jth component. Furthermore if ind is an index function such then ind(j) is the index of the jth largest of the argument values then $b_j = y_{ind(j)}$. This allows us to have the following equivalent representations of the OWA aggregation operator

$$F(y_1, ..., y_n) = \sum_{j=1}^{n} w_j b_j = \sum_{j=1}^{n} w_j y_{ind(j)} = W^T B$$

Thus we it is the inner product of the weighting vector and the ordered argument vector.

The OWA operator is parameterized by the choice of weights. If $w_i = 1$ and $w_j = 0$ for $j \neq 1$ then $F(y_1, ..., y_n) = y_{ind(1)}$. If $w_n = 1$ and $w_j = 0$ for $j \neq n$ then $F(y_1, ..., y_n) = y_{ind(n)}$. If $w_j = \frac{1}{n}$ then $F(y_1, ..., y_n) = \frac{1}{n} \sum_{i=1}^{n} y_i$, it is the simple average. More generally if most of the weights are associated with w_j's that have smaller indexes, then the aggregation is giving preference to the arguments with the
bigger values, the $b_j = y_{ind(j)}$ with the lower j. Conversely if most of the weight are associated with w_j's that have higher indices, then the aggregation is giving preference to the argument with the smaller values, the $b_j = y_{ind(j)}$ with the larger j.

Thus the OWA operator describes a family of averaging operators parameterized by its weighting vector W. In [4] a measure was introduced to characterize an OWA operator with regard to its preference for bigger or smaller elements in the argument. This measure called the **attitudinal character** is defined as

$$A\text{-}C(W) = \frac{1}{n-1} \sum_{i=1}^{n} (n-j) w_j$$

It takes its maximal value of one for the case where $w_1 = 1$ and its minimum value of zero when $w_n = 1$. For the case when all $w_j = \frac{1}{n}$ it assumes the value 0.5. Generally if the weights tend to be at the top of the W vector A-C(W) gets closer to one while if the weights tend to be near the bottom A-C(W) get close to zero.

It is interesting to note the attitudinal character itself is an OWA aggregator of a special argument collection using the W In particular

$$A\text{-}C(W) = F(\frac{n-1}{n-1}, \frac{n-2}{n-1}, \frac{n-3}{n-1}, ..., \frac{n-n}{n-1})$$

It is the OWA aggregation of $y_j = \frac{n-j}{n-1}$.

Another measure [4] associated with an OWA vector is the measure of dispersion or entropy

$$E(W) = = - \sum_{j=1}^{n} w_i \ln(w_i)$$

It can be easily shown that the OWA operator is a mean operator, it is monotonic, commutative and bounded, $\text{Min}_i[y_i] \le F(y_1, y_2, ..., y_n) \le \text{Max}_i[y_i]$. It is also idempotent, if all $y_i = c$ then $F(y_1, y_2, ..., y_n) = c$. Typically mean operators are not associative. This means that there is no preordained way of extending these operators from dimension n to n + 1.

Thus an interesting problem associated with OWA operators is providing *consistent* ways of defining the weights associated with aggregations of different dimensions. Three predominate approaches to addressing this issue can be noted. One approach, introduced in [10], is to provide some function, called a BUM function, 1 and use this to generate the weights. A BUM function is a monotonic mapping f: [0, 1] → [0, 1] satisfying **1** f(0) = 0, **2.** f(1) = 1, **3.** f(x) ≥ f(y) if x > y. Using this BUM function consistent weights can be obtained by setting for all j

$$w_j = f(\frac{j}{n}) - f(\frac{j-1}{n})$$

where n is the dimension of the aggregation. Using this BUM function we are able to generate weights in a consistent manner for aggregations of different dimensions.

Another general approach, initiated by the work of O'Hagan [11], focuses on determining the weights so that the attitudinal character remains the same as we change the dimensionality. A third approach is to directly express the process of assigning the weights in such a way that it can be easily extended to any number of arguments. Finding the simple average is an example of this. Obtaining the Max or Min of the argument also falls in the category. We have essentially described the process of determining the weight in a manner independent of the number of arguments.

In the proceeding we defined the OWA operator as

$$F_W(y_1, ..., y_n) = \sum_{j=1}^{n} w_j b_j = W^T B$$ where $b_j = y_{\text{ind}(j)}$. Here ind(j) is the index

of the jth largest argument value. This is essentially taking an average of the argument values where the weights associated with the different arguments are determined by the ordered position of the argument with respect to their value.

An equivalent view of this operator is the following. Let M: {1, ..., n} → {1, ..., n} be a one to one and onto (bijective) mapping such that M(i) is the ordered position of the argument y_i. Using this notation we have

$$F(y_1, ..., y_n) = \sum_{i=1}^{n} w_{M(i)} y_i$$. Hence M is a mapping that associates with each

argument a unique integer determined by is ordered argument position. We note that a unique relationship exists between M and ind, $ind(j) = M^{-1}(j)$. Using this

$$F_W(y_1, \ldots\ldots, y_n) = \sum_{j=1}^{n} w_j \, b_j \text{ where } b_j = y_{M^{-1}(j)}.$$

We now introduce a generalization of the OWA operator called the Induced OWA (IOWA) operator [12]. Behind this generalization is the realization that the process of ordering the arguments does not necessarily have to be guided by an ordering based on the value of the arguments to still result in a mean aggregation of the arguments. The assignment of weights to arguments can be induced in other ways. More formally $W^T B$ where B is any vector of containing all the arguments in any order is still a mean operator.

Let us describe the IOWA operator. As in the case of OWA operator with the IOWA operator we have a weighting vector W whose components w_i sum to one and lie in the unit interval. However in the case of the IOWA operator the input are tuples, (y_i, v_i), where y_i is called the argument variable and v_i is called the order inducing variable. Here we are still interested in getting a mean aggregation of the y_i but the process of associating the weights with the argument values, ordering the elements in B, is determined by the v_i.

Here we denote $M_V: I \to J$ where $I = J = \{1, 2, ..., n\}$. M_V is a mapping that associates with each $i \in I$ a unique value $j \in J$ determined by the value v_i. We note M_V is a bijective mapping, one to one and onto,.

Using this notation we have

$$F_W((y_1, v_1), ..., (y_n, v_n)) = \sum_{i=1}^{n} w_{M_V(i)} \, y_i$$

or alternatively

$$F_W((y_1, v_1), ..., (y_n, v_n)) = \sum_{j=1}^{n} w_j \, y_{M_V^{-1}(j)}$$

We can then let $b_j = y_{M_V^{-1}(j)}$ and express this aggregation as $F_W((y_1, v_1), ..., (y_n, v_n)) = W^T B_v$. We emphasize that since the w_j are positive and sum to one we will still getting a mean of the y_i.

An example when v is a variable taking values from the real line would be to the mapping $M_V: I \to J$ be such that $M_V(i) = j$ if the v_i has the jth largest value for the order inducing variable. However the ordering can be inverse, we can have $m_V(i) = j$ if the ith tuple has the $n - j + 1$ largest value for the order inducing variable. Furthermore the order inducing variable v need not take its values from the real line. It can take its values from any space that has an ordering. Thus the v_i can be words that can be ordered such as *small, medium* and *large*.

A particularly interesting application of the IOWA aggregation is called the *last best* model that was introduced in [12]. Assume we have n experts who predict the value of today's closing value of the ABC stock. Let y_i be the prediction of the ith expect. Our objective is to average these to get an aggregate prediction. In the *last best* model we use an ordering based on the expert's previous day's performance to assign the OWA weights to the different experts. In particular let Δ_i be absolute error in the ith expert's performance yesterday. That is $\Delta_i = |b_i - \hat{b}|$ where b_i was experts guess yesterday and $\hat{b}$ was the actual value that was attained by ABC yesterday. In this case the input pairs to the IOWA aggregation are $((y_i, \Delta_i), ..., (y_n, \Delta_n))$. In the best yesterday model we induce the weight assignment using $M_\Delta: I \rightarrow J$. In particular we order the experts in increasing order of Δ_i, then assign $m_V(i) = j$ where i has the jth smallest of the Δ_i values.

5 Using OWA Operators in Time Series

An important special case of the IOWA aggregation is the case where the arguments are a sequence of values, $(x_1, ..., x_n)$, and the order inducing variable is the position in the sequence. Here we are aggregating $F((1, x_1), (2, x_2), ..., (n, x_n))$ with order inducing variable $v_i = i$. If we choose M_V to be such that $M_V(i) = v_i = i$ then we get

$$\sum_{j=1}^{n} w_{M_{V(i)}} x_i = \sum_{i=1}^{n} w_i x_i$$

where the w_i are a collection of OWA weights, they lie in the unit interval and sum to one. This then allows us to simultaneously view this aggregation process as a time series smoothing problem and OWA aggregation. This multiple view allows to bring tools from both perspectives to the problem of smoothing. In particular we shall be able to draw upon the large body of work on families of OWA operators to suggest families for the aggregation of time series data.

An interesting and useful relationship can be shown between the OWA view and the time series view. In the framework of the OWA operator we defined the attitudinal character associated with a weighting vector W as

$$A\text{-}C(W) = \frac{1}{n-1} \sum_{j=1}^{n} w_j(n-j)$$

In the time series smoothing framework we introduced the idea of the average age of the data used which is expressed as

$$\overline{AGE}(W) = \sum_{j=1}^{n} w_j \, Age(j) = \sum_{j=1}^{n} w_j(n-j)$$

what can be easily seen is their relationship

$$\overline{AGE}(W) = (n-1) \, A\text{-}C(W)$$

Thus the average age of the data is equal to the attitudinal character times n - 1. In particular if W and $\tilde{W}$ are two OWA weighting vectors such that A-C(W) ≤ A-C($\tilde{W}$) then $\overline{AGE}(W) \le \overline{AGE}(\tilde{W})$. As we previously noted in time series we are interested in obtaining aggregations in which we have a small average age. This can be seen to correspond to OWA weighting vectors with A-C values tending toward zero. In the framework of the ordinary OWA operator we are looking for weighting vectors that are "and-like".

Another interesting correspondence exists between the measure of dispersion or entropy in the OWA framework, $E(W) = -\sum_{j=1}^{n} w_j \ln(w_j)$ and the measure of variance in the time series framework $H(W) = -\sum_{j=1}^{n} w_j^2$, these two measures are cointensive in that they are formulating the same intuitive notion. We first note that both are invariant with respect to the indexing of the w_j. They both attain their minimum value when $w_j = \frac{1}{n}$ for all j and both obtain their maximum value when $w_k = 1$ for some arbitrary k. They both are expandable [13, 14], if an element $w_{n+1} = 0$ is added to the collection [w_1, ..., w_n] both H(W) and -E(W) are unchanged.

A key feature of both these is what we shall refer as a **preference for equalitarism**. Let w_1, ..., w_n be a collection of positive weights which sum to one. If one of the weights is less than $\frac{1}{n}$ then there must be at least one other weight greater than $\frac{1}{n}$. Let w_1 be a weight less than $\frac{1}{n}$ and let w_2 be a weight greater than $\frac{1}{n}$. Let Δ be a value such that $\Delta \le Min[|w_1 - \frac{1}{n}|, |w_2 - \frac{1}{n}|]$. Let v_1, ..., v_n be a new set of weights such that

$$v_1 = w_1 + \Delta$$

$$v_2 = w_2 - \Delta$$

$$v_j = w_j \text{ for } j > 2$$

It can be shown that H(V) ≤ H(W) and E(V) ≤ E(W), both of these have increased.

6 Linear Decaying Weights

We now describe a parameterized family of weights for the OWA aggregation and discuss their potential role in the smoothing framework, we refer to these as Linearly Decaying (LD) weights. In the following we denote our observations as $x_1, ..., x_n$ where x_n is the most recent. We associate with the LD weights a single window–like parameter, an integer m such that $1 < m \leq n$. Using this parameter we obtain the LD weights as

$$w_j = \frac{j - (n - m)}{T} \qquad n - m + 1 \leq j \leq n$$

$$w_j = 0 \qquad j \leq n - m$$

where $T = \sum_{j=1}^{m} j = \frac{m(m+1)}{2}$.

In the case where m = 5 we have T = 5 + 4 + 3 + 2 + 1 = 15 and our weights are

$$w_n = 5/15, \ w_{n-1} = 4/15, \ w_{n-2} = 3/15, \ w_{n-3} = 2/15, \ w_{n-4} = 1/15$$

$$w_j = \quad 0 \quad \text{for all others}$$

For the case where m = 10 we get T = ((11)(10))/2 = 55 and hence

$$w_n = 10/55, \ w_{n-1} = 9/55, \ w_{n-2} = 8/55, \ w_{n-3} = 7/55, \ w_{n-4} = 6/55,$$
$$w_{n-5} = 5/55, \ w_{n-6} = 4/55, \ w_{n-7} = 3/55, \ w_{n-8} = 2/55, \ w_{n-9} = 1/55$$
$$w_j = 0 \text{ for all others.}$$

In figure #2 we clearly see the linearly decaying nature of the weights

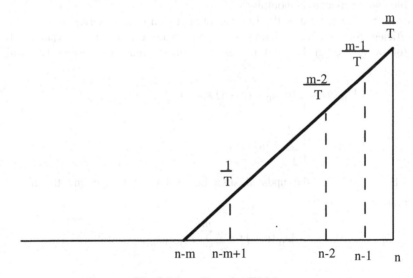

Fig. 2. Linear Decaying Weights

The average age of the data using LD weights is

$$\overline{AGE} = \sum_{j=1}^{n} w_j(n-j) = \frac{1}{T} \sum_{j=n-m+1}^{n} (j-(n-m))(n-j) = \frac{m-1}{3}$$

The average age depends linearly on the parameter m, as m increases the average age increases.

We can also calculate the associated expected variance in this case of LD weights

$$H(W) = \sum_{j=1}^{m} w_j^2 = \sum_{j=1}^{m} \frac{j^2}{T^2} = \frac{1}{T^2} \frac{4}{6} (m)(m+1)(1+2m)$$

$$H(W) = \frac{4(1+2m)}{6(m)(m+1)}$$

Here we see as the value of m increases the value of $H(W)$ decreases.

Let us compare the flexibility of this LD approach with the moving average. We observe that we get the same average age when $\frac{m-1}{3} = \frac{P-1}{2}$ and hence $P = \frac{2m+1}{3}$. With these values the variance of the moving average method is $H(MA) = \frac{1}{P} = \frac{3}{2m+1}$. The variance of the LD weights approach is $H(LD) = \frac{2}{3} \frac{(1+2m)}{(m)(m+1)}$. If we take the difference we get $H(LD) - H(MA) \leq 0$. Thus the LD weights are more flexible than the moving average. Since the moving average has the same flexibility as the exponential smoothing the LD weights are more flexible than exponential smoothing.

An interesting feature of the LD weight approach to time series smoothing is the efficient form for the updation when going from n to n + 1 observations. In the following we let $LD_m(n)$ be the smoothing value at n using LD with parameters m

$$LD_m(n) = \frac{1}{T} (m\, x_n + (m-1)\, x_{n-1} + \cdots\cdots + x_{n-m+1})$$

$$LD_m(n) = \frac{1}{T} \sum_{j=1}^{m} j\, x_{n-m+j}$$

Let $LD_m(n + 1)$ be the updated smoothed value after receiving the n + 1 observation

$$LD_m(n+1) = \frac{1}{T} \sum_{j=1}^{m} j\, x_{n+1-m+j}$$

Note here we use $x_{n+1-m+j}$ instead of x_{n-m+j}. If we expand this we get

$$LD_m(n + 1) = \frac{1}{T} (m\, x_{n+1} + (m - 1)\, x_n + \ldots\ldots + x_{n-m+2}$$

From the preceding we see that

$$LD_m(n + 1) = LD_m(n) + \frac{m}{T} x_{n+1} - \frac{1}{T} (x_n + x_{n-1} + \ldots\ldots + x_{n-m+1})$$

$$LD_m(n + 1) = LD_m(n) + \frac{2}{m+1} a_{n+1} - \frac{2}{m+1}\left(\frac{1}{m} \sum_{j=1}^{m} a_{n-(m-j)}\right)$$

However we note that $\dfrac{1}{m} \displaystyle\sum_{j=1}^{m} a_{n-(m-j)} = \dfrac{1}{m}(x_n + x_{n-1} + \ldots\ldots + x_{n-m+1})$ is

simply the moving average with window m and time n which we denote the $MA_m(n)$. Using this notation we have

$$LD_m(n + 1) = LD_m(n) + \frac{2}{m + 1}[x_{n + 1} - MA_m(n)]$$

Thus we see the updation is simply obtained by adding to the last value of LD_m the difference between the new observation and the average of the m previous observations divided by one half m + 1.

Thus this provides a simple formulation for calculating $a_{n + 1}$ from a_n

$$a_{n + 1} = a_n + \frac{2}{m + 1}[x_{n + 1} - MA_m(n)]$$

In a particular application the determination of the value for m can be made by the standard methods used in time series. Either the parameter m can be chosen a priori or learned by finding the value of m that minimizes the mean square error between the predicted (smoothed value $LD_m(n)$) and the observations $x_{n + 1}$.

7 Squarely Decaying Weights

We consider another closely related weighting scheme for time series data smoothing based on Squarely Decaying (SD) weights. Here again we let our single parameter, the integer m, be the window size. However in this case our weights are

$$w_n = \frac{m^2}{T_2}, \quad w_{n-1} = \frac{(m - 1)^2}{T_2}, \quad w_{n-2} = \frac{(m - 2)^2}{T_2}, \quad \ldots\ldots\ldots, w_{n-(m+1)} = \frac{1}{T_2}$$

$$w_j = 0 \qquad\qquad \text{for } j \leq n - m$$

where $T_2 = \displaystyle\sum_{j=1}^{m} j^2 = (\frac{1}{6})(m)(m + 1)(1 + 2m)$

We can more successfully express these weights as

$$w_j = \frac{(j - (n - m))^2}{T_2} \quad \text{for } j = n - m + 1 \text{ to } n$$

$$w_j = 0 \quad \text{for } j \leq n - m$$

In this case average age as a function of m is

$$\overline{AGE} = \frac{1}{2} \frac{(m^2 - m)}{(1 + 2m)}$$

If we denote $\frac{1}{2} \frac{(m^2 - m)}{(1 + 2m)}$ as $\overline{AGE_2}$ and if we denote the average age for the linear

decaying weights $\frac{m - 1}{3}$ as $\overline{AGE_1}$ we see that

$$\overline{AGE_1} - \overline{AGE_2} = \frac{m - 1}{3} - \frac{1}{2} \frac{m^2 - m}{2m + 1} = \frac{2(m - 1)(2m + 1) - 3m^2 + 3m}{6(2m + 1)} \geq 0$$

$\overline{AGE_2}$ is always less then $\overline{AGE_1}$.

Let us now consider the variance in this case of squarely decaying weights

$$H(W) = \sum_{j=1}^{m} w_j^2 = \sum_{j=1}^{m} \frac{j^4}{(T_2)^2} = \frac{1}{(T_2)^2} \frac{1}{30}(m)(m + 1)(1 + 2m)(-1 + 3m + 3m^2)$$

$$H(W) = \frac{36}{30} \frac{(m)(m + 1)(1 + 2m)(-1 + 3m + 3m^2)}{(m)^2(m + 1)^2(1 + 2m)^2} = \frac{6}{5} \frac{(3m^2 + 3m - 1)}{(m)(m + 1)(1 + 2m)}$$

Let us now compare this with the moving average. For moving average with window P we get the same average age as the squarely decaying weights when

$$\frac{P - 1}{2} = \frac{1}{2} \frac{(m^2 - m)}{(1 + 2m)}. \text{ In this case } P = \frac{m^2 + m + 1}{2m + 1}.$$

Let us compare the variance in this for the square decay weights W_s we showed

$$H(W_s) = \frac{6}{5} \frac{(3m^2 + 3m - 1)}{(m)(m + 1)(1 + 2m)}$$

For the moving average we have $H(W_{MA}) = \frac{1}{P} = \frac{2m + 1}{m^2 + m + 1}$. Calculating the

difference we get

$$H(W_s) - H(W_{MA}) \leq 0$$

Thus for a given average age the square decaying weights have a smaller variance then the moving average.

More generally we can consider the case of weights with two parameters m and K with

$$w_j = \frac{(j - (n - m))^K}{T_K} \qquad \text{for } n - m + 1 \le j \le n$$

$$w_j = 0 \qquad\qquad\qquad j \le n - m$$

where $T_K = \sum_{j=1}^{m} j^K$ and $K \ge 0$

We observe when $K = 0$ we get the ordinary moving average and when $K \to \square$ we get the case where $w_n = 1$ and all others $w_j = 0$.

We note that except in the cases where $K = 0$, 1 or ∞ the updated smoothed value is not easily expressed in terms of the previous value. However given the current computation power the process of calculating smoothed value after each iteration is not difficult. In particular since the parameters w_j are already determined, we just require storing the last $m - 1$ readings plus the new reading and the calculation is linear.

More generally we see that we have a two parameter class smoothing operators. Here the parameters are $K > 0$ and m is a positive integers. We can learn the parameters from the data.

8 Using Inverse Sum Weights

We now briefly consider another data smoothing model based on using an OWA weighting vector introduced by Ahn in [15, 16]. Consider the OWA weights of dimension m defined as

$$w_j = \frac{1}{m} \sum_{i=j}^{m} \frac{1}{i} \qquad \text{for } j = 1 \text{ to } m.$$

In particular we note that

$$w_1 = \frac{1}{m}(1 + \frac{1}{2} + \frac{1}{3} + \ldots + \frac{1}{m})$$

$$w_2 = \frac{1}{m}(\frac{1}{2} + \frac{1}{3} + \ldots + \frac{1}{m})$$

$$w_{m-i} =$$

$$w_{m-1} = w_m = \frac{1}{m}(\frac{1}{m})$$

We observe that for these weights $w_j = w_{j+1} + (\frac{1}{j})\frac{1}{m}$. We shall refer to these weights as **Inverse Sum (IS)** weights.

It can be shown [15, 16] that the degree of orness obtained using these weights is $A\text{-}C(W) = \frac{1}{m-1} \sum_{i=1}^{m} (m - j)w_j = \frac{3}{4}$. We emphasize that this is fixed independent of the dimension m. Since $\overline{AGE(W)} = (n-1) \, A\text{-}C(W)$ then $\overline{AGE(W)} = \dfrac{m-1}{4}$.

Thus the average age is linearly related to our window parameter m.

We now observe the form of the smoothing updation using these

$$a_n = \sum_{j=1}^{m} w_j \, a_{n-j+1}$$

$$a_{n+1} = \sum_{j=1}^{m} w_j \, a_{n-j+2}$$

$$a_{n+1} = w_1 \, x_{n+1} + \sum_{j=2}^{m} w_j \, x_{n-j+2} = w_1 \, x_{n+1} + \sum_{j=1}^{m-1} w_{j+1} \, x_{n-j+1}$$

$$a_{n+1} = w_1 \, x_{n+1} + \sum_{j=1}^{m-1} (w_j - \frac{1}{m}\frac{1}{j}) x_{n-j+1} + (\frac{1}{m^2} - \frac{1}{m^2}) x_{n-m+1}$$

$$a_{n+1} = w_1 \, x_{n+1} + a_n - \sum_{j=1}^{m} \frac{1}{jm} x_{n+1-j}$$

$$a_{n+1} = a_n + \frac{1}{m}(\sum_{j=1}^{m} \frac{1}{j}(x_{n+1} - x_{n+1-j})$$

Thus we have this very nice form for the updation. If we get Δ_j be the difference between the new observation and the observation j back then

$$a_{n+1} = a_n + \frac{1}{m}(\sum_{j=1}^{m} \frac{1}{j} \Delta_j)$$

Here m the window size is our only parameter.

9 Truncated Exponential Smoothing

In some cases classes of probability density functions provide sources interesting weighting functions that can be used in data smoothing. Here we consider the use the exponential type probability density function. This type of probability density function is define by

$$f(x) = \begin{cases} \lambda e^{-\lambda x} & x \geq 0 \\ 0 & x < 0 \end{cases}$$

where $\lambda \geq 0$ is its parameter. It is well known here that $\int_a^b f(x)dx = \int_a^b \lambda e^{-\lambda x} dx = e^{-a\lambda} - e^{-b\lambda}$ and specifically $\int_0^\infty f(x)dx = 1$.

Consider now using this to generate our weights in the smoothing operation. For simplicity in the following we shall let y_j be the data reading j back from the present reading. Here y_0 would be our most recent observation. Using an aggregation of the last m readings we get as our smoothed value $\bar{a} = \sum_{j=0}^{m-1} w_j y_j$.

Now let

$$u_j = \int_j^{j+1} \lambda e^{-\lambda x} dx = e^{-j\lambda} - e^{-(j+1)\lambda} = e^{-j\lambda}(1 - e^{-\lambda})$$

Consider now using as our weights $w_j = \dfrac{u_j}{K_m}$ where $K_m = \sum_{j=0}^{m-1} u_j$. We observe that $w_j \geq w_{j+1}$ and hence the weights are in decreasing order, the more recent the data the more the associated weight. Furthermore $\sum_{j=1}^{m-1} w_j = 1$. We also observe that $K_m = \sum_{j=0}^{m-1} u_j = 1 - e^{-\lambda m}$

Using these weights we get

$$\bar{a} = \sum_{j=0}^{m-1} w_j y_j = \frac{1}{K_m} \sum_{j=0}^{m-1} e^{-j\lambda}(1 - e^{-\lambda}) y_j$$

Letting $\alpha = (1 - e^{-\lambda})$ we get $\bar{a} = \dfrac{1}{K_m} \sum_{j=0}^{m-1} \alpha e^{-j\lambda} y_j$.

Consider now the updation process. Let y_{-1} be the next reading and let $\bar{\bar{a}}$ be the new smoothed value. We see that

$$\bar{\bar{a}} = w_0 y_{-1} + \sum_{j=0}^{m-2} w_{j+1} y_j = \frac{u_0}{K_m} y_{-1} + \frac{1}{K_m} \sum_{j=0}^{m-2} u_{j+1} y_j$$

Since $u_j = \alpha e^{-j\lambda}$ then $u_{j+1} = \alpha e^{-(j+1)\lambda} = e^{-\lambda} u_j$ and after some calculation we can show

$$\bar{\bar{a}} = \frac{u_0}{K_m} y_{-1} + \frac{1}{K_m} \sum_{j=0}^{m-2} u_j e^{-\lambda} y_j$$

$$\overline{\overline{a}} = \frac{u_0}{K_m}y_{-1} + \frac{1}{K_m}\sum_{j=0}^{m-2} u_j e^{-\lambda}y_j + \frac{e^{-\lambda}u_{m-1}y_m}{K_m} - \frac{e^{-\lambda}u_{m-1}y_1}{K_m}$$

$$\overline{\overline{a}} = e^{-\lambda}\,\overline{a} + w_0\,y_{-1} - e^{-\lambda}\,w_{m-1}y_{m-1}$$

At this point we shall make a change in notation in the above which allows an easier calculation. We denote $\overline{a} = a_t$, $\overline{\overline{a}} = a_{t+1}$, $y_{-1} = x_{t+1}$ and $y_{m-1} = x_{t-m+1}$ this gives us

$$a_{t+1} = e^{-\lambda}\,a_t + w_0\,x_{t+1} - e^{-\lambda}\,w_{m-1}\,x_{t-m+1}$$

$$a_{t+1} = e^{-\lambda}\,a_t + \frac{u_0}{K_m}\,x_{t+1} - e^{-\lambda}\,\frac{u_{m-1}}{K_m}\,x_{t-m+1}$$

Since $u_0 = (1 - e^{-\lambda})$ and $u_{m-1} = e^{-(m-1)\lambda}(1 - e^{-\lambda})$ and $K_m = 1 - e^{-\lambda m}$ and we have denoted $(1 - e^{-\lambda}) = \alpha$ then we get $a_{t+1} = (1 - \alpha)\,a_t + \dfrac{\alpha}{1 - e^{-\lambda m}}(x_{t+1} - e^{-\lambda m}x_{t-m+1})$. Since $1 - e^{-\lambda} = \alpha$ then $e^{-\lambda} = 1 - \alpha$ and hence $e^{-\lambda m} = (1 - \alpha)^m = \overline{\alpha}^{\,m}$. From this we see

$$a_{t+1} = \overline{\alpha}a_t + \frac{\alpha}{1 - \overline{\alpha}^{\,m}}(x_{t+1} - \overline{\alpha}^{\,m} x_{t-m+1})$$

This begins to look like a truncated type exponential smoothing. Actually we see that if $m \to \infty$ then $\overline{\alpha}^{\,m} \to 0$ and we get the classical exponential smoothing formula

$$a_{t+1} = \overline{\alpha}\,a_t + \alpha\,x_{t+1}$$

We also observe that as λ gets bigger $e^{-\lambda}$ gets smaller and α gets bigger.

Here then we have a smoothing method with two parameters, λ and m. For which when $m \to \infty$ we get the classical exponential smoothing.

10 Conclusion

We discussed the process of predicting future values of a variable from past observations. This process, often referred to as times series smoothing, involve an aggregation of the past observations to predict the future values. Our objective here was to use recent advances in computational intelligence to suggest new and better approaches for performing the necessary aggregations. We first looked at some special features associated with the types of aggregations needed in times series smoothing. We showed how these requirements impact on our choice of weights in the aggregations. We then noted the connection between the method of aggregation used in times series smoothing and that used in the intelligent type aggregation method known as the Ordered Weighted Averaging (OWA) operator. We then took advantage of this connection to allow us to simultaneously view the problem from a times series smoothing perspective and OWA aggregation operations perspective. Using this multiple view we drew upon the large body of work on families of OWA operators to suggest families for the aggregation of time series data. A particularly notable result of this linkage was the introduction of the use of linear decaying weights for time series data smoothing

Acknowledgement. This work has been supported by a Multidisciplinary University Research Initiative (MURI) grant (Number W911NF-09-1-0392) for "Unified Research on Network-based Hard/Soft Information Fusion", issued by the US Army Research Office (ARO) under the program management of Dr. John Lavery. This work has also been supported by an ONR grant for "Human Behavior Modeling Using Fuzzy and Soft Technologies", award number N000141010121. We gratefully appreciate this support.

References

[1] Brown, R.G.: Smoothing, Forecasting and Prediction of Discrete Time Series. Prentice-Hall, Englewood Cliffs (1963)

[2] Box, G.E.P., Jenkins, G.M., Reinsel, G.C.: Time Series Analysis: Forecasting and Control. John Wiley, New York (2008)

[3] Newbold, P.: Statistics for Business and Economics. Prentice Hall, Upper Saddle River (1995)

[4] Yager, R.R.: On ordered weighted averaging aggregation operators in multi-criteria decision making. IEEE Transactions on Systems, Man and Cybernetics 18, 183–190 (1988)

[5] Yager, R.R., Kacprzyk, J.: The Ordered Weighted Averaging Operators: Theory and Applications. Kluwer, Norwell (1997)

[6] Yager, R.R.: Time series smoothing and OWA aggregation. IEEE Transactions on Fuzzy Systems 16, 994–1007 (2008)

[7] Beliakov, G., Pradera, A., Calvo, T.: Aggregation Functions: A Guide for Practitioners. Springer, Heidelberg (2007)

[8] Dubois, D., Prade, H.: A review of fuzzy sets aggregation connectives. Information Sciences 36, 85–121 (1985)

[9] Hyndman, R., Koehler, A.B., Ord, J.K., Snyder, R.D.: Forecasting with Exponential Smoothing: The State Space Approach. Springer, Berlin (2008)

[10] Yager, R.R.: Quantifier guided aggregation using OWA operators. International Journal of Intelligent Systems 11, 49–73 (1996)

[11] O'Hagan, M.: Using maximum entropy-ordered weighted averaging to construct a fuzzy neuron. In: Proceedings 24th Annual IEEE Asilomar Conf. on Signals, Systems and Computers, Pacific Grove, CA, pp. 618–623 (1990)

[12] Yager, R.R., Filev, D.P.: Induced ordered weighted averaging operators. IEEE Transaction on Systems, Man and Cybernetics 29, 141–150 (1999)

[13] Klir, G.J.: Uncertainty and Information. John Wiley & Sons, New York (2006)

[14] Yager, R.R.: Expansible measures of specificity. International Journal of General Systems (to appear)

[15] Ahn, B.S.: The uncertain OWA aggregation with weighting functions having a constant level of orness. International Journal of Intelligent Systems 21, 469–483 (2006)

[16] Ahn, B.S.: On the properties of OWA operator weights functions with constant level of orness. IEEE Transactions on Fuzzy Systems 14, 511–515 (2006)

Chapter 4
Financial Fuzzy Time Series Models Based on Ordered Fuzzy Numbers

Adam Marszałek and Tadeusz Burczyński

Abstract. The purpose of this chapter is to present an original concept of financial fuzzy time series models based on financial data in the form of Japanese Candlestick Charts. In this approach the Japanese Candlesticks are modeled using Ordered Fuzzy Numbers (OFN) called further Ordered Fuzzy Candlesticks (OFC). The use of ordered fuzzy numbers allows modeling uncertainty associated with financial data. Thanks to well-defined arithmetic of ordered fuzzy numbers, one can construct models of fuzzy time series, such as e.g. an autoregressive process, where all input values are OFC, while the coefficients and output values are arbitrary OFN, in the form of classical equations, without using rule-based systems. Finally, several applications of these models for modeling and forecasting selected financial time series are presented.

1 Introduction

It is hard to disagree with opinion that among all different sources of data, the financial market is the most uncertain. The main reason is the fact that huge amount of information is reflected in the financial market. What more, we can say that everything that happens in the world (e.g. in economy, politics) has an effect on quotations of financial instruments. On the other hand, how the information influence the market is decided by investors by taking a long or short position in the market.

Adam Marszałek · Tadeusz Burczyński
Cracow University of Technology, Institute of Computer Science,
Computational Intelligence Department, Warszawska 24, 31-155 Cracow, Poland
e-mail: amarszalek@pk.edu.pl

Tadeusz Burczyński
Silesian University of Technology,
Department for Strength of Materials and Computational Mechanics,
Konarskiego 18A, 44-100 Gliwice, Poland
e-mail: tadeusz.burczynski@polsl.pl

W. Pedrycz & S.-M. Chen (Eds.): Time Series Analysis, Model. & Applications, ISRL 47, pp. 77–95.
DOI: 10.1007/978-3-642-33439-9_4   © Springer-Verlag Berlin Heidelberg 2013

The investors can be simple divided into two groups. The first group of investors decides using fundamental analysis, while the second group decides on a basis technical analysis. Both groups must make a subjective assessment of macroeconomic factors and signals of technical analysis, respectively, so the human factor is a cause of uncertainty as well.

The second group of investors very often uses price charts analysis to make decisions. The price charts (e.g. Japanese Candlestick chart) are used to illustrate movements in the price of a financial instrument over time. Notice, that using the price chart, a large part of the information about the process is lost, e.g. using Japanese Candlestick chart with one hour frequency, for one hour, we know only four prices, while in this time the price must have changed hundreds of times.

In this paper we propose fuzzy logic (i.e. ordered fuzzy numbers), to model uncertainty associated with financial data and reduce the size of lost information. Further, we show how the concept (OFC) can be used to build models of financial time series.

2   Financial Data

In this work as a financial data we mean the quotations of financial instruments (e.g. stock prices or currency pair). Making investment decisions based on observation of each single quotation is very difficult or even impossible, when price changes tens times a minute.

In practice, quotations of financial instruments are represented using price charts [12]. The open-high-low-close chart (also OHLC chart, or simply bar chart) and Japanese Candlestick are most often used in technical analysis. Both types of charts are presented in Figs. 1 and 2, respectively.

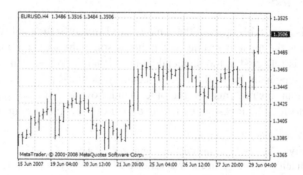

Fig. 1 Open-High-Low-Close chart of EUR/USD, four hour frequency

Each bar represents the range of price movement over a given time interval. In both types of charts, bars are described by only four prices from given time period: first (open), highest, lowest and last (close) price at a given time interval. In addition, Japanese Candlestick has a body, whose color illustrates the relationship between

Fig. 2 Japanese Candlestick chart of EUR/USD, four hour frequency

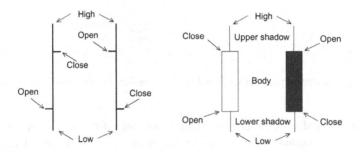

Fig. 3 The long and short OHLC bar, and long and short Japanese Candlestick

the opening and closing price. If the Candlestick closed higher than it opened, the body is white or unfilled, else the body is black. The formation of OHLC bar and Japanese Candlestick are shown in Fig. 3. More details about the Japanese Candlesticks and trading techniques based on them can be found in [13].

3 Ordered Fuzzy Numbers

One of many ways of uncertainty modeling is an approach based on fuzzy logic. Fuzzy data analysis requires also fuzzy arithmetic. Applications of classical fuzzy numbers (sets) [17, 18] or so-called (L, R)-numbers with two shape functions L and R [1] lead to some drawbacks that concern properties of fuzzy algebraic operations, as well as produce unexpected and uncontrollable results when using these operations in an iterative way [16, 17]. In the series of papers [4, 5, 6, 7, 8], W. Kosiński et al. introduced and developed main concepts of the space of ordered fuzzy numbers (OFN), whose arithmetic eliminates these drawbacks.

3.1 Definition of Ordered Fuzzy Number

The concept of membership functions has been weakened by requiring a mere *membership relation*. Consequently, *an ordered fuzzy number A* is identified with an ordered pair of continuous real functions defined on the interval $[0,1]$, i.e. $A = (f,g)$ with $f,g\colon [0,1] \to \mathbb{R}$.

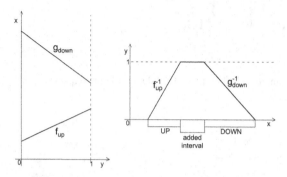

Fig. 4 Graphical interpretation of OFN and an OFN presented as fuzzy number in classical meaning

Functions f and g are called the *up* and *down*-parts of the fuzzy number A, respectively. The continuity of both parts implies their images are bounded intervals, say *UP* and *DOWN*, respectively. In general, the functions f and g need not be invertible, and only continuity is required. If we assume, however, that these functions are monotonous, i.e., invertible, and add the constant function of x on the interval $[1_A^-, 1_A^+]$ with the value equal to 1, we might define the membership function

$$\mu(x) = \begin{cases} f^{-1}(x) & \text{if } x \in [f(0), f(1)], \\ g^{-1}(x) & \text{if } x \in [g(1), g(0)], \\ 1 & \text{if } x \in [1_A^-, 1_A^+], \end{cases} \tag{1}$$

if f is increasing and g is decreasing, and such that $f \le g$ (pointwise). In this way, the obtained membership function $\mu(x)$, $x \in \mathbb{R}$ represents a mathematical object which resembles a convex fuzzy number in the classical sense. The ordered fuzzy number and ordered fuzzy number as a fuzzy number in classical meaning are presented in Fig. 4.

Let us note that a pair of continuous functions (f,g) determines different ordered fuzzy number than the pair (g,f). It follows from the fact that we are dealing with an ordered pair of functions. In this way, we specified an extra feature to this object, named the *orientation*. In graphical interpretation of the ordered fuzzy number, orientation is presented by arrow. Depending on the orientation, the ordered fuzzy numbers can be divided into two types: *a positive orientation*, if the direction of ordered fuzzy number is consistent with the direction of the axis Ox and *a negative orientation*, if the direction of the ordered fuzzy number is opposite to the direction of the axis Ox, as shown in Fig. 5.

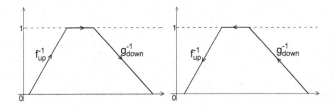

Fig. 5 Positively and negatively oriented OFN

3.2 Operations

The basic arithmetic operations on ordered fuzzy numbers are defined as the pairwise operations of their elements.

Let $A = (f_A, g_A)$, $B = (f_B, g_B)$ and $C = (f_C, g_C)$ are mathematical objects called ordered fuzzy numbers. The sum $C = A + B$, subtraction $C = A - B$, product $C = A \cdot B$, and division $C = A \div B$ are defined by formula

$$f_C(y) = f_A(y) * f_B(y), \qquad g_C(y) = g_A(y) * g_B(y) \qquad (2)$$

where $*$ works for $+$, $-$, $\cdot$ and $\div$, respectively, and where $C = A \div B$ is defined, if the functions $|f_B|$ and $|g_B|$ are bigger than zero. In a similar way, if we want to multiply an ordered fuzzy number A by a scalar $\lambda \in \mathbb{R}$, then the product $C = \lambda \cdot A$ is defined by formula

$$f_C(y) = \lambda \cdot f_A(y), \qquad g_C(y) = \lambda \cdot g_A(y) \qquad (3)$$

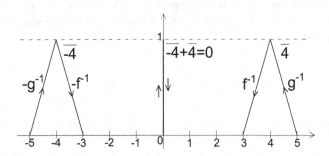

Fig. 6 Sum of two opposite ordered fuzzy numbers

Notice that the subtraction of B is the same as the addition of the opposite of B, i.e. the number $(-1) \cdot B$. If we will do $B + (-1) \cdot B$ we get a numeric zero, i.e., an ordered fuzzy number represented by the pair of constant functions equal to zero. In a similar way, the inverse $1/B$ of an ordered fuzzy number B is defined as an ordered fuzzy number such that the product $B \cdot (1/B)$ gives a number, i.e.,

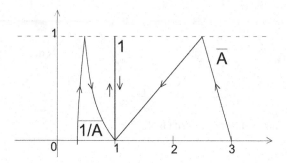

Fig. 7 Product of ordered fuzzy number and its inverse

an ordered fuzzy number represented by the pair of constant functions equal to one. This is presented in Figs. 6 and 7, respectively.

The existence of neutral elements of addition and multiplication is the most important advantage for our further consideration. This fact causes that not always the result of an arithmetic operation is a fuzzy number with a larger support. This allows to build fuzzy models based on ordered fuzzy numbers in the form of the classical equations without losing the accuracy.

3.3 Defuzzification of Ordered Fuzzy Number

Let $\mathcal{O}$ be a universe of all ordered fuzzy numbers. $\mathcal{O}$ can be identified with $\mathscr{C}^0([0,1]) \times \mathscr{C}^0([0,1])$, hence the space $\mathcal{O}$ is a Banach space [7]. A class of defuzzification operators of ordered fuzzy numbers can be defined, as a linear and continuous functionals on the Banach space $\mathcal{O}$, thanks to the general representation theorem (of Banach-Kakutami-Riesz) they are uniquely determined by a pair of Radon measures (v_1, v_2) on $[0,1]$, as

$$Def(A) = \int_0^1 f_A dv_1 + \int_0^1 g_A dv_2 \qquad (4)$$

where $Def(A)$ is the value of a defuzzification operator at the ordered fuzzy number $A = (f_A, g_A)$.

The above formula gives a continuum of defuzzification operators, both linear and nonlinear, which map ordered fuzzy numbers into reals. For example, the standard defuzzification procedure in terms of the area under membership relation can be defined. It is realized by a linear combinations of two Lebesgue measures of $[0,1]$.

4 Ordered Fuzzy Candlesticks

The aim of our research is to find a new tool for modeling of financial data. We want to make it so easy as classical Japanese Candlesticks for observing by investors. At the same time to allow for modeling of uncertainty associated with financial data and also keep more information about the prices than Japanese Candlestick. Ordered fuzzy numbers presented previously, in a simple way, satisfy our requirements.

Generally, in this approach, further as Japanese Candlestick is identified with ordered fuzzy number and it is called *Ordered Fuzzy Candlestick* (OFC). The general idea is presented in Fig. 8. Notice, that the orientation of the ordered fuzzy number shows whether the ordered fuzzy candlestick is long or short. While the information about movements in the price are contained in the shape of the f and g functions. In the following sections we will show how the ordered fuzzy candlestick can be constructed.

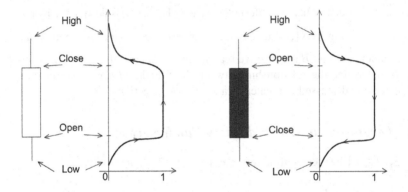

Fig. 8 The Japanese Candlesticks presented as a ordered fuzzy numbers

4.1 Proposal of Global Definition of Ordered Fuzzy Candlestick

Let $\{X_t : t \in T\}$ be a given time series and $T = \{1, 2, \ldots, n\}$. The ordered fuzzy candlestick is defined as an ordered fuzzy number $C = (f, g)$ which satisfies the following properties 1 - 4 or 5 - 8.

Long Candlestick

1. $X_1 \leq X_n$

2. $f : [0, 1] \to \mathbb{R}$ is continuous and increasing on $[0, 1]$

3. $g : [0, 1] \to \mathbb{R}$ is continuous and decreasing on $[0, 1]$

4. $S_1 < S_2$, $f(1) = S_1$, $f(0) = \min_{t \in T} X_t - C_1$, $g(1) = S_2$ and $g(0)$ such that

$$\frac{\int_0^1 g(y)dy - S_2}{A} = \frac{S_1 - \int_0^1 f(y)dy}{B} \qquad (5)$$

Short Candlestick

5. $X_1 > X_n$

6. $f: [0,1] \to \mathbb{R}$ is continuous and decreasing on $[0,1]$

7. $g: [0,1] \to \mathbb{R}$ is continuous and increasing on $[0,1]$

8. $S_1 < S_2$, $f(1) = S_2$, $f(0) = \max_{t \in T} X_t + C_2$, $g(1) = S_1$ and $g(0)$ such that

$$\frac{\int_0^1 f(y)dy - S_2}{A} = \frac{S_1 - \int_0^1 g(y)dy}{B} \qquad (6)$$

The center of ordered fuzzy candlestick (i.e. added interval) is designated by parameters $S_1, S_2 \in \left[\min_{t \in T}, \max_{t \in T}\right]$, while C_1 and C_2 are arbitrary nonnegative real numbers. The parameters A and B are positive real numbers, and together with equations (5) and (6) determine the relationship between the function f and g. A selection of parameters are discussed in greater detail in the next section.

4.2 Parameters of Ordered Fuzzy Candlesticks

Let $\{X_t : t \in T\}$ be a given time series and $T = \{1,2,\ldots,n\}$.

Parameters S_1 and S_2
For to designate the center of the ordered fuzzy candlestick, we can use the average of time series X_t. There are many types of average, the most popular ones are

Simple Average

$$SA = \frac{1}{n}(X_1 + X_2 + \cdots + X_n) \qquad (7)$$

Linear Weighted Average

$$LWA = \frac{X_1 + 2X_2 + \cdots + nX_n}{1 + 2 + \cdots + n} \qquad (8)$$

Exponential Average

$$EA = \frac{(1-\alpha)^{n-1}X_1 + (1-\alpha)^{n-2}X_2 + \cdots + (1-\alpha)X_{n-1} + X_n}{(1-\alpha)^{n-1} + (1-\alpha)^{n-2} + \cdots + (1-\alpha) + 1}, \quad \alpha = \frac{2}{n+1} \qquad (9)$$

Consequently we propose the following

$$S_1, S_2 \in \{SA, LWA, EA\} \text{ such that } S_1 \leq S_2$$

Determination of parameters A and B
For parameters A and B the following formula is proposed

$$A = 1 + S^{+S_2} \quad \text{and} \quad B = 1 + S^{-S_1}$$

where S^{+S_2} and S^{-S_1} means that one of the sums from numerator in the formulas (7), (8) or (9), calculated only for $X_t \geq S_2$ and $X_t \leq S_1$, respectively. These parameters shows how much the movement is concentrated above and below parameters S_1 and S_2, respectively. If formula (8) or (9) is selected then we assume that the more recent time series values are more important than the past ones, which is a natural assumption in financial processes.

Parameters C_1 and C_2
The parameters C_1 and C_2 are defined as a standard deviation of X_t

$$C_1 = C_2 = \sigma_{X_t}$$

4.3 Special Types of Ordered Fuzzy Candlesticks

In this section, some simple types of ordered fuzzy candlesticks are presented.

Trapezoid OFC
Suppose that f and g are linear functions in form

$$f(y) = (f(1) - f(0))y + f(0) \tag{10}$$

$$g(y) = (g(1) - g(0))y + g(0) \tag{11}$$

then the ordered fuzzy candlestick $C = (f, g)$ is called *a trapezoid OFC*, especially if $S_1 = S_2$ then also can be called *a Triangular OFC*.

Let X_t be a given time series. Suppose that $X_1 \leq X_n$ then we have

$$f(y) = (S_1 - \min X_t + C_1)y + \min X_t - C_1 \tag{12}$$

$$g(y) = (S_2 - g(0))y + g(0) \tag{13}$$

where

$$g(0) = \frac{A}{B}(S_1 - \min X_t + C_1) + S_2 \tag{14}$$

Whereas if $X_1 > X_n$ then we have

$$f(y) = (S_2 - \max X_t + C_2)y + \max X_t + C_2 \tag{15}$$

$$g(y) = (S_1 - g(0))y + g(0) \tag{16}$$

where

$$g(0) = \frac{B}{A}(S_2 - \max X_t - C_2) + S_1 \tag{17}$$

Gaussian OFC

The ordered fuzzy candlestick $C = (f,g)$ where the membership relation has a shape similar to the Gaussian function is called *a Gaussian OFC*. It means that f and g are given by functions

$$f(y) = f(z) = \sigma_f \sqrt{-2\ln(z)} + m_f \tag{18}$$

$$g(y) = g(z) = \sigma_g \sqrt{-2\ln(z)} + m_g \tag{19}$$

where e.g. $z = 0.99y + 0.01$.

Let X_t be a given time series. Suppose that $X_1 \le X_n$ then we have

$$f(z) = \sigma_f \sqrt{-2\ln(z)} + m_f \text{ where } m_f = S_1, \ \sigma_f = \frac{\min X_t - C_1 - S_1}{\sqrt{-2\ln(0.01)}} \le 0 \tag{20}$$

$$g(z) = \sigma_g \sqrt{-2\ln(z)} + m_g \text{ where } m_g = S_2, \ \sigma_g = -\frac{A}{B}\sigma_f \tag{21}$$

Whereas if $X_1 > X_n$ then we have

$$f(z) = \sigma_f \sqrt{-2\ln(z)} + m_f \text{ where } m_f = S_2, \ \sigma_f = \frac{\max X_t + C_1 - S_2}{\sqrt{-2\ln(0.01)}} \ge 0 \tag{22}$$

$$g(z) = \sigma_g \sqrt{-2\ln(z)} + m_g \text{ where } m_g = S_1, \ \sigma_g = -\frac{B}{A}\sigma_f \tag{23}$$

4.4 Experimental Studies

Let X_t be a given time series of quotations of EUR/USD for the 1-hour period ending 09.01.2011 at 7pm (236 ticks). The time series X_t and its histogram are presented in Fig. 9. For time series X_t we have $X_0 = X_{235} = 1.2894$, so this Japanese Candlestick has no body. It is so-called Doji Candlestick. Assume that $S_1 = EA = 1.28972$, $S_2 = SA = 1.28986$ and $C_1 = C_2 = \sigma_{X_t} = 2.23e^{-7}$. The exponential average was used in the calculation of parameters A and B of the ordered fuzzy candlesticks, so we have $A = 60.32825$ and $B = 70.83852$. The classical Japanese Candlestick, Trapezoid OFC and Gaussian OFC for time series X_t are presented in Fig. 10.

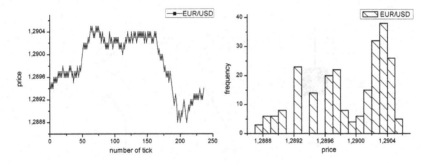

Fig. 9 The tick chart and histogram of EUR/USD

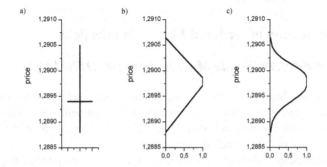

Fig. 10 Types of Candlesticks for time series X_t: a) classical Japanese Candlestick, b) Trapezoidal OFC, c) Gaussian OFC

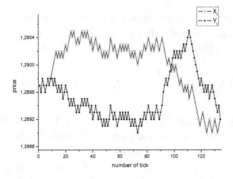

Fig. 11 The tick charts of the time series X_t and Y_t

Now, the two different time series X_t and Y_t are presented in Fig. 11. Both have the same Japanese Candlestick (see Fig. 12a), because the main prices (i.e. OHLC) are the same. However, the ordered fuzzy candlesticks for time series X_t and Y_t presented in Figs. 12b and 12c are different. Therefore, we can conclude that the ordered fuzzy candlesticks effectively contain more information than classical Japanese Candlesticks.

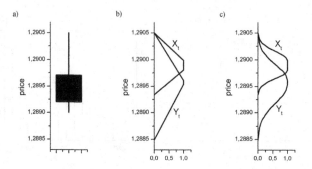

Fig. 12 Types of Candlesticks for time series X_t and Y_t: a) Classical Japanese Candlestick, b) Trapezoid OFC, c) Gaussian OFC

5 An Application of Ordered Fuzzy Candlesticks

5.1 Ordered Fuzzy Simple Moving Average (OFSMA(s))

Ordered fuzzy candlesticks can be used e.g. to construct a fuzzy version of simple technical indicators (i.e. a indicators that require only arithmetic operations such as addition, subtraction and multiplication by a scalar). The Simple Moving Average is presented as an example of technical indicator.

The classical Simple Moving Average with order s at a time period t is given by formula

$$SMA_t(S) = \frac{1}{s}(X_t + X_{t-1} + \cdots + X_{t-s+1}) \tag{24}$$

where X_t is the observation (real) at a time period t (e.g. closing prices) [12].

Now, the Ordered Fuzzy Simple Moving Average with order s at a time period t is also given by formula (24) but the observations X_t are OFC, i.e.

$$OFSMA_t(S) = \frac{1}{s}(\bar{X}_t + \bar{X}_{t-1} + \cdots + \bar{X}_{t-s+1}) \tag{25}$$

where $\bar{X}_t$ is the ordered fuzzy candlestick at a time period t. The process of fuzzification of the other simple technical indicators can be done by analogy.

Notice, if A and B are Trapezoidal (Gaussian) ordered fuzzy candlesticks and $\lambda \in \mathbb{R}$ then ordered fuzzy candlesticks $C = A + B$, $C = A - B$ and $C = \lambda \cdot A$ are Trapezoidal (Gaussian) OFC as well. Moreover, if their functions are in the form of following expressions

$$\phi(y; a, b) = ay + b, \quad \text{for Trapezoid OFC} \tag{26}$$

$$\psi(y; \sigma, m) = \psi(z; \sigma, m) = \sigma\sqrt{-2\ln(z)} + m, \quad \text{for Gaussian OFC} \tag{27}$$

then we have

$$\phi(y;a_1,b_1) \pm \phi(y;a_2,b_2) = \phi(y;a_1 \pm a_2, b_1 \pm b_2)$$
$$\psi(y;\sigma_1,m_1) \pm \psi(y;\sigma_2,m_2) = \psi(y;\sigma_1 \pm \sigma_2, m_1 \pm m_2) \tag{28}$$

$$\lambda \cdot \phi(y;a,b) = \phi(y;\lambda \cdot a, \lambda \cdot b)$$
$$\lambda \cdot \psi(y;\sigma,m) = \psi(y;\lambda \cdot \sigma, \lambda \cdot m) \tag{29}$$

This causes that the numerical implementation of these operations is much simpler.

Empirical Results

The practical case study was performed on data from FOREX market. The data covering the period of 93 hours from 5pm of 09.01.2011 till 2pm of 14.01.2011 of quotations of EUR/USD. The data set included 65376 ticks and is presented in Fig. 13. The classical Japanese Candlestick chart of 1 hour frequency for the set data is shown in Fig. 14. The result of fuzzification of each Candlestick by Gaussian OFC is presented in Fig. 15 by a triangle symbols. The triangles correspond to the value of the function f and g for values 0, 0.5 and 1. Moreover, if an OFC is long then the triangles are pointing straight up, otherwise down.

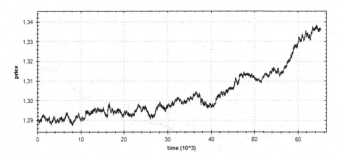

Fig. 13 Tick chart of the data set

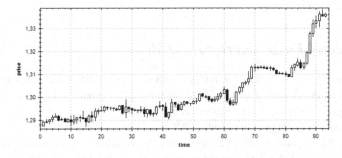

Fig. 14 Tick chart and Japanese Candlestick chart of the data set

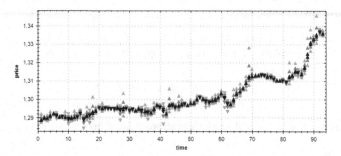

Fig. 15 The ordered fuzzy candlestick chart of the data

Fig. 16 shows results of realization of classical (line with xcross symbol) and ordered fuzzy (triangle symbols) simple moving average with order equal to 7 for the data set. Fig. 16 also shows the ordered fuzzy simple moving average defuzzification by center of gravity operator (line with circle symbol). In technical analysis the moving average indicator usually is used to define the current trend. Notice that the ordered fuzzy moving average determines the current trend by orientation of ordered fuzzy candlesticks, if orientation is positive then trend is long else trend is short.

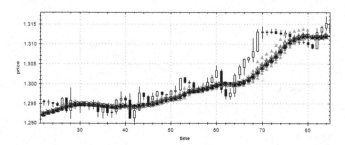

Fig. 16 The Japanese Candlestick chart of the data set, realization of a classical and ordered fuzzy simple moving average

5.2 Ordered Fuzzy Autoregressive Model

In a similar way as it is shown in the previous section we construct fuzzy financial time series models based on ordered fuzzy numbers and candlesticks. In this section, the autoregressive process is presented as an example.

An classical autoregressive model $(AR(p))$ is one where the current value of a variable, depends upon only the values that the variable took in previous periods plus an error term [15]. The presented approach, an ordered fuzzy autoregressive model of order p, denoted as $OFAR(p)$, in natural way is fully fuzzy $AR(p)$ and can be expressed as

$$\bar{X}_t = \bar{\alpha}_0 + \sum_{i=1}^{p} \bar{\alpha}_i \bar{X}_{t-i} + \bar{\varepsilon}_t \qquad (30)$$

where $\bar{X}_{t-i}$ are the ordered fuzzy candlesticks at a time period t, $\bar{\alpha}_i$ are fuzzy coefficients given by arbitrary ordered fuzzy numbers and $\bar{\varepsilon}_t$ is an error term.

Fuzzy Coefficients and Their Estimation
The assumption that fuzzy coefficients of $OFAR$ are arbitrary ordered fuzzy numbers requires the ability to approximate all possible shapes of functions f and g, and the ability to perform arithmetic operations on them.

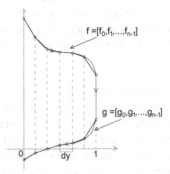

Fig. 17 Discretization of ordered fuzzy numbers

Furthermore, if we multiply the functions of the same class (e.g. linear) we get a function of another class, so the output of $OFAR$ must be represented by arbitrary ordered fuzzy numbers as well. The simplest of solutions, discretization and approximation using the linear function is proposed by us and presented in Fig. 17. Then the arithmetic operations are performed on individual points.

The Least Squares Method is proposed for estimation fuzzy parameters $\bar{\alpha}_i$ in $OFAR(p)$ model. Rearranging the terms in (30) we obtain

$$\bar{\varepsilon}_t = \bar{X}_t - \left(\bar{\alpha}_0 + \sum_{i=1}^{p} \bar{\alpha}_i \bar{X}_{t-i} \right) \qquad (31)$$

From a least-square perspective, the problem of estimation then becomes

$$\min \sum_t \bar{\varepsilon}_t^2 = \min \sum_t \left(\bar{X}_t - \bar{\alpha}_0 - \sum_{i=1}^{p} \bar{\alpha}_i \bar{X}_{t-i} \right)^2 \qquad (32)$$

However, the error term $\bar{\varepsilon}_t$ is the ordered fuzzy number so we do not know what equation (32) mean. Therefore, the least-square method is defined using a distance measure. The measure of the distance between two ordered fuzzy numbers is expressed by formula

$$d(A,B) = d((f_A, g_A), (f_B, g_B)) = \|f_A - f_B\|_{L^2} + \|g_A - g_B\|_{L^2} \qquad (33)$$

where $\| \cdot \|$ is a metric induced by the L^2-norm. Hence, the least-square method for $OFAR(p)$ is to minimize the following objective function

$$E = \sum_t d\left(\bar{X}_t, \bar{\alpha}_0 + \sum_{i=1}^{p} \bar{\alpha}_i \bar{X}_{-i}\right) \qquad (34)$$

So-defined function does not guarantee that received coefficients will be ordered fuzzy numbers, so we have to control coefficients in the course of estimation.

Empirical Results

For the case study the empirical data was the same as in Section 5.1. but was divided into two sets, the first 80 candlesticks are used for estimation, while the next 13 candlesticks are used to evaluate the quality of prediction. The empirical results of several types of ordered fuzzy autoregressive processes are presented.

Model 1

First, in Fig. 18 we can see the realization of a classical autoregressive process with order 4, where the variables X_t are selected prices. On the left side we can see $AR(4)$ of close prices, while on the right side we can see $AR(4)$ of average of OHLC prices. Estimation of $AR(4)$ processes was performed in statistical applications Eviews. We can notice, that the ordered fuzzy autoregressive process is natural generalization of the classical autoregressive process in the space of ordered fuzzy numbers. Assume that all coefficients and input values are numbers (i.e. ordered fuzzy numbers, where functions f and g are equal and constant), then the processes $OFAR$ and AR are equivalent (i.e. give the same results). For the set data, it is presented in Fig. 19.

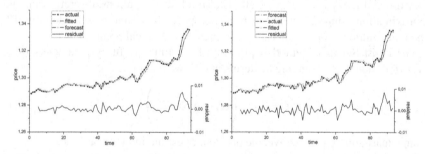

Fig. 18 Realization and static forecast of classical $AR(4)$ processes of close prices and average of OHLC prices, respectively

Model 2

Now, assume that the coefficients still are numbers, while input values are ordered fuzzy candlesticks. Then $OFAR$ can be identified with the vector autoregressive model (VAR) and we can use Eviews for estimation coefficients. The realization $OFAR(4)$ are presented in Fig. 20. In Fig. 20 are shown also defuzzification values of $OFAR(4)$ received by the center of gravity operator (black line).

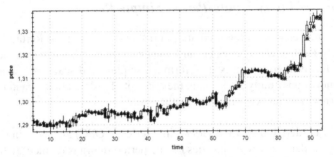

Fig. 19 Realization and static forecast of $OFAR(4)$ (triangle symbols) and classical $AR(4)$ (plus and star symbols)

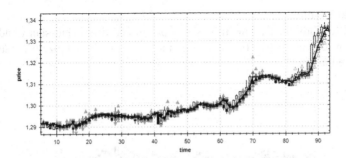

Fig. 20 Realization and static forecast of $OFAR(4)$ with assumption from Model 2 (triangle symbols) with Gaussian OFC and defuzzification values of $OFAR(4)$ (black line)

Model 3

Finally, assume that the coefficients are ordered fuzzy numbers and input values are ordered fuzzy candlesticks. In this case the realization of $OFAR(4)$ are presented in Fig. 21. In Fig. 21 are shown also defuzzification values of $OFAR(4)$ (black line).

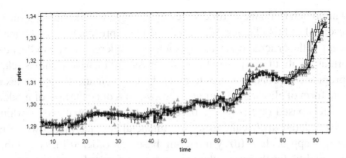

Fig. 21 Realization and static forecast of $OFAR(4)$ with assumption from Model 3 (triangle symbols) with Gaussian OFC and defuzzification values of $OFAR(4)$ (black line)

5.3 Ordered Fuzzy Single-Period Simple Return

In financial studies more often used are returns, instead of prices. Return series are easier to handle than price series because they have more attractive statistical properties and for average investors they form a complete and scale-free summary of the investment opportunity [15]. Using the concept of ordered fuzzy candlestick the fuzzy return series can be defined in a natural way.

Let $\bar{X}_t$ be a ordered fuzzy time series (time series of OFC) given by time series of prices. Then ordered fuzzy time series of one period return is defined by following formula

$$\bar{R}_t = \frac{\bar{X}_t - \bar{X}_{t-1}}{\bar{X}_{t-1}} \qquad (35)$$

Empirical Results

For the case study we take the time series of ordered fuzzy Gaussian Candlestick obtained in section 5.1 (see Fig. 15). The time series of ordered fuzzy simple return is presented in Fig. 22.

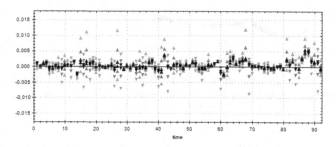

Fig. 22 Ordered fuzzy one period return series

6 Conclusions

The novel approach to financial time series modeling based on ordered fuzzy numbers is presented in this chapter. We described the representation of financial data using concept of the ordered fuzzy candlestick. The ordered fuzzy candlestick keeps more information about the prices than the classical Japanese Candlestick. Moreover, the proposed approach enables to build the fuzzy financial time series models in the simple form of classical equations. It allows to reduce the size of models compared to models based on fuzzy rule-based systems. It is too early to evaluate the usefulness of ordered fuzzy candlesticks in financial engineering, however one can expect that this approach to fuzzy modeling based on ordered fuzzy numbers will bring a new quality. Furthermore, the time series of ordered fuzzy return presented in section 5.3 can be used in the most interesting area of financial modeling, i.e. modeling of volatility. Results of further experiments to validate this approach will be reported on in the future.

References

1. Dubois, D., Prade, H.: Operations on fuzzy numbers. Int. J. System Science 9, 576–578 (1978)
2. Kao, C., Chyu, C.-L.: Least-squares estimates in fuzzy regression analysis. European Journal of Operational Research 148, 426–435 (2003)
3. Kosiński, W., Piechór, K., Prokopowicz, K., Tyburem, K.: On algorithmic approach to operations on fuzzy numbers. In: Burczyński, T., Cholewa, W. (eds.) Methods of Artificial Intelligence in Mechanics and Mechanical Engineering, pp. 95–98. PACM, Gliwice (2001)
4. Kosiński, W., Prokopowicz, P., Ślęzak, D.: Drawback of fuzzy arithmetic - New intuitions and propositions. In: Burczyński, T., Cholewa, W., Moczulski, W. (eds.) Proc. Methods of Artificial Intelligence, pp. 231–237. PACM, Gliwice (2002)
5. Kosiński, W., Prokopowicz, P., Ślęzak, D.: On algebraic operations on fuzzy numbers. In: Klopotek, M., Wierzchoń, S.T., Trojanowski, K. (eds.) Intelligent Information Processing and Web Mining, Proc. Int. Symp. IIS: IIPWM 2003, Zakopane, Poland, pp. 353–362. Physica Verlag, Heidelberg (2003)
6. Kosiński, W., Prokopowicz, P., Ślęzak, D.: Ordered fuzzy numbers. Bull. Polish Acad. Sci., Ser. Sci. Math. 51(3), 327–338 (2003)
7. Kosiński, W., Prokopowicz, P.: Algebra of fuzzy numbers. Matematyka Stosowana. Matematyka dla Spoeczestwa 5(46), 37–63 (2004) (in Polish)
8. Kosiński, W.: On soft computing and modelling. Image Processing Communications 11(1), 71–82 (2006)
9. Lee, C.L., Liu, A., Chen, W.: Pattern Discovery of Fuzzy Time Series for Financial Prediction. IEEE Trans. on Knowledge and Data Engineering 18(5) (2006)
10. Łachwa, A.: Fuzzy World of Sets, Numbers, Relations, Fazts, Rules and Decisions. EXIT, Warsaw (2001) (in Polish)
11. Łęski, J.: Neuro-fuzzy systems. WNT, Warsaw (2008) (in Polish)
12. Murphy, J.J.: Technical Analysis of the Financial Markets. New York Institute of Finance, New York (1999)
13. Nison, S.: Japanese Candlestick Charting Techniques. New York Institute of Finance, New York (1991)
14. Tanaka, H., Uejima, S., Asia, K.: Linear regression analysis with Fuzzy model. IEEE Trans. Systems Man. Cybernet. 12, 903–907 (1982)
15. Tsay, R.S.: Analysis of Financial Time Series, 2nd edn. John Wiley & Sons, Inc., Hoboken (2005)
16. Wagenknecht, M.: On the approximate treatment of fuzzy arithmetics by inclusion, linear regression and information content estimation. In: Chojcan, J., Łęski, J. (eds.) Fuzzy Sets and Their Applications, pp. 291–310. Silesian University of Technology Press, Gliwice (2001)
17. Wagenknecht, M., Hampel, R., Schneider, V.: Computational aspects of fuzzy arithmetic based on Archimedean t-norms. Fuzzy Sets Syst. 123(1), 49–62 (2001)
18. Zadeh, L.A.: The concept of a linguistic variable and its application to approximate reasoning, Part I. Inf. Sci. 8(3), 199–249 (1975)

Chapter 5
Stochastic-Fuzzy Knowledge-Based Approach to Temporal Data Modeling

Anna Walaszek-Babiszewska and Katarzyna Rudnik

Abstract. In the chapter an advanced fuzzy modeling method has been presented which can be useful in temporal data analysis. The method joints fuzzy and probabilistic approaches. The notions of the stochastic process with fuzzy states, and linguistic random variable have been defined to create a knowledge representation of the SISO and MISO dynamic systems. As the basic description of the stochastic process with fuzzy states observed at fixed moments, the joint probability distribution of n linguistic random variables has been assumed. The joint, conditional and marginal probability distributions of the stochastic process with fuzzy states valuate weights of particular rules of the knowledge rule base. Also, the probability distributions determine the probabilistic structure of the particular steps of the tested process. A mean fuzzy conclusion (prediction) can be calculated by the proposed inference procedure.

The implemented knowledge-based system, which creates the knowledge base with optimal number of elementary rules, has been also presented. The optimization method uses a fast algorithm to find fuzzy association rules as a process of automatic knowledge base extraction.

Two examples illustrate the presented methods of the knowledge base extraction from different numeric time series.

1 Introduction

In the topic literature there are many different approaches to time series modeling. The main distinguish can be made between statistical and fuzzy methods. Statistical methods are well known in econometrics and in control theory areas and there are many identification methods of the models. The key role in ordering of the statistical methods in time series modeling has played the work by Box and

Anna Walaszek-Babiszewska · Katarzyna Rudnik
Opole University of Technology, Opole, Poland
e-mail: {a.walaszek-babiszewska,k.rudnik}@po.opole.pl
        a.walaszekbabiszewska@gmail.com

W. Pedrycz & S.-M. Chen (Eds.): Time Series Analysis, Model. & Applications, ISRL 47, pp. 97–118.
DOI: 10.1007/978-3-642-33439-9_5   © Springer-Verlag Berlin Heidelberg 2013

Jenkins [2]. Also fuzzy approaches or more general, neuro-fuzzy and genetic-fuzzy approaches to time series modeling have a long history and a very large literature. Certain review of trends in fuzzy models and identification methods the interested reader can find e.g. in works [11] and [8].

It is often assumed, that temporal data collected from many objects of human activities constitute realizations of stochastic processes. Since the complete description of the stochastic process needs calculations of the series of nD probability distributions [3, 5, 12], many types of models have been invented, which are sufficient under the specific assumptions. Time series models are well known as the models of the specific realizations of time-discrete stochastic processes. In the fuzzy systems theory, the fuzzy representations of time-discrete stochastic processes are known in forms of the linguistic rule-based models, as well as, the Takagi-Sugeno-Kang (TSK) fuzzy models with equations at the consequent parts of rules [8, 10, 19].

In the chapter we present the method of temporal data analysis, which joints fuzzy and probabilistic approaches. The notions of the stochastic process with fuzzy states [18], and linguistic random variable have been defined. As the basic description of the stochastic process with fuzzy states observed at fixed moments $t_1, t_2, ..., t_n$, the joint probability distribution of n linguistic random variables has been assumed. The joint, conditional and marginal probability distributions of the stochastic process with fuzzy states determine respective weights of particular rules of the knowledge base. Also, the probability distributions are used to determine the probabilistic structure of the particular steps of the tested process and to calculate a mean fuzzy conclusion (prediction) by the proposed inference procedure.

We also present the implementation of the knowledge-based system [13], which creates the probabilistic-fuzzy knowledge base with the optimal number of elementary rules. The optimization method uses a fast algorithm to find fuzzy association rules as a process of automatic knowledge base extraction.

Exemplary calculations are presented with results derived by using the implemented knowledge-based system and chosen numeric time series.

2 Stochastic Process with Fuzzy States

2.1 Introduction

According to the theory of stochastic processes, a family of time dependent random variables (dependent on a real parameter t), denoted as

$$\{X(t, \omega),\ X \in \chi, t \in T, \omega \in \Omega\}, \tag{1}$$

is defined as stochastic process (shortly written as $X(t)$), where $\chi \subseteq R$ is a domain of the process values, $T \subset R$ is a domain of parameter t, and Ω is an elementary events domain.

For each given t, $X(t) = X_t$ is a random variable and $F_t(x) = P\{X_t < x\}$ is a distribution function of X_t.

For any given set of parameter values, $\{t_1, t_2, ..., t_n\}$, stochastic process $X(t)$ is determined by n-D probability distribution function

$$F_{t_1, t_2, ..., t_n}(x_1, x_2, ..., x_n) = P\{X(t_1) < x_1, ..., X(t_n) < x_n\}$$

$$= P\{X_{t_1} < x_1, ..., X_{t_n} < x_n\}. \qquad (2)$$

Stochastic process is fully determined by a family of all n-D probability distribution functions, where $n=1,2,...$ For any given elementary event $\omega' \in \Omega$, the function $x(t) = X(t, \omega')$ is a realization (trajectory) of the stochastic process $X(t)$ [3, 5, 12].

2.2 One Dimensional Probability Distribution of the Stochastic Process with Fuzzy States

Let $X(t)$ denotes a stochastic process, a family of time dependent random variables, taking its values in $\chi \subset R$, $t \in T \subset R$. Let $(\chi, \mathcal{B}, p)$ be a probability space, where $\mathcal{B}$ is a σ-field of Borel sets in $\chi \subset R$ and p is a probability measure over $(\chi, \mathcal{B})$.

Let us determine, in the domain of the stochastic process values χ, a linguistic variable which is generated by the process $X(t)$, at fixed t. The linguistic variable is given by quintuple $< X_t, L(X), \chi, G, M >$, where X_t is the name of the variable and $L(X) = \{LX_i\}$, $i=1,2,...,I$ is a collection of its linguistic values. The semantic rule M assigns fuzzy event A_i, $i=1,2,...,I$ to every meaning of LX_i, $i=1,2,...,I$ [21]. Let also, membership functions $\mu_{A_i}(x): \chi \to [0,1]$ be Borel measurable and meet

$$\sum_{i=1}^{I} \mu_{A_i}(x) = 1, \ \forall x \in \chi. \qquad (3)$$

Then, the collection of linguistic values $L(X) = \{LX_i\}$, $i=1,2,...,I$ and the collection of corresponding fuzzy sets A_i, $i=1,2,...,I$ defined over χ, will be called the *linguistic (fuzzy) states of the stochastic process $X(t)$*.

According to Zadeh's definitions from [20], fuzzy states A_i, $i=1,2,...,I$ of the stochastic process constitute *fuzzy events* in the probability space $(\chi, \mathcal{B}, p)$. Probability of the occurrence the fuzzy state A_i, can be calculated by the following Lebesgue-Stietljes' integral

$$P(A_i) = \int_{x \subseteq \chi} \mu_{A_i}(x) dp, \qquad (4)$$

if the integral exists [20]. The existence of the integral (4) results from the assumption that $\mu_{A_i}(x)$ is a Borel measurable function. If the universal set is a countable collection, $X=\{x_n\}$, $n=1,2,...$, and the probability function is determined for discrete process values $P(X = x_n) = p_n$, such that $\sum_n p_n =1$, then the probability of fuzzy event $A_i = \sum_n (\mu_{A_i}(x_n)/x_n)$, denoted as $P(A_i)$, is defined as

$$P(A_i) = \sum_n \mu_{A_i}(x_n)p_n . \tag{5}$$

One dimensional probability distribution of linguistic values (fuzzy states) of the stochastic process X(t), for any fixed value t, can be defined as a set of probabilities of fuzzy events

$$P(X_t) = \{P(A_i)\}, i = 1,2,...,I , \tag{6}$$

where $P(A_i)$, i=1,2,...,I are determined according to (4) or (5) and the following relationships must be fulfilled [16]:

$$0 \leq P(A_i) \leq 1, i=1,2,...,I ; \quad \sum_{i=1}^{I} P(A_i) = 1 . \tag{7}$$

2.3   nD Probability Distribution of the Stochastic Process with Fuzzy States

One-dimensional probability of the stochastic process is an efficient description for the special type of stochastic processes, so called 'white noise processes'.

To determine a probability description of the stochastic process with fuzzy states for two fixed moments t_1, t_2, let us take into account two random variables $(X(t_1), X(t_2))$, determined in the probability space $(X^2, \mathcal{B}, p)$, where $\chi^2 \subseteq R^2$. Two linguistic random variables (linguistic random vector) (X_{t_1}, X_{t_2}) generated by stochastic process values in χ^2, can be defined. The simultaneous linguistic values $LX_i \times LX_j$, i,j=1,2,...,I and the corresponding collection of fuzzy events $\{A_i \times A_j\}_{i,j=1,...,I}$ can be determined over χ^2 by membership functions $\mu_{A_i \times A_j}(u): \chi^2 \rightarrow [0,1]$, i,j=1,2,...,I. The membership functions $\mu_{A_i \times A_j}(u)$ in the linguistic vector domain, χ^2, should be Borel measurable and fulfill the following relationship:

$$\sum_{i=1}^{I}\sum_{j=1}^{I} \mu_{A_i \times A_j}(u) = 1, \ \forall (u) \in \chi^2 . \tag{8}$$

Then, fuzzy sets $A_i \times A_j$ determined in χ^2 are the *simultaneous fuzzy events* and probability $P(A_i \times A_j)$, is defined according to (4), as follows

$$P(A_i \times A_j) = \int_{u \in \chi^2} \mu_{A_i \times A_j}(u) dp , \tag{9}$$

where

$$\mu_{A_i \times A_j}(u) = T(\mu_{A_i}(x), \mu_{A_j}(x)), \tag{10}$$

in particular

$$\mu_{A_i \times A_j}(u) = \mu_{A_i}(x)\mu_{A_j}(x). \tag{11}$$

If universe χ^2 is a finite set, $\chi^2 = \{(x_k, x_l)\}$, $k=1,...,K$, $l=1,...,L$, then the *probability of simultaneous fuzzy event* $A_i \times A_j$ is determined, according to

$$P(A_i \times A_j) = \sum_{(x_k, x_l) \in \chi^2} p(x_k, x_l)\mu_{A_i \times A_j}(x_k, x_l), \tag{12}$$

where $\{p(x_k, x_l)\}_{k=1,2,...k;l=1,2,...,L}$ is a probability function of the discrete random vector variable $(X(t_1), X(t_2))$, at two fixed moments t_1, t_2.

   The *joint 2D probability distribution of the linguistic values (fuzzy states) of the stochastic process* $X(t)$ is determined by the collection of probabilities of *simultaneous fuzzy events* $A_i \times A_j$

$$P(X_{t_1}, X_{t_2}) = \{P(A_i \times A_j)\}_{\cdot i, j=1,2,...,I} , \tag{13}$$

if the following relationships are fulfilled

$$0 \le P(A_i \times A_j) \le 1, \ \forall i, j = 1,...,I \ \text{ and } \ \sum_{i=1}^{I}\sum_{j=1}^{I} P(A_i \times A_j) = 1. \tag{14}$$

To determine the *nD probability distribution* of the stochastic process with fuzzy states, assume first, that stochastic process $X(t)$, for a set of moments $t_1,...,t_n$ is represented by a random vector $(X(t_1),..., X(t_n))$ and $(\chi^n, \mathcal{B}, p)$ is a probability space. Let the linguistic variables

$$< X_{t_1}, L(X), \chi, G, M >,...,< X_{t_n}, L(X), \chi, G, M > \tag{15}$$

be generated by the stochastic process in the domain X. The same sets of the linguistic values

$$L(X_{t_1}) = ... = L(X_{t_n}) = L(X) = \{LX_i\}_{i=1,...,I} , \tag{16}$$

for particular linguistic variables are represented by fuzzy sets A_i, $i=1,...,I$, with membership functions, $\mu_{A_i}(x): \chi \to [0,1]$.

Let the random linguistic vector variable whose name is determined by a vector $(X_{t_1},...,X_{t_n})$ *takes simultaneous linguistic values*

$$L(X)^n = \{LX_{i_1} \times LX_{i_2} \times ... \times LX_{i_n}\}, \ \forall i_1,...,i_n = 1,...,I , \tag{17}$$

whose meanings are represented by the collection of *simultaneous fuzzy events* (*fuzzy states*)

$$\{ (A_{i_1} \times ... \times A_{i_n}) \}, \ \forall i_1,...,i_n = 1,...,I . \tag{18}$$

Fuzzy events (18) are determined on χ^n by membership functions $\mu_{A_{1i} \times ... \times A_{i_n}}(u)$, $u \in \chi^n$, which are Borel measurable and fulfill the relationship

$$\sum_{i_1=1}^{I} ... \sum_{i_n=1}^{I} \mu_{A_{i_1} \times ... \times A_{i_n}}(u) = 1, \forall u \in \chi^n. \tag{19}$$

Let also probabilities of the simultaneous fuzzy events (18), calculated according to (4) or (5), respectively, exist and fulfill the relationships

$$0 \le P(A_{i_1} \times ... \times A_{i_n}) \le 1, \ \forall i_1,...,i_n = 1,...,I ; \tag{20}$$

$$\sum_{i_1=1}^{I} ... \sum_{i_n=1}^{I} P(A_{i_1} \times ... \times A_{i_n}) = 1. \tag{21}$$

Then, *nD joint probability distribution of linguistic values (fuzzy states) of the stochastic process X(t) at moments* $t_1,...,t_n$ is a probability distribution of linguistic vector variable $(X_{t_1},...,X_{t_n})$, determined by the following collection of probabilities of the simultaneous fuzzy events [16]

$$P(X_{t_1},...,X_{t_n}) = \{P(A_{i_1} \times ... \times A_{i_n})\}_{i_1=1,...,I;...;i_n=1,...,I} . \tag{22}$$

In the nD joint probability distribution of linguistic values of the stochastic process $X(t)$ we can distinguish rD, $r<n$ marginal probability distributions, e.g.

$$P(X_{t_1},...,X_{t_{n-1}}) = \{\sum_{i_n=1}^{I} P(A_{i_1} \times ... \times A_{i_{n-1}} \times A_{i_n})\}_{i_1=1,...,I;...;i_{n-1}=1,...,I} , \tag{23}$$

as well as, conditional probability distributions. Generally, nD joint probability distribution of linguistic values of the stochastic process $X(t)$ can be expressed by using marginal and conditional probability distributions in a way

$$P(X_{t_n}, X_{t_{n-1}}, ..., X_{t_2}, X_{t_1}) = P(X_{t_n} / X_{t_{n-1}}, ..., X_{t_2}, X_{t_1}) P(X_{t_{n-1}}, ..., X_{t_2}, X_{t_1}) =$$
$$P(X_{t_n} / X_{t_{n-1}}, ..., X_{t_2}, X_{t_1}) P(X_{t_{n-1}} / X_{t_{n-2}}, ..., X_{t_2}, X_{t_1}) .. P(X_{t_2} / X_{t_1}) P(X_{t_1}). \tag{24}$$

2.4 Fuzzy Mean Value of the Stochastic Process with Fuzzy States

Let the stochastic process $X(t)$ takes its linguistic values $L(X)=\{LX_i\}$, $i=1,2,...,I$, which are represented by fuzzy events A_i, $i=1,2,...,I$ in χ. Let the probability distribution of the fuzzy states, $P(X_t) = \{P(A_i)\}, i = 1,2,..., I$ exists. Then, a *fuzzy mean value of the stochastic process with fuzzy states*, denoted as $\overline{A}(X)$, is a fuzzy set determined as

$$\overline{A}(X) = \sum_{i=1}^{I} A_i P(A_i), \quad \forall x \in \chi, \tag{25}$$

and the membership function is calculated as follows:

$$\mu_{\overline{A}}(x) = \sum_{i=1}^{I} \mu_{A_i}(x) P(A_i), \quad \forall x \in \chi. \tag{26}$$

3 Fuzzy Knowledge Base of the Stochastic Systems

Fuzzy rule based models of dynamic systems are being used not only when knowledge about the real system functioning is incomplete but also when the fuzzy rule based model has to approximate the real system characteristics when the system is too complex or nonlinear. Those models are well known and they are described in the subject literature, eg. in [8, 19]. They are often connected with algorithms of clustering or evolving algorithms.

The novelty in the propositions implemented into model known in subject literature through this work, is the model validation by the probability distributions determined by empirical data.

Defining the fuzzy knowledge base for stochastic environment, it is necessary to make some assumptions about the possibility of existing multidimensional probability distributions of stochastic processes realizations observed in long time intervals. Usually, we assume also ergodicity and stationarity of the processes.

3.1 Fuzzy SISO Model of the Stochastic Process

Let $X(t)$ be a stochastic process with fuzzy states, as it was shown in paragraph 2. Assuming that the process was observed at two fixed moments $t_1, t_2 \in T$; $t_2 > t_1$, the process realizations have been used to calculate $2D$ empirical probability distributions of fuzzy states.

The fuzzy knowledge representation of the stochastic process is a collection of the following weighted file rules, in the form [18]:

$$\forall A_i \in L(X), i=1,...,I$$

$$R^{(i)}: w_i[If\ (X_{t_1}\ is A_i)]\ Then\ (X_{t_2}\ is\ A_1)w_{1/i}$$

$$- - - - - - - - - - - -$$

$$Also\,(X_{t_2}\ is\ A_j)w_{j/i} \qquad\qquad , \qquad (27)$$

$$- - - - - - - - - - -$$

$$Also\,(X_{t_2}\ is\ A_J)w_{J/i}$$

where $A_i, A_j \in L(X), i.j = 1,2,...,I$ denote the fuzzy states of the process, and weights

$$w_i = P(X_{t_1} = A_i), i=1,2,...,I \qquad (28)$$

are the probabilities of fuzzy events at the antecedents of the rules (marginal probability distribution), and weights

$$w_{j/i} = P[(X_{t_2} = A_j)/(X_{t_1} = A_i)], j=1,2,...,I; \ i=const \qquad (29)$$

are the conditional probabilities of the fuzzy events at the consequent part of the rules (conditional probability distribution). According to the probability distribution features, the following relationships are fulfilled

$$\sum_{i=1,...,I} w_i = 1 \ \sum_{j=1,...,J} w_{j/i} = 1.$$

The model can be also presented as a collection of the elementary weighted rules

$$\forall A_i \in L(X), \ \forall A_j \in L(X), i,j=1,2,...,I$$

$$R^{(i,j)}: w_{ij}[If\ (X_{t_1}\ is\ A_i)\ Then\ (X_{t_2}\ is\ A_j)], \qquad (30)$$

where

$$w_{ij} = P(X_{t_1}, X_{t_2}) = P(A_i \times A_j), i,j=1,2,...,I \qquad (31)$$

is a joint probability of fuzzy events in the rule (joint probability distribution) and the following relationship must be fulfilled

$$\sum_{i=1,\dots,I} \sum_{j=1,\dots,J} w_{ij} = 1.$$

The above propositions of the knowledge representation contain weights: w_i, $w_{j/i}$, w_{ij}, which stand for the frequency of the occurrence the fuzzy events in particular parts of rules. The weights, real numbers from the interval [0, 1], do not change logic values of the sentences.

Assuming stationarity of the process with fuzzy states, the prediction of the process can be determined by means of approximate reasoning.

3.2 Inference Procedure (Prediction Procedure) from the SISO Model of the Stochastic Process

For the logic analysis we take into account the following fuzzy relation representing file rule (30)

$$R^{(i)} : A_i \Rightarrow (A_1 \cup \dots \cup A_j \cup \dots \cup A_I), \tag{32}$$

which can be described by membership function

$$\mu_{R^{(i)}}(x_{t_1}, x_{t_2}) = \mu_{A_i \Rightarrow (A_1 \cup \dots \cup A_I)}(x_{t_1}, x_{t_2}), i=1,\dots,I. \tag{33}$$

To consider the *prediction procedure*, which is based on well-known procedure of approximate reasoning (e.g. in [8, 10, 19]), let us assume the crisp value of the stochastic process at moment t_1, $X(t_1) = x_{t_1}^*$.

Then the level of activation of the elementary rule is determined as

$$\tau_i = \mu_{A_i}(x_{t_1}^*), i=1,\dots,I \tag{34}$$

and the fuzzy value of the conclusion $(X_{t_2} \text{ is } A'_{j/i})$, computed e.g. based on Larsen's rule of reasoning, is a fuzzy set $A'_{j/i}$, determined by its membership function

$$\mu_{A'_{j/i}}(x_{t_2}) = \tau_i \mu_{A_j}(x_{t_2}), j=1,\dots,I; i=\text{const}. \tag{35}$$

The fuzzy conditional expected value (fuzzy conditional mean value) of the output of *i-th* rule $E\{(X_{t_2} \text{ is } \varphi(A_j))/[X(t_1) \text{ is } A_i]\} = A'_i$, stands for the aggregated outputs of elementary rules, $j=1,\dots,J$, according to the formula [16]

$$\mu_{A'_i}(x_{t_2}) = \sum_j w_{j/i} \mu_{A'_{j/i}}(x_{t_2}). \tag{36}$$

The   fuzzy   expected   value   of   the   prediction, $E\{(X_{t_2}\,is\,\varphi(A_j))/[X(t_1)=x_{t_1}*]\}=A'$, computed as the aggregated outputs of all active *i-th* rules, is determined by the formula [16]

$$\mu_{A'}(x_{t_2})=\sum_i w_i \mu_{A_i'}(x_{t_2})=\sum_i w_i \tau_i \sum_j w_{j/i}\mu_{A_j}(x_{t_2}).\tag{37}$$

The prediction according to the generalized Mamdani-Assilian's type interpretation of fuzzy models gives us the following conclusion

$$\mu_{A'_{j/i}}(x_{t_2})=T(\tau_i,\mu_{A_j}(x_{t_2})),\,j=1,\ldots,I;\,i=\text{const}.\tag{38}$$

Prediction determined by using the logic type interpretation of fuzzy models, gives us the following relationships, instead of (35) or (38):

$$\mu_{A'_{j/i}}(x_{t_2})=I(\tau_i,\mu_{A_j}(x_{t_2})),\,j=1,\ldots,I;\,i=const,\tag{39}$$

where T denotes a *t*-norm and I means the implication operator.

The scheme of the prediction procedure from the SISO model is presented in Fig. 1.

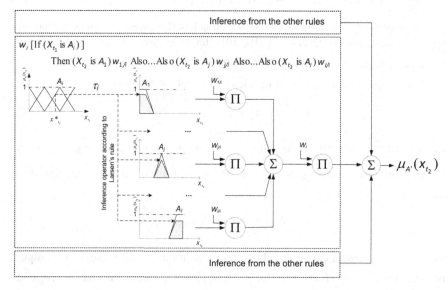

Fig. 1 Scheme of the prediction procedure from the SISO fuzzy model of the stochastic process

3.3 Fuzzy MISO Model of the Long Memory Stochastic Process

Let the stochastic process $X(t)$ with fuzzy states be determined, as it has been shown in paragraph 2. To create the representation of knowledge base in the form of fuzzy *If...Then* rules, the linguistic random variables have been determined with their collection of linguistic values, $\left\{ LX_{k,i_k} \right\}$, the same for each variable. The meanings of linguistic values are represented by fuzzy sets, A_{k,i_k}, $k=1,...,n$; $i_k=1,...,I$ in χ, with the Borel measurable membership functions $\mu_{A_{k,i_k}}(x): \chi \rightarrow [0,1]$, $k=1,...,n$; $i_k=1,...,I$.

The fuzzy sets divide the space χ^n of values of samples into n-D fuzzy areas: $(A_{1,i_1} \times ... \times A_{n,i_n})$, $i_1=1,2,...,I$;...; $i_n=1,2,...,I$.

Assuming that the process was observed at fixed moments $t_1, t_2,..., t_n \in T$; $t_n > t_{n-1} > ... > t_1$, the process realizations have been used to calculate the following empirical probability distributions of fuzzy states:

- nD joined probability distribution of the linguistic random vector variable $\left(X_{t_1},..., X_{t_n} \right)$

$$P(X_{t_1},..., X_{t_n}) = \{ P(A_{1,i_1} \times ... \times A_{n,i_n}) \}_{i_1=1,...,I;...;i_n=1,...,I}, \quad (40)$$

- marginal $(n-1)D$ probability distribution of the linguistic random vector variable (of the antecedent fuzzy events)

$$w_j = P(X_{t_1},..., X_{t_{n-1}}) = \{ P(A_{1,i_1} \times ... \times A_{n-1,i_{n-1}}) \}_{i_1=1,...,I;...;i_{n-1}=1,...,I}, \quad (41)$$

- conditional probability distribution (of the consequent fuzzy events)

$$w_{i_n/j} = P(X_{t_n} / X_{t_1},..., X_{t_{n-1}}) = \{ P[A_{n,i_n} /(A_{1,i_1} \times ... \times A_{n-1,i_{n-1}})] \} \quad (42)$$

$$i_n = 1,2,...,I; i_1, i_2,..., i_{n-1} = const.$$

The MISO fuzzy model, as the knowledge representation of the stochastic process, has the form of the collection $\left\{ R^{(j)} \right\}_{j=1,2,...,J}$ of weighted file rules [16]:

$$\forall A_{1,i_1} \in L(X_{t_1}), \; \forall A_{2,i_2} \in L(X_{t_2}),..., \forall A_{n,i_n} \in L(X_{t_n}),$$

$$i_1, i_2,..., i_n = 1,2,..., I;$$

$$R^{(j)}: w_j [If\, (X_{t_1}\, is\, A_{1,i_1})\, And(X_{t_2}\, is\, A_{2,i_2})\, And...And(X_{t_{n-1}}\, is\, A_{n-1,i_{n-1}})]$$

$$Then(X_{t_n}\, is\, A_{n,1})w_{1/j}$$

$$- - - - - - - - - - - -$$

$$Also(X_{t_n}\, is\, A_{n,i_n})w_{i_n/j} \qquad\qquad (43)$$

$$- - - - - - - - - - - -$$

$$Also(X_{t_n}\, is\, A_{n,I})w_{I/j}$$

$j=1,2,\dots J$; J - number of file rules. The number of file rules, J, depends on the number of fuzzy $(n-1)D$ areas in the input space X^{n-1}, with non-zero probabilities w_i; it can be even $J = I^{n-1}$.

3.4 Prediction Procedure from the MISO Fuzzy Model of the Stochastic Process

Assuming, that the process with fuzzy states is a stationary process, that is, the nD joint probability distribution does not depend on time, we can use the created knowledge representation for the prediction of the process. The conclusion, fuzzy or numeric, determined by means of approximate reasoning represents the prediction of the process. The input data can be fuzzy or numeric in their character.

Let us consider the prediction procedure, assuming crisp data of observations of the process, $X(t_1) = x^*_{t_1}, X(t_2) = x^*_{t_2},\dots, X(t_{n-1}) = x^*_{t_{n-1}}$. Then, the level of activation of j-th rule is determined by the t-norm of membership functions of fuzzy sets in antecedents as follows [8, 10, 19]:

$$\tau_j = T_1\left(\mu_{A_{1,i_1}}(x^*_{t_1}), \mu_{A_{2,i_2}}(x^*_{t_2}),\dots, \mu_{A_{n-1,i_{n-1}}}(x^*_{t_{n-1}})\right), j=1,\dots,J. \tag{44}$$

If the values of the process at moments $t_1, t_2,\dots, t_{n-1}$ are expressed by fuzzy numbers (linguistic values), that is

$$(X_{t_1} is\ A'_{1,i_1})\ And\ (X_{t_2} is\ A'_{2,i_2})\ And\ \dots And\ (X_{t_{n-1}} is\ A'_{n-1,i_{n-1}}),$$

where A'_{k,i_k} are given by the membership functions $\mu_{A'_{k,i_k}}(x_{t_k}): \chi \to [0,1]$, $k=1,2,\dots,n-1$, then the level of activation of j-th rule is expressed as [8, 10, 19]:

$$\tau_j = T_1\left\{\left\{\sup_{x\in\chi}[\mu_{A'_{1,i_1}}(x_{t_1})\wedge\mu_{A_{1,i_1}}(x_{t_1})]\right\},\dots,\left\{\sup_{x\in\chi}[\mu_{A'_{n-1,i_{n-1}}}(x_{t_{n-1}})\wedge\mu_{A_{n-1,i_{n-1}}}(x_{t_{n-1}})]\right\}\right\}. \tag{45}$$

The *fuzzy conclusion*, $X_{t_n} is\ A'_{n,i_n}$, from i_n-th consequent part of the rule can be determined by one of the ways [8, 10, 19]:

• according to Mamdani-Assilian's rule of inference

$$\mu_{A'_{n,i_n}}(x_{t_n}) = \tau_j \wedge \mu_{A_{n,i_n}}(x_{t_n}), \tag{46}$$

• according to Larsen's rule

$$\mu_{A'_{n,i_n}}(x_{t_n}) = \tau_j \mu_{A_{n,i_n}}(x_{t_n}), \tag{47}$$

- according to generalized Mamdani-Assilian's type of interpretation

$$\mu_{A'_{n,i_n}}(x) = T_2\left(\tau_j, \mu_{A_{n,i_n}}(x)\right) \tag{48}$$

- according to the logic interpretation

$$\mu_{A'_{n,i_n}}(x_{t_n}) = I\left(\tau_j, \mu_{A_{n,i_n}}(x_{t_n})\right). \tag{49}$$

The fuzzy conditional expected value of the conclusion A'_{n,i_n}

$$E\left[\left(X_{t_n}\text{ is }A'_{n,i_n}\right)/\left(X_{t_1}\text{ is }A'_{1,i_1}\right)\cap...\cap\left(X_{t_1}\text{ is }A'_{1,i_1}\right)\right] = A'_{n/j} \tag{50}$$

is the aggregated value (weighted sum) of conclusions from particular i_n-th outputs, $i_n=1,...,I$ (calculated according to one of relationships (46) - (49)) and the conditional probabilities of fuzzy events in i_n-th consequents, as follows

$$\mu_{A'_{n/j}}(x_{t_n}) = \sum_{i_n} w_{i_n/j}\mu_{A'_{n,i_n}}(x_{t_n}). \tag{51}$$

Taking into account all j-th active rules, the *fuzzy conditional expected value of the prediction, A_n'*,

$$A_n' = E\left\{\left(X_{t_n}\text{ is }A'_{n/j}\right)/\left(X(t_1) = x^*_{t_1}, X(t_2) = x^*_{t_2},..., X(t_{n-1}) = x^*_{t_{n-1}}\right)\right\} \tag{52}$$

can be calculated as the aggregated value (weighted sum) of conclusions from particular j-th file rules, $A'_{n/j}$, (51), and joint probabilities of fuzzy events in particular antecedents, as follows

$$\mu_{A'_n}(x_{t_n}) = \sum_j w_j\mu_{A'_{n/j}}(x_{t_n}) = \sum_j w_j\sum_{i_n} w_{i_n/j}\mu_{A'_{n,i_n}}(x_{t_n}). \tag{53}$$

Subscript n in A_n' shows, that fuzzy conclusion is the fuzzy value of the linguistic random variable X_{t_n}.

The discussed prediction procedure from the MISO model is presented in Fig. 2.

The numerical value of the prediction, x^*_n, can be determined as the centroid of A_n' calculated e.g. by the COA method

$$x^*_n = \frac{\int\limits_{x\in\chi} x\mu_{A'_n}(x)dx}{\int\limits_{x\in\chi} \mu_{A'_n}(x)dx}. \tag{54}$$

The fuzzy value of the prediction $(X_{t_n} \text{ is } A_n')$ is a function of fuzzy propositions $(X_{t_k} \text{ is } A'_{k,i_k})$, or numerical propositions $X(t_k) = x^*_{t_k}$, for $k=1,2,\ldots,n$-1, on the input of the system, as well as the chosen procedures of inference.

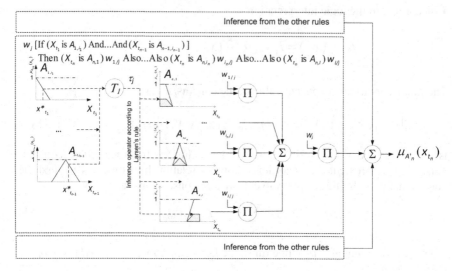

Fig. 2 Scheme of the prediction procedure from the MISO fuzzy model of the stochastic process

3.5 Probability of Fuzzy Predictions

We can also determine the probability of the fuzzy conclusions, fuzzy predictions, derived from the stochastic-fuzzy rule bases of the SISO and the MISO model.

Since the fuzzy conclusion (prediction), determined during the reasoning procedures, is given by its membership function in a domain of the output variable, and the probability distribution $p(x)$ has been determined based on data, then, probability of the fuzzy prediction can be determined by the following formula

$$P\left((X_{t_n} \text{ is } A_n')/(X(t_1) = x^*_{t_1},\ldots,X(t_{n-1}) = x^*_{t_{n-1}})\right)= \int_{x\in\chi}\mu_{A_n'}(x)p(x)dx . \quad (55)$$

4   Conception of the Knowledge-Based Inference System

The knowledge-based systems are usually composed of the following parts [8, 10, 19]:

- knowledge base in the form of if-then rules (43), that contains information essential to solve a given problem,

- fuzzification block that transforms quantitative data into qualitative data represented by fuzzy sets on the bases of membership grades entered in the database,
- inference block that utilizes the database and the implemented aggregation methods and final inference (reasoning) to solve specialized problems,
- defuzzification block that calculates the crisp value (defuzzified value) at the system output on the bases of the resulting membership grades.

The implemented knowledge base contains database and rule base. The database contains information defined by experts on a given application field containing linguistic values of the variables accounted in the rule base and definitions of fuzzy sets identified with these values. On the other hand, knowledge base contains a set of linguistic rules created on the grounds of a modified algorithm generating fuzzy association rules. The algorithm makes it possible to adjust the model to measurement data. The characteristic form of the rules, exposing an empirical probability distribution of fuzzy events enables a simple interpretation of the knowledge contained in the model and additional analysis of the considered problem.

The inference mechanism with multiple inputs and a single output enables the calculation of the membership function of the conclusion, on the bases of the crisp input data, and, in consequence, the defuzzified value of the model output. For the system with the rule base in the form of (43), there are many possible ways of obtaining crisp output results. In this conception of the knowledge-based inference system, we consider the methods presented in chapter 3.4.

4.1   Methods of Fuzzy Knowledge Discovery

The *if-then* rules that constitute the knowledge bases of the fuzzy system may be defined in two ways:

- as logical rules constituting subjective definitions created by experts on the grounds of experience and knowledge of the investigated phenomenon,
- as physical rules constituting objective knowledge models defined on the grounds of observations and natural research into the analyzed process (object) and its regularities.

In the case of fuzzy modeling there were initially logics rules, yet, in consideration of machine learning a hybrid of rules was gradually implemented according to which initial assumptions concerning fuzzy sets and the associated rules are defined following the experts' conviction, whereas other parameters are adjusted to measurement data. The objective of automatic data discovery is to obtain the smallest set of *if-then* rules enabling as accurate representation of the modelled object or phenomenon as possible.

Methods of knowledge discovery for fuzzy systems of Mamdani type include [10, 19]:

- Wang-Mendel method,
- Nozaki-Ishibuchi-Tanaki method,
- Sugeno-Yasukawa method,
- template-based method of modelling fuzzy systems.

In order to obtain databases for fuzzy systems, data mining methods have also been applied.

Data mining, considered as the main stage in knowledge discovery [4] is focused on non-trivial algorithms of searching "hidden", so far unknown and potentially required information [6] and its records in the form of mathematical expressions and models. Some of the data mining methods identify zones in the space of system variables, which, consequently, create fuzzy events in the rules. This may be accomplished by searching algorithm clusters or covering algorithms, also called separate and conquer algorithms. Other methods, for example: fuzzy association rules, are based on constant division for each attribute (fuzzy grid) and each grid element is regarded as a potential component of the rule. As far as the first approach is concerned, each identified rule has its own fuzzy sets [17]. Therefore, from the point of view of rules interpretation, the second approach seems more applicable [9].

4.2 Association Rules as Ways of Fuzzy Knowledge Discovery

Irrespective of automatic knowledge discovery, rules of the fuzzy model are obtained on the bases of their optimal adjustment to experimental data. In view of this, the generation of the rules may be understood as a search for rules with high occurrence frequency, where, the frequency parameter influences the optimal rules adjustment. In such case, fuzzy rules may be analyzed as the co-existence of fuzzy variable values in experimental data, i.e.: fuzzy association rules.

The issue of association rules was first discussed in [1]. Nowadays it is one of the most common data mining methods. In a formal approach, the association rules have the form of the following implications:

$$X \Rightarrow Y \, (s,c), \qquad\qquad (56)$$

where X and Y are separable variable sets (attributes) in the classic approach to mathematical sets, often referred to as: X – conditioning values set, Y- conditioned values set.

Considering the fuzzy rules of association for the MISO model (43), the following may be derived:

$$X \Rightarrow Y :$$

$$(X_{t_1} \text{ is } A_{1,i_1}) And (X_{t_2} \text{ is } A_{2,i_2}) And...And (X_{t_{n-1}} \text{ is } A_{n-1,i_{n-1}}) \Rightarrow X_{t_n} \text{ is } A_{n,1} (s,c) \qquad (57)$$

where A_{k,i_k} , $k=1,...,n;$ $i_k=1,...,I$ denote the fuzzy states of the process.

Each association rule is connected with two statistical measures that determine the validity and power of the rule: *support* (s) – probability of the simultaneous incidence of set ($X \cap Y$) in the set collection and *confidence* (c) – also called credibility which is conditional probability ($P(Y \mid X)$). The issue of discovering fuzzy association rules involves finding, in a given database, all support and trust values that are higher than the association rules the support and trust of which are higher than the defined minimal values of support and trust given by users.

The first application of the association rules was in basket analysis. However, taking into account the fact that rules may include variables that are derived from diverse variables expressed in a natural language, the ranges of the application of the discussed method may be extended to forecasting, decision-making, planning, control etc. In the inference system with stochastic-fuzzy knowledge base, we proposed to use the idea of fuzzy association rules to knowledge discovery.

In the topic literature we can find many algorithms of creating the association rules and modifications [1, 7] but they generate association rules only in non-fuzzy version. To knowledge discovery in the form of (43) two algorithms have been proposed [13, 14, 17]. One is based on the *Apriori* algorithm and the second algorithm uses the *FP-Growth* assumption. In these algorithms the so called frequent fuzzy set is a set of which the probability of the occurrence is bigger than the value of the assumed minimal support s. Thus, the inputs of the proposed algorithm are: set of measurements used for model identification, predefined database (linguistic values of variables considered in the model and definitions of fuzzy sets identified with the values), and the threshold value of minimal support (s). Threshold value of the minimal confidence (c) is not in use. The output of the algorithms is a rule base of a probabilistic-fuzzy knowledge representation. Fig. 3. presents the results of comparison of the generating time of the probabilistic-fuzzy knowledge base as the function of the minimal support value for the modified Apriori and FP-Growth algorithms. The chart presents the advantage of modified FP-Growth algorithm.

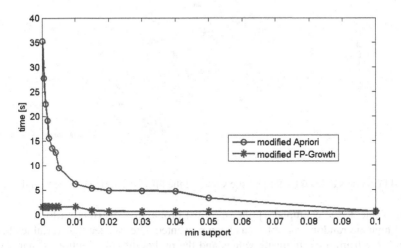

Fig. 3 The time of generating the probabilistic-fuzzy knowledge base as the function of the minimal support value for the modified Apriori and FP-Growth algorithms (3 input variables, one output variable, 5 fuzzy sets for each variable, near 500 learning data)

5 Exemplary Calculations

The created in the work [13] inference system has been applied to predict the values of time varying variables determining the natural phenomena such as wind speed and ash contents (incombustible matter) in row coal. Both variables are very difficult for prediction because of many random rates influence on the measurements results.

The set of 11 000 measurements of the wind speed $X(t)=\{v(t)\}$ $t=1,2,\ldots n$ were recorded at 1-minute samplings. The averages of measurements from 4 steps were researched. First 2000 measurements were treated as learning data, the remaining ones – test data. The forecasts of wind speed $v(t)$ have been made on the grounds of the last three measurements of wind speed denoted as $v(t\text{-}3)$, $v(t\text{-}2)$, $v(t\text{-}1)$. For each variable, in the space of process values, 9 fuzzy sets have been defined, with the linguistic values describing the wind speed, as: "very light", "light", "mild", "moderate", "fairly strong", "strong", "very strong", "squally", "very squally", assuming 45 disjoint intervals of the variables values. Exemplary values of the membership functions for variable $v(t\text{-}3)$ are shown in Fig. 4.

The membership grades for other variables have been analogically defined.

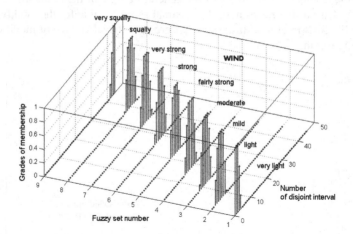

Fig. 4 Fuzzy sets defined for 9 linguistic values of the linguistic variable 'speed wind'

In Table 1. the exemplary joint empirical probability distribution for two chosen linguistic random variables has been presented. We can see that variables take their three from nine linguistic values and the probability distribution is 'narrow', concentrated only over the few linguistic values.

Table 1 Exemplary joint probability distribution of two linguistic random variables (X_t, X_{t-1}) under the conditions: $((X_{t-3}$ *is moderate) and* $(X_{t-2}$ *is moderate))*

Assumption: wind X(t-3) is 'moderate' and wind X(t-2) is 'moderate'

X(t-1) \ X(t)	very light	light	mild	moderate	fairly strong	strong
very light	0	0	0	0	0	0
light	0	0	0	0	0	0
mild	0	0	0,0145	0,0196	0	0
moderate	0	0	0,0223	0,0934	0,0180	0
fairly strong	0	0	0	0,0153	0,0156	0
strong	0	0	0	0	0	0

The optimal model structure is derived at the minimal support value, equal to $s=0.001$, then, the root mean square error for the learning data is 0.5514 m/sec, whereas for the testing data it is 0.6434 m/sec. The model consists of 92 elementary rules (47 file rules). The most important file rules are:

R1: (0.1337) IF ($X(t$-3) IS 'moderate') AND ($X(t$-2) IS 'moderate') AND ($X(t$-1) IS 'moderate')
       THEN ($X(t)$ IS 'moderate') (0.6989)
       ALSO ($X(t)$ IS 'mild') (0.1665)
       ALSO ($X(t)$ IS 'fairly strong') (0.1346)
R2: (0.0973) IF ($X(t$-3) IS 'fairly strong') AND ($X(t$-2) IS 'fairly strong')
    AND ($X(t$-1) IS 'fairly strong')
           THEN ($X(t)$ IS 'fairly strong') (0.6827)
           ALSO ($X(t)$ IS 'moderate') (0.2253)
           ALSO ($X(t)$ IS 'strong') (0.0920)
R3: (0.0749) IF ($X(t$-3) IS 'mild') AND ($X(t$-2) IS 'mild') AND ($X(t$-1) IS 'mild')
           THEN ($X(t)$ IS 'mild') (0.6683)
           ALSO ($X(t)$ IS 'moderate') (0.2131)
           ALSO ($X(t)$ IS 'light') (0.1186)

The results of predicted numeric values of the wind speed and measured data have been presented in Fig. 5.

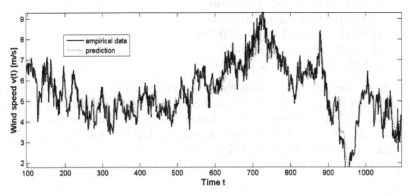

Fig. 5 Comparison of the prediction values and empirical data of wind speed for testing data

The next application of the tested inference system in dynamic system modelling will be shown on the example of coal parameters analysis. Measurements of some coal parameters, as contents of particular density fractions of grains, ash or sulfur contents are the bases of quality control at coal preparation plants and power stations. Sampling research of grain materials are used by technological process engineers to approximate needed coal parameters or characteristics. Sample taking is a random process, according to respective scheme of randomness. In the other hand, experts of technology often express some values of measurements in linguistic categories. These are the reasons that knowledge base of the variation of empirical data concerning coal parameters is created with regards fuzziness and randomness.

The interested reader can find more details on probabilistic-fuzzy modelling characteristics of grain materials in the work [15].

In the example the coegsisting of two variables: the content of light grains fraction, X, and the ash content in that fraction, Y, in the time series has been analyzed. Spaces of considerations of both variables have been divided into 40 disjoint intervals and 7 triangular fuzzy sets have been defined as the representations of linguistic values: {"very small", "small", "medium small", "medium", "medium large", "large", "very large"}, for both variables.

Derived knowledge base for that dynamic system has a form of 779 file rules. The most important rules are presented below:

R1: (0.0487) IF ($X(t$-2) IS 'med. large') AND ($Y(t$-2) IS 'small') AND ($X(t$-1) IS 'med. large') AND ($Y(t$-1) IS 'small') THEN ($Y(t)$ IS 'small') (0.5277)

        ALSO ($Y(t)$ IS 'v. small') (0.2004)
        ALSO ($Y(t)$ IS 'med. small') (0.1954)
        ALSO ($Y(t)$ IS 'medium') (0.0398)
        ALSO ($Y(t)$ IS 'med. large') (0.0257)
        ALSO ($Y(t)$ IS 'large') (0.0102)
        ALSO ($Y(t)$ IS 'v. large') (0.0009)

R2: (0.0247) IF ($X(t$-2) IS 'medium') AND ($Y(t$-2) IS 'small') AND ($X(t$-1) IS 'med. large') AND ($Y(t$-1) IS 'small') THEN ($Y(t)$ IS 'small') (0.4673)

        ALSO ($Y(t)$ IS 'med. small') (0.2454)
        ALSO ($Y(t)$ IS 'v. small') (0.1478)
        ALSO ($Y(t)$ IS 'medium') (0.1293)
        ALSO ($Y(t)$ IS 'med. large') (0.0075)
        ALSO ($Y(t)$ IS 'large') (0.0018)
       ALSO ($Y(t)$ IS 'v. large') (0.0009)

In Table 2. the probability distribution of two linguistic random variables (Y_t, Y_{t-1}), under the conditions: ((X_{t-2} is medium large) and (Y_{t-2} is small) and (X_{t-1} is medium large)), has been presented. It is easy to observe that occurrence of any linguistic value of any variable is possible with a probability grater then zero. This is different distribution then in the first example.

Table 2 Exemplary probability distribution of two linguistic random variables (Y_t, Y_{t-1}) under the conditions: ((X_{t-2} *is medium large*) *and* (Y_{t-2} *is small*) *and* (X_{t-1} *is medium large*))

Assumption: $X(t-2)$ is 'med. large' and $Y(t-2)$ is 'small' and $X(t-1)$ is 'med. large'

$Y(t-1)$ \ $Y(t)$	'v. small'	'small'	'medium small'	'medium'	'medium large'	'large'	'v. large'
'v. small'	0,0047	0,0087	0,0028	0,0002	0,0004	0,0010	0,0002
'small'	0,0097	0,0257	0,0095	0,0019	0,0012	0,0005	4,37E-05
'med. small'	0,0029	0,0063	0,0041	0,0005	0,0002	0,0001	0
'medium'	0,0009	0,0021	0,0008	0,0005	0,0003	9,29E-05	0
'med. large'	0,0003	0,0003	0,0002	0,0001	9,985E-06	0	0
'large'	2,07E-05	0,0005	5,119E-05	0	0	0	0
'v. large'	2,099E-06	9,38E-05	8,362E-05	0	0	0	0

The computed marginal and conditional probability distributions have been used in the prediction procedure. The optimal structure of the model has been derived at the complete probability distribution, by using both Larsen's inference rule and Fodor's t-norm as a representation of the logic AND (see T_1 in chapter 3.4.). Than the root mean square for training data was equal to 0.87, and for testing data 1.85.

6 Conclusions

The use of fuzzy logics in the knowledge-based system makes it possible to express incomplete and uncertain information in a natural language, typical for expression and cognition of human beings. In addition, the application of the probability of events expressed in linguistic categories enables the adjustment of the model on the grounds of numerical information derived from the data stored in the course of the operation of a given real processes. The created model becomes easier for interpretation by its users, what is very important in the strategic decision-making situations, as well as, in diagnostic systems.

References

[1] Agrawal, R., Imielinski, T., Swami, A.: A Mining association rules between sets of items in large databases. In: ACM Sigmod Intern. Conf. on Management of Data, Washington D.C., pp. 207–216 (May 1993)
[2] Box, G.E.P., Jenkins, G.M.: Time Series Analysis; Forecasting and Control. Holden-Day, San Francisco (1976)
[3] Doob, J.L.: Stochastic Processes. Wiley, New York (1953)
[4] Fayyad, U., Piatetsky-Shapiro, G., Smyth, P.: From data mining to knowledge discovery in databases. AI Magazine, 37–54 (1996)

[5] Fisz, M.: Probability and Statistics Theory. PWN, Warsaw (1967) (in Polish)
[6] Frawley, W., Piatetsky-Shapiro, G., Matheus, C.: Knowledge Discovery in Databases: An Overview. AI Magazine, 57–70 (1992)
[7] Han, J., Pei, H., Yin, Y.: Mining Frequent Patterns without Candidate Generation. In: Proc. Conf. Management of Data, SIGMOD 2000, Dallas, TX. ACM Press, New York (2000)
[8] Hellendoorn, H., Driankov, D. (eds.): Fuzzy Model identification, Selected Approaches. Springer, Berlin (1997)
[9] Hüllermeier, E.: Fuzzy methods in machine learning and data mining: status and prospects. Fuzzy Sets and System 156(3), 387–406 (2005)
[10] Łęski, J.: Neuro-Fuzzy Systems. WNT, Warsaw (2008) (in Polish)
[11] Oh, S.-K., Pedrycz, W., Park, K.-J.: Identification of fuzzy systems by means of genetic optimization and data granulation. Journal of Intelligent & Fuzzy Systems 18, 31–41 (2007)
[12] Papoulis, A.: Probability, Random Variables, and Stochastic Processes. WNT, Warsaw (1972) (Polish edition)
[13] Rudnik, K.: Conception and implementation of the inference system with the probabilistic-fuzzy knowledge base. Dissertation, Opole University of Technology, Opole, Poland (2011) (in Polish)
[14] Rudnik, K., Walaszek-Babiszewska, A.: Probabilistic-fuzzy knowledge-based system for managerial applications. Management and Production Engineering Review 3 (in press 2012)
[15] Walaszek-Babiszewska, A.: Fuzzy probability for modelling of particle preparation processes. In: Proc. 4th Int. Conf. Intelligence Processing and Manufacturing of Materials (IPMM 2003), Sendai, Japan (2003)
[16] Walaszek-Babiszewska, A.: IF-THEN Linguistic Fuzzy Model of a Discrete Stochastic System. In: Cader, A., Rutkowski, L., Tadeusiewicz, R., Żurada, J. (eds.) Artificial Intelligence and Soft Computing, pp. 169–174. Academic Publishing House EXIT, Warsaw (2006)
[17] Walaszek-Babiszewska, A., Błaszczyk, K.: A modified Apriori algorithm to generate rules for inference system with probabilistic-fuzzy knowledge base. In: Proc. 7th Int. Workshop on Advanced Control and Diagnosis, November 19-20, CD-ROM, Zielona Góra, Poland (2009)
[18] Walaszek-Babiszewska, A.: Fuzzy Modeling in Stochastic Environment; Theory, knowledge bases, examples. LAP LAMBERT Academic Publishing, Saarbrücken (2011)
[19] Yager, R.R., Filev, D.: Essentials of Fuzzy Modeling and Control. John Wiley and Sons (1994)
[20] Zadeh, L.A.: Probability Measures of Fuzzy Events. Journal of Mathematical Analysis and Applications 23(2), 421–427 (1969)
[21] Zadeh, L.A.: The concept of a linguistic variable and its application to approximate reasoning. Part I. Information Sciences 8, 199–249 (1975); Part II. 8, 301–357, Part III. 9, 43–80

Chapter 6
A Novel Choquet Integral Composition Forecasting Model for Time Series Data Based on Completed Extensional L-Measure

Hsiang-Chuan Liu

Abstract. In this study, based on the Choquet integral with respect to complete extensional L-measure and M-density, a novel composition forecasting model which composed the time series model , the exponential smoothing model and GM(1,1) forecasting model was proposed.   For evaluating this improved composition forecasting model, an experiment with the data of the grain production in Jilin during 1952 to 2007 by using the sequential mean square error was conducted. Based on the M-density and N- density, the performances of Choquet integral composition forecasting model with the completed extensional L-measure, extensional L-measure, L-measure, Lambda-measure and P-measure, respectively, a ridge regression composition forecasting model and a multiple linear regression composition forecasting model and the traditional linear weighted composition forecasting model were compared. The experimental results showed that the Choquet integral composition forecasting model with respect to the completed extensional L-measure and M-density outperforms other ones. Furthermore, for each fuzzy measure, including the completed extensional L-measure, extensional L-measure, L-measure, Lambda-measure and P-measure, respectively, the Choquet integral composition forecasting model based on M-density is better than the one based on N-density.

Keywords: Choquet integral, composition forecasting model, M-density, completed extensional L-measure.

Hsiang-Chuan Liu
Department of Biomedical Informatics, Asia University,
500, Lioufeng Rd. Wufeng, Taichung 41354, Taiwan, R.O.C.
Graduate Institute of Educational Measurement and Statistics,
National Taichung University of Education
140 Min-Shen Road, Taichung 40306 Taiwan R.O.C
e-mail: lhc@asia.edu.tw

W. Pedrycz & S.-M. Chen (Eds.): Time Series Analysis, Model. & Applications, ISRL 47, pp. 119–137.
DOI: 10.1007/978-3-642-33439-9_6   © Springer-Verlag Berlin Heidelberg 2013

1   Introduction

The composition forecasting model is first considered in the work of Bates and Granger (1969) [1]. They are now in a widespread use in many areas, especially in economic field. Zhang Wang and Gao (2008) [2] applied the linear composition forecasting model which composed the time series model, the second-order exponential smoothing model and GM(1,1) forecasting model in the Agricultural Economy Research, the GM(1,1) is one of the most frequently used grey forecasting model, it is a time series forecasting model, encompassing a group of differential equations adapted for parameter variance, rather than a first order differential equation [3-4]. In our previous works [5-9], we extended the work of Zhang, Wang, and Gao by proposing some nonlinear composition forecasting model which also composed the time series model, the second-order exponential smoothing model and GM(1,1) forecasting model by using the ridge regression model [5] and the theory of Choquet integral with respect to some fuzzy measures, including Sugeno's λ-measure [13], Zadeh's P-measure [14] and authors' fuzzy measures, L-measure, extensional L-measure and completed extensional L-measure [6-12]. Since the first two well-known fuzzy measures are univalent measures, each of them has just one feasible fuzzy measure satisfying the conditions of its own definition, but the others proposed by our previous works are multivalent fuzzy measures, all of them have infinitely feasible fuzzy measures satisfying the conditions of their own definition. The fuzzy measure based Choquet integral composition forecasting models are supervised methods, by comparing the mean square errors between the estimated values and the corresponding true values, each of our multivalent fuzzy measures based forecasting models has more opportunity to find the better feasible fuzzy measure, the performances of them are always better than the one based on the univalent fuzzy measures, λ-measure and P-measure. In addition, the author has proved that the P-measure is a special case of the L-measure [7], we know that all of the extended multivalent fuzzy measures of L-measure are at lest as good as their special case P-measure. However, the λ-measure is not a special case of the L-measure, so the improved L-measure, called extensional L-measure, was proposed to contain the λ-measure as a special case [7]. And then, all of the P-measure, λ-measure and L-measure are special cases of the extensional L-measure. However, the extensional L-measure does not attend the largest fuzzy measure B-measure, it is not a completed fuzzy measure, for overcoming this drawback, an improved extensional L-measure, called completed extensional L-measure was proposed, all of other above-mentioned fuzzy measures proposed are the special cases of it. The real data experiment showed that the extensional L-measure Choquet integral based composition forecasting model is the best one. On the other hand, all of above mentioned Choquet integral composition forecasting models with some different fuzzy measures are based on N-density. From the definition of Choquet integral and fuzzy measures, we know that the Choquet integral can be viewed as a function of its fuzzy measure, and the fuzzy measure can be viewed as a function of its fuzzy density function,   therefore, the performance of any Choquet integral is predominate by its fuzzy measure, and the performance of any fuzzy measure is predominate by its fuzzy density function, in other words, the performance of any Choquet integral is predominate by its fuzzy density function. Since the older fuzzy density function N-density is based on the linear correlation coefficient, the new fuzzy density function M-density based on the mean square error is non-linear, the relations among the composition forecasting model and three given forecasting models are non-linear as well, hence, in the same Choquet integral with respect to the same fuzzy measure, the performance of the non-linear fuzzy density functions is always better than the linear fuzzy density functions.

In this paper, a novel fuzzy measure, called the completed extensional L-measure, and the new fuzzy density function, M-density, are considered. Based on the M-density and the proposed completed extensional L-measure, a novel composition forecasting model is also considered. For comparing the forecasting efficiency of two fuzzy densities M-density and N-density, is also considered.

2   The Composition Forecasting Model

In this paper, for evaluating the forecasting validation of forecasting model to sequential data, the sequential mean square error is used, its formal definition is listed as follows.

***Definition 1.*   Sequential Mean Square Error (SMSE) [9-10]**

If θ_{t+j} is the realized value of target variable at time $(t+j)$, $\hat{\theta}_{t+j|t}$ is the forecasted value of target variable at time $(t+j)$ based on training data set from time 1 to time t,

and
$$SMSE\left(\hat{\theta}_t^{(h)}\right) = \frac{1}{h}\sum_{j=1}^{h}\left(\hat{\theta}_{t+j|t+j-1} - \hat{\theta}_{t+j}\right)^2 \tag{1}$$

then $SMSE\left(\hat{\theta}_t^{(h)}\right)$ is called the sequential mean square error (SMSE) of the h forecasted values of target variable from time $(t+1)$ to time $(t+h)$ based on training data set from time 1 to time t. The composition forecasting model or combination forecasting model can be defined as follows.

***Definition 2.*   Composition Forecasting Model [9-10]**
(i) Let y_t be the realized value of target variable at time t.
(ii) Let $x_{t,1}, x_{t,2}, ..., x_{t,m}$ be a set of m competing predictors of y_t, $\hat{y}_t$ be a function f of $x_{t,1}, x_{t,2}, ..., x_{t,m}$ with some parameters, denoted as

$$\hat{y}_t = f\left(x_{t,1}, x_{t,2}, ..., x_{t,m}\right) \tag{2}$$

(iii) Let $x_{t+j|t,k}$ be the forecasted values of y_t by competing predictor k at time $(t+j)$ based on training data set from time 1 to time t, and for the same function f as above,

Let
$$\hat{y}_{t+j|t} = f\left(x_{t+j,1}, x_{t+j,2}, \dots, x_{t+j,m}\right) \tag{3}$$

(iv) Let
$$SMSE\left(\hat{y}_t^{(h)}\right) = \frac{1}{h} \sum_{j=1}^{h} \left(\hat{y}_{t+j|t+j-1} - y_{t+j}\right)^2 \tag{4}$$

$$SMSE\left(x_{t,k}^{(h)}\right) = \frac{1}{h} \sum_{j=1}^{h} \left(x_{t+j,k} - y_{t+j}\right)^2 \tag{5}$$

For current time t and the future h times, if

$$SMSE\left(\hat{y}_t^{(h)}\right) \leq \min_{1 \leq k \leq m} SMSE\left(x_{t,k}^{(h)}\right) \tag{6}$$

then $\hat{y}_t$ is called a composition forecasting model for the future h times of $x_{t,1}, x_{t,2}, \dots, x_{t,m}$ or, in brief, a composition forecasting model of $x_{t,1}, x_{t,2}, \dots, x_{t,m}$.

Definition 3.   Linear Combination Forecasting Model [9-10]

For given parameters $\beta_k \in R, \sum_{k=1}^{m} \beta_k = 1$, let

$$\hat{y}_t = \sum_{k=1}^{m} \beta_k x_{t,k} \tag{7}$$

If $\hat{y}_t$ is a composite forecasting model of $x_{t,1}, x_{t,2}, \dots, x_{t,m}$ then $\hat{y}_t$ is called a linear combination forecasting model or linear composition forecasting model, otherwise, it is called a non-linear combination forecasting model or non-linear composition forecasting model.

Definition 4.   Ridge Regression Composition Forecasting Model [5,9,10]

(i) Let $\underline{y}_t = \left(y_1, y_2, \dots, y_t\right)^T$ be realized data vector of target variable from time 1 to time t, $\underline{x}_{t,k} = \left(x_{1,k}, x_{2,k}, \dots, x_{t,k}\right)^T$ be a forecasted value vector of competing predictor k of target variable y_t from time 1 to time t.

(ii) Let X_t be a forecasted value matrix of m competing predictors of target variable y_t from time 1 to time t.

(iii) Let
$$\underline{\hat{y}}_t = \left(\hat{y}_1, \hat{y}_2, \dots, \hat{y}_t\right)^T \tag{8}$$

$$f\left(X_{t}\right)=f\left(\underline{x}_{t,1},\underline{x}_{t,2},...,\underline{x}_{t,m}\right) \tag{9}$$

(iv) Let $\quad \underline{\beta}_{t}^{(r)}=\left(\beta_{t,1}^{(r)},\beta_{t,2}^{(r)},...,\beta_{t,m}^{(r)}\right)^{T}=\left(X_{t}^{T}X_{t}+rI_{m}\right)^{-1}X_{t}^{T}\underline{y}_{t} \tag{10}$

$$\hat{\underline{y}}_{t}=f\left(X_{t}\right)=X_{t}\underline{\beta}_{t}^{(r)} \tag{11}$$

Then $\qquad \hat{\underline{y}}_{t+j|t}=f\left(X_{t+j}\right)=X_{t+j}\underline{\beta}_{t}^{(r)} \tag{12}$

$$\hat{y}_{t+j|t}=f\left(x_{t+j,1},x_{t+j,2},...,x_{t+j,m}\right)$$
$$=\left[x_{t+j,1},x_{t+j,2},...,x_{t+j,m}\right]\underline{\beta}_{t}^{(r)}=\sum_{k=1}^{m}\beta_{t,k}^{(r)}x_{t+j,k} \tag{13}$$

For current time t and the future h times, if

$$SMSE\left(\hat{y}_{t}^{(h)}\right)\leq\min_{1\leq k\leq m}SMSE\left(x_{t,k}^{(h)}\right) \tag{14}$$

And ridge coefficient $r=0$ then $\hat{y}_{t}$ is called a multiple linear regression combination forecasting model of $x_{t,1},x_{t,2},...,x_{t,m}$. If formula (14) is satisfied and $r>0$, then $\hat{y}_{t}$ is called a ridge regression composition forecasting model of $x_{t,1},x_{t,2},...,x_{t,m}$. Note that Hoerl, Kenard, and Baldwin (1975) suggested that the ridge coefficient of ridge regression is

$$r=\frac{m\hat{\sigma}^{2}}{\underline{\beta}_{t}^{T}\underline{\beta}},\quad \hat{\sigma}^{2}=\frac{1}{t}\sum_{i=1}^{t}\left(y_{i}-\hat{y}_{t}\right)^{2} \tag{15}$$

3   Choquet Integral Composition Forecasting Model

3.1   Fuzzy Measures [6-13]

Definition 5.   Fuzzy Measure [6-13]
A fuzzy measure μ on a finite set X is a set function $\mu:2^{X}\rightarrow[0,1]$ satisfying the following axioms:

$$\mu(\phi)=0,\mu(X)=1 \qquad \text{(boundary conditions)} \tag{16}$$

$$A\subseteq B\Rightarrow\mu(A)\leq\mu(B) \qquad \text{(monotonicity)} \tag{17}$$

3.2   *Fuzzy Density Function [6-10]*

Definition 6.   **Fuzzy Density Function, Density [6-10]**
(i) A fuzzy density function of a fuzzy measure μ on a finite set X is a function $d : X \rightarrow [0,1]$ satisfying:

$$d(x) = \mu(\{x\}), x \in X \qquad (18)$$

$d(x)$ is called the density of singleton x.

(ii) A fuzzy density function is called a normalized fuzzy density function or a density if it satisfying

$$\sum_{x \in X} d(x) = 1 \qquad (19)$$

Definition 7.   **Standard Fuzzy Measure [6-10]**
A fuzzy measure is called a standard fuzzy measure, if its fuzzy density function is a normalized fuzzy density function.

Definition 8.   **N-density [8-10]**
Let μ be a fuzzy measure on a finite set $X = \{x_1, x_2, ..., x_n\}$, y_i be global response of subject i and $f_i(x_j)$ be the evaluation of subject i for singleton x_j, satisfying:

$$0 < f_i(x_j) < 1, i = 1, 2, ..., N, \quad j = 1, 2, ..., n \qquad (20)$$

If
$$d_N(x_j) = \frac{r(f(x_j))}{\sum_{j=1}^{n} r(f(x_j))}, j = 1, 2, ..., n \qquad (21)$$

Where $r(f(x_j))$ is the linear regression coefficient of y_i on $f(x_j)$ satisfying

$$r(f(x_j)) = \frac{S_{y,x_j}}{S_y S_{x_j}} \geq 0 \qquad (22)$$

$$S_y^2 = \frac{1}{N} \sum_{i=1}^{N} \left(y_i - \frac{1}{N} \sum_{i=1}^{N} y_i \right)^2 \qquad (23)$$

$$S_{x_j}^2 = \frac{1}{N} \sum_{i=1}^{N} \left[f_i(x_j) - \frac{1}{N} \sum_{i=1}^{N} f_i(x_j) \right]^2 \qquad (24)$$

$$S_{y,x_j} = \frac{1}{N}\sum_{i=1}^{N}\left(y_i - \frac{1}{N}\sum_{i=1}^{N}y_i\right)\left[f_i(x_j) - \frac{1}{N}\sum_{i=1}^{N}f_i(x_j)\right] \qquad (25)$$

then the function $d_N : X \rightarrow [0,1]$ satisfying $\mu(\{x\}) = d_N(x), \forall x \in X$ is a fuzzy density function, called N-density of μ.

Note that

(i) N-density is a normalized fuzzy density function.

(ii) N-density is a linear fuzzy density function based on linear correlation coefficients

3.3   M-Density [10]

We know that any linear function can be viewed as a special case of some corresponding non-linear function, In this paper, a non-linear fuzzy density function based on Mean Square Error, denoted M-density, is proposed, its formal definition is introduced as follows:

Definition 9.   M-density

Let μ be a fuzzy measure on a finite set $X = \{x_1, x_2, ..., x_n\}$, y_i be global response of subject i and $f_i(x_j)$ be the evaluation of subject i for singleton x_j, satisfying:

$$0 < f_i(x_j) < 1, i = 1, 2, ..., N, \quad j = 1, 2, ..., n \qquad (26)$$

If

$$d_M(x_j) = \frac{\left[MSE(x_j)\right]^{-1}}{\sum_{j=1}^{n}\left[MSE(x_j)\right]^{-1}}, j = 1, 2, ..., n \qquad (27)$$

Where

$$MSE(x_j) = \frac{1}{N}\sum_{i=1}^{N}\left(y_i - f_i(x_j)\right)^2 \qquad (28)$$

then the function $d_M : X \rightarrow [0,1]$ satisfying $\mu(\{x\}) = d_M(x), \forall x \in X$ is a fuzzy density function, and called M-density of μ.

3.4   Classification of Fuzzy Measures [6-10]

Definition 10. Additive measure, sub-additive measure and supper- additive measure

(i)   A fuzzy measure μ is called an sub-additive measure, if

$$\forall A, B \subset X, A \cap B = \phi \Rightarrow g_\mu(A \cup B) < g_\mu(A) + g_\mu(B) \qquad (29)$$

(ii)   A fuzzy measure μ is called an additive measure, if

$$\forall A, B \subset X, A \cap B = \phi \Rightarrow g_\mu (A \cup B) = g_\mu (A) + g_\mu (B) \tag{30}$$

(iii)   A fuzzy measure μ is called a supper-additive measure, if

$$\forall A, B \subset X, A \cap B = \phi \Rightarrow g_\mu (A \cup B) > g_\mu (A) + g_\mu (B) \tag{31}$$

(iv)   A fuzzy measure is called a mixed fuzzy measure, if is not a Additive measure, sub-additive measure and supper- additive measure.

Theorem 1. Let d be a given fuzzy density function of an additive measure, A-measure, then its measure function $g_A : 2^X \rightarrow [0,1]$ satisfies

$$\forall E \subset X \Rightarrow g_A (E) = \sum_{x \in E} d(x) \tag{32}$$

3.4 λ-Measure [13]

Definition 10. λ-measure [13]
For a given fuzzy density function d on a finite set X, $|X| = n$, a measure is called λ-measure, if its measure function, $g_\lambda : 2^X \rightarrow [0,1]$, satisfying:

(i) $$g_\lambda (\phi) = 0, \, g_\lambda (X) = 1 \tag{33}$$

(ii) $$A, B \in 2^X, A \cap B = \phi, A \cup B \neq X$$
$$\Rightarrow g_\lambda (A \cup B) = g_\lambda (A) + g_\lambda (B) + \lambda g_\lambda (A) g_\lambda (B) \tag{34}$$

(iii) $$\prod_{i=1}^{n} \left[1 + \lambda d(x_i) \right] = \lambda + 1 > 0, \; d(x_i) = g_\lambda (\{x_i\}) \tag{35}$$

Theorem 2. Let d be a given fuzzy density function on a finite set X, $|X| = n$,
Under the condition of λ-measure, the equation (35) determines the parameter λ uniquely:

(i) $$\sum_{x \in X} d(x) > 1 \Rightarrow \lambda < 0, \lambda\text{-measure is a sub-additive measure} \tag{36}$$

(ii) $$\sum_{x \in X} d(x) = 1 \Rightarrow \lambda = 0, \; \lambda\text{-measure is an additive measure} \tag{37}$$

(iii) $$\sum_{x \in X} d(x) < 1 \Rightarrow \lambda > 0, \; \lambda\text{-measure is a supper-additive measure} \tag{38}$$

Note that

(i) λ-measure has just one feasible fuzzy measure satisfies the conditions of its own definition.

(ii)   In equation (35), the value of $d(x_i)$ is decided first, and then to find the solution of the measure parameter λ, and $\prod_{i=1}^{n}\left[1+\lambda d(x_i)\right]$ can be viewed as a function of its fuzzy density $d(x_i)$. Therefore, we can say that λ-measure is predominate by its fuzzy density function.

(iii) λ-measure can not be a mixed fuzzy measure.

3.5   P-Measure [14]

Definition 11.   **P-measure [14]**
For a given fuzzy density function d on a finite set X, $|X| = n$, a measure is called P-measure, if its measure function, $g_P : 2^X \to [0,1]$, satisfying:

(i) $$g_P(\phi) = 0,\ g_P(X) = 1 \tag{39}$$

(ii) $$^\forall A \in 2^X \Rightarrow g_P(A) = \max_{x \in A} d(x) = \max_{x \in A} g_P(\{x\}) \tag{40}$$

Theorem 3. **P-measure is always a sub-additive measure [6-10]**

Note that since the maximum of any finite set is unique, hence, P-measure has just one feasible fuzzy measure satisfies the conditions of its own definition.

3.6   Multivalent Fuzzy Measure [6-10]

Definition 12.   **Univalent fuzzy measure, multivalent fuzzy measure [4-8]**
A fuzzy measure is called a univalent fuzzy measure, if it has just one feasible fuzzy measure satisfies the conditions of its own definition, otherwise, it is called a multivalent fuzzy measure.

Note that both λ-measure and P-measure are univalent fuzzy measures.

3.7   L-Measure [6-10]

In my previous work [4], a multivalent fuzzy measure was proposed, which is called L-measure, since my last name is Liu. Its formal definition is as follows

Definition 13.   **L-measure [6-10]**
For a given fuzzy density function d on a finite set X, $|X| = n$, a measure is called L-measure, if its measure function, $g_L : 2^X \to [0,1]$, satisfying:

(i)
$$g_L(\phi) = 0, \; g_L(X) = 1 \tag{41}$$

(ii) $L \in [0,\infty), X \neq A \subset X \Rightarrow g_L = \max_{x \in A} d(x) + \dfrac{\left(|A|-1\right)L\sum\limits_{x \in A}d(x)\left[1-\max\limits_{x \in A}d(x)\right]}{\left[n-|A|+L\left(|A|-1\right)\right]\sum\limits_{x \in X}d(x)} \tag{42}$

Theorem 4. Important Properties of L-measure [6]

(i) For any $L \in [0,\infty)$, L-measure is a multivalent fuzzy measure, in other words, L-measure has infinite fuzzy measure solutions.

(ii) L-measure is an increasing function on L.

(iii) If $L = 0$ then L-measure is just the P-measure.

(iv) L-measure may be a mixed fuzzy measure

Note that

(i)   P-measure is a special case of L-measure

(ii)   L-measure does not contain additive measure and λ-measure, in other words, additive measure and λ-measure are not special cases of L-measure.

3.8   *Extensional L-Measure [7]*

For overcoming the drawback of L-measure, an improving multivalent fuzzy measure which containing additive measure and λ-measure., called extensional L-measure, was proposed by my next previous paper [7],   Its formal definition is as follows;

Definition 14.   **Extensional L-measure, L_E-measure [7]**

For a given fuzzy density function d on a finite set X, $|X| = n$, a measure is called extensional L-measure, if its measure function, $g_{L_E} : 2^X \to [0,1]$, satisfying:

(i)
$$g_{L_E}(\phi) = 0, \; g_{L_E}(X) = 1 \tag{43}$$

$L \in [-1,\infty), A \subset X$

(ii) $\Rightarrow g_{L_E}(A) = \begin{cases} (1+L)\sum\limits_{x \in A}d(x) - L\max\limits_{x \in A}d(x) & , L \in [-1,0] \\[2em] \sum\limits_{x \in A}d(x) + \dfrac{\left(|A|-1\right)L\sum\limits_{x \in A}d(x)\left[1-\sum\limits_{x \in A}d(x)\right]}{\left[n-|A|+L\left(|A|-1\right)\right]\sum\limits_{x \in X}d(x)} & , L \in (0,\infty) \end{cases} \tag{44}$

Theorem 5. Important Properties of L_E –measure [7]

(i) For any $L \in [-1, \infty)$, L_E -measure is a multivalent fuzzy measure, in other words, L_E-measure has infinite fuzzy measure solutions.

(ii) L_E-measure is an increasing function on L.

(iii) if $L = -1$ then L_E-measure is just the P-measure.

(iv) if $L = 0$ then L_E-measure is just the additive measure.

(v) if $L = 0$ and $\sum_{x \in X} d(x) = 1$, then L_E-measure is just the λ-measure.

(vi) if $-1 < L < 0$ then L_E-measure is a supper-additive measure.

(vii) if $L > 0$ then L_E-measure is a sub-additive measure

Note that additive measure, λ-measure and P-measure are two special cases of L_E-measure.

3.9   B-Measure [7]

For considering to extend the extensional L-measure, a special fuzzy measure was proposed by my previous work as below;

Definition 15.   B-measure [7]
For a given fuzzy density function d, a B-measure, g_B, is a measure on a finite set $X, |X| = n$, satisfying:

$$\forall A \subset X \Rightarrow g_B(A) = \begin{cases} \sum_{x \in A} d(x) & if\ |A| \le 1 \\ 1 & if\ |A| > 1 \end{cases} \tag{45}$$

Theorem 6.   Any B-measure is a supper-additive measure.

3.10   Comparison of Two Fuzzy Measures [7-10]

Definition 16. Comparison of two fuzzy measures [7-10]
For a given fuzzy density function, $d(x)$, on a finite set, X,   let μ_1 and μ_2 be two fuzzy measures on X,

(i)   If $g_{\mu_1}(A) = g_{\mu_2}(A), \forall A \subset X,$, then we say that μ_1-measure is equal to μ_2-measure, denoted as

$$\mu_1 - measure = \mu_2 - masure \tag{46}$$

(ii)   If $g_{\mu_1}(A) < g_{\mu_2}(A), \forall A \subset X, 1 < |A| < |X|$ then we say that μ_1-measure is less than μ_2-measure, or μ_2-measure is larger than μ_1-measure, denoted as

$$\mu_1 - measure < \mu_2 - masure \tag{47}$$

(iii)   If $g_{\mu_1}(A) \le g_{\mu_2}(A), \forall A \subset X, 1 < |A| < |X|$, then we say that μ_1-measure is not larger than μ_2-measure, or μ_2-measure is not smaller than μ_1-measure, denoted as

$$\mu_1 - measure \le \mu_2 - masure \tag{48}$$

Theorem 7.   For any given fuzzy density function, if $\mu-measure$ is a fuzzy measure, then we have

$$P-measure \le \mu as-meaure \le B-measure \tag{49}$$

In other words, for any given fuzzy density function, the P-measure is the smallest fuzzy measure, and the B-measure is the largest fuzzy measure.

3.11   Completed Fuzzy Measure

Definition 17. Completed fuzzy measure [8]
If the measure function of a multivalent fuzzy measure has continuously infinite fuzzy measure solutions, and both P-measure and B-measure are its limit fuzzy measure solutions, then this multivalent fuzzy measure is called a completed fuzzy measure.

Note that both the L –measure and L_E –measure are not completed fuzzy measures, since

$$\lim_{L \to \infty} \frac{(|A|-1)L \sum_{x \in A} d(x)}{\left[n-|A|+(|A|-1)L\right] \sum_{x \in X} d(x)} = \frac{\sum_{x \in A} d(x)}{\sum_{x \in X} d(x)} \ne 1 \text{, the B-measure is not a limit fuzzy}$$

measure of the L –measure and L_E –measure

3.12   Completed Extensional L-Measure

Definition 18. Completed extensional L-measue, L_{CE} –measure
For a given fuzzy density function d on a finite set X, $|X| = n$, a measure is called extensional L-measure, if its measure function, $g_{L_{CE}} : 2^X \to [0,1]$, satisfying:

(i) $$g_{L_{CE}}(\phi)=0, g_{L_{CE}}(X)=1 \tag{50}$$

$$L \in [-1, \infty), A \subset X$$

(ii)

$$\Rightarrow g_{L_{CE}}(A) = \begin{cases} (1+L)\sum_{x \in A} d(x) - L \max_{x \in A} d(x) & , L \in [-1, 0] \\ \sum_{x \in A} d(x) + \dfrac{(|A|-1)L\sum_{x \in A} d(x)\left[1 - \sum_{x \in A} d(x)\right]}{\left[n - |A|\right]\sum_{x \in X} d(x) + L(|A|-1)\sum_{x \in A} d(x)} & , L \in (0, \infty) \end{cases} \tag{51}$$

Theorem 7. Important Properties of L_{CE} –measure [7]

(i) For any $L \in [-1, \infty)$, L_{CE} -measure is a multivalent fuzzy measure, in other words, L_{CE} -measure has infinite fuzzy measure solutions.

(ii) L_{CE} -measure is an increasing function on L.

(iii) if $L = -1$ then L_{CE} -measure is just the P-measure.

(iv) if $L = 0$ then L_{CE} -measure is just the additive measure.

(v) if $L = 0$ and $\sum_{x \in X} d(x) = 1$, then L_{CE} -measure is just the λ-measure.

(vi) if $-1 < L < 0$ then L_{CE} -measure is a sub-additive measure.

(vii) if $L > 0$ then L_{CE} -measure is a supper-additive measure

(viii) $L \rightarrow \infty$ then L_{CE} -measure is a B- measure

(ix) L_{CE} -measure is a completed fuzzy measure.

Note that additive measure, λ-measure, P-measure and B-measure are special cases of L_{CE} -measure.

3.13   Choquet Integral

Definition 19.   **Choquet Integral [9-10]**

Let μ be a fuzzy measure on a finite set $X = \{x_1, x_2, ..., x_m\}$. The Choquet integral of $f_i : X \rightarrow R_+$ with respect to μ for individual i is denoted by

$$\int_C f_i d\mu = \sum_{j=1}^{m} \left[f_i\left(x_{(j)}\right) - f_i\left(x_{(j-1)}\right) \right] \mu\left(A_{(j)}^i\right), \quad i = 1, 2, ..., N \tag{52}$$

where $f_i\left(x_{(0)}\right) = 0$, $f_i\left(x_{(j)}\right)$ indicates that the indices have been permuted so that

$$0 \le f_i\left(x_{(1)}\right) \le f_i\left(x_{(2)}\right) \le ... \le f_i\left(x_{(m)}\right), A_{(j)} = \left\{x_{(j)}, x_{(j+1)}, ..., x_{(m)}\right\} \tag{53}$$

Note that from Definition 19, for given integrand $f_i : X \rightarrow R_+$, the Choquet integral can be viewed as a function of the fuzzy measure μ -measure, in other words, the value of Choquet integral is predominate by its fuzzy measure.

Theorem 8. If a λ-measure is a standard fuzzy measure on $X = \{x_1, x_2, ..., x_m\}$, and $d : X \to [0,1]$ is its fuzzy density function, then the Choquet integral of $f_i : X \to R_+$ with respect to λ for individual i satisfying

$$\int_C f_i d\lambda = \sum_{j=1}^{m} d(x_j) f_i(x_j), \quad i = 1, 2, ..., N \tag{54}$$

3.14   Choquet Integral Composition Forecasting Model

Definition 20.   **Choquet Integral Composition Forecasting Model [8]**
(i) Let y_t be the realized value of target variable at time t,

(ii) Let $X = \{x_1, x_2, ..., x_m\}$ be the set of m competing predictors,

(iii) Let $f_t : X \to R_+$, $f_t(x_1), f_t(x_2), ..., f_t(x_m)$ be m forecasting values of y_t by competing predictors $x_1, x_2, ..., x_m$ at time t.

If μ is a fuzzy measure on X, $\alpha, \beta \in R$ satisfying

$$(\hat{\alpha}, \hat{\beta}) = \arg \min_{\alpha, \beta} \left[\sum_{t=1}^{N} \left(y_i - \alpha - \beta \int_C f_t dg_\mu \right) \right] \tag{55}$$

$$\hat{\alpha} = \frac{1}{N} \sum_{t=1}^{N} y_t - \hat{\beta} \frac{1}{N} \sum_{t=1}^{N} \int f_t dg_\mu, \quad \hat{\beta} = \frac{S_{yf}}{S_{ff}} \tag{56}$$

$$\hat{\alpha} = \frac{1}{N} \sum_{t=1}^{N} y_t - \hat{\beta} \frac{1}{N} \sum_{t=1}^{N} \int f_t dg_\mu \tag{57}$$

$$S_{yf} = \frac{\sum_{t=1}^{N} \left[y_i - \frac{1}{N} \sum_{t=1}^{N} y_t \right] \left[\int f_t dg_\mu - \frac{1}{N} \sum_{t=1}^{N} \int f_t dg_\mu \right]}{N-1} \tag{58}$$

then $\hat{y}_t = \hat{\alpha} + \hat{\beta} \int f_t dg_\mu, t = 1, 2, ..., N$ is called the Choquet integral regression composition forecasting estimator of y_t, and this model is also called the Choquet integral regression composition forecasting model with respect to μ-measure.

Theorem 9. If a λ-measure is a standard fuzzy measure then Choquet integral regression composition forecasting model with respect to λ-measure is just a linear combination forecasting model.

4   Experiments and Results

A real data of the grain production with 3 kinds of forecasted values of the time series model, the exponential smoothing model and GM(1,1) forecasting model, respectively, in Jilin during 1952 to 2007 from the paper of Zhang, Wang and Gao [2],was listed in Table 2. For evaluating the proposed new density based composition forecasting model, an experiment with the above-mentioned data by using sequential mean square error was conducted.

We arrange the first 50 years grain production and their 3 kinds of forecasted values as the training set and the rest data as the forecasting set. And the following N-density and M-density of all fuzzy measures were used

$$N\text{-density:} \qquad \{0.3331, \quad 0.3343, \quad 0.3326\} \tag{59}$$

$$M\text{-density:} \qquad \{0.2770, \quad 0.3813, \quad 0.3417\} \tag{60}$$

The performances of Choquet integral composition forecasting model with extensional L-measure, L-measure, λ-measure and P-measure, respectively, a ridge regression composition forecasting model and a multiple linear regression composition forecasting model and the traditional linear weighted composition forecasting model were compared. The result is listed in Table 1.

Table 1 SMSEs of 2 densities for 7 composition forecasting models

Composition forecasting Models		SMSE	
		N-density	M-density
	L_{CE}-measure	13149.64	13217.31
	L_E-measure	13939.84	13398.29
Choquet integral regression	L-measure	14147.83	13751.60
	λ-measure	21576.38	19831.86
	P-measure	16734.88	16465.98
Ridge regression		18041.92	
Multiple linear regression		24438.29	

Table 1 shows that the M-density based Choquet integral composition forecasting model with respect to L_{CE}-measure outperforms other composition forecasting models. Furthermore, for each fuzzy measure, including the L_{CE}-measure, L_E-measure, L-measure, λ-measure and P-measure, the M-density based Choquet integral composition forecasting model is better than the N-density based.

5   Conclusion

In this paper, a new density, M-density, was proposed. Based on M-density, a novel composition forecasting model was also proposed. For comparing the forecasting efficiency of this new density with the well-known density, N-density, a real data experiment was conducted. The performances of Choquet integral composition forecasting model with the completed extensional L-measure, extensional L-measure, λ-measure and P-measure, by using M-density and N-density, respectively, a ridge regression composition forecasting model and a multiple linear regression composition forecasting model and the traditional linear weighted composition forecasting model were compared. Experimental result showed that for each fuzzy measure, including the L_{CE}-measure, L_E-measure, L-measure, λ-measure and P-measure, the M-density based Choquet integral composition forecasting model is better than the N-density based, and the M-density based Choquet integral composition forecasting model outperforms all of other composition forecasting models.

Acknowledgment. This study is partially supported by the grant of National Science Council of Taiwan Government (NSC 100-2511-S-468-001).

References

1. Bates, J.M., Granger, C.W.J.: The Combination of Forecasts. Operations Research Quarterly 4, 451–468 (1969)
2. Zhang, H.-Q., Wang, B., Gao, L.-B.: Application of Composition Forecasting Model in the Agricultural Economy Research. Journal of Anhui Agri. Sci. 36(22), 9779–9782 (2008)
3. Hsu, C.-C., Chen, C.-Y.: Applications of improved grey prediction model for power demand forecasting. Energy Conversion and Management 44, 2241–2249 (2003)
4. Kayacan, E., Ulutas, B., Kaynak, O.: Grey system theory-based models in time series prediction. Expert Systems with Applications 37, 1784–1789, (2010)
5. Hoerl, A.E., Kenard, R.W., Baldwin, K.F.: Ridge regression: Some simulation. Communications in Statistics 4(2), 105–123 (1975)
6. Liu, H.-C., Tu, Y.-C., Lin, W.-C., Chen, C.C.: Choquet integral regression model based on L-Measure and γ-Support. In: Proceedings of 2008 International Conference on Wavelet Analysis and Pattern Recognition (2008)
7. Liu, H.-C.: Extensional L-Measure Based on any Given Fuzzy Measure and its Application. In: Proceedings of 2009 CACS International Automatic Control Conference, November 27-29, pp. 224–229. National Taipei University of Technology, Taipei Taiwan (2009)
8. Liu, H.-C.: A theoretical approach to the completed L-fuzzy measure. In: Proceedings of 2009 International Institute of Applied Statistics Studies (IIASS), 2nd Conference, Qindao, China, July 24-29 (2009)
9. Liu, H.-C., Ou, S.-L., Cheng, Y.-T., Ou, Y.-C., Yu, Y.-K.: A Novel Composition Forecasting Model Based on Choquet Integral with Respect to Extensional L-Measure. In: Proceedings of the 19th National Conference on Fuzzy Theory and Its Applications (2011)

10. Liu, H.-C., Ou, S.-L., Tsai, H.-C., Ou, Y.-C., Yu, Y.-K.: A Novel Choquet Integral Composition Forecasting Model Based on M-Density. In: Pan, J.-S., Chen, S.-M., Nguyen, N.T. (eds.) ACIIDS 2012, Part I. LNCS, vol. 7196, pp. 167–176. Springer, Heidelberg (2012)
11. Choquet, G.: Theory of capacities. Annales de l'Institut Fourier 5, 131–295 (1953)
12. Wang, Z., Klir, G.J.: Fuzzy Measure Theory. Plenum Press, New York (1992)
13. Sugeno, M.: Theory of fuzzy integrals and its applications. Unpublished doctoral dissertation, Tokyo Institute of Technology, Tokyo, Japan (1974)
14. Zadeh, L.A.: Fuzzy Sets as a Basis for Theory of Possibility. Fuzzy Sets and Systems 1, 3–28 (1978)

Appendix

Table 2 SMSEs of 2 densities for 6 composition forecasting models

Years	Y	X_1	X_2	X_3	X_4
1952	613.20	490.67	518.60	399.51	472.45
1953	561.45	549.73	570.84	414.09	511.35
1954	530.95	542.83	586.41	429.20	524.94
1955	556.53	549.57	584.31	444.86	530.10
1956	493.64	582.69	591.12	461.09	542.51
1957	429.35	598.64	570.80	477.91	538.81
1958	528.84	610.69	531.14	495.35	524.37
1959	526.60	633.88	540.11	513.43	537.85
1960	394.70	655.07	544.78	532.16	549.04
1961	398.55	672.97	497.45	551.58	531.58
1962	437.16	694.53	465.26	571.71	523.02
1963	501.67	617.26	457.04	592.57	519.94
1964	491.80	738.99	475.53	614.19	547.93
1965	525.10	761.94	484.57	636.61	563.02
1966	597.60	786.18	503.23	659.84	583.82
1967	647.74	810.67	543.38	683.91	616.78
1968	622.15	835.87	589.95	708.87	653.65
1969	498.70	862.13	612.17	734.74	677.54
1970	738.80	889.12	580.22	761.55	672.02
1971	713.05	916.86	647.93	789.33	721.78

Table 2 (*continued*)

1972	556.99	945.56	684.75	818.14	754.99
1973	783.00	975.15	650.45	847.99	749.50
1974	858.15	1005.63	711.01	878.93	796.71
1975	906.50	1037.08	780.79	911.01	849.50
1976	755.50	1069.53	846.59	944.25	900.60
1977	728.35	1102.98	833.96	978.70	909.05
1978	914.70	1137.47	813.29	1014.40	913.61
1979	903.34	1173.05	867.92	1051.40	960.20
1980	859.60	1209.74	900.44	1089.80	995.23
1981	921.91	1247.58	905.13	1129.60	1015.54
1982	1000.04	1286.60	930.45	1170.80	1047.82
1983	1477.98	1326.84	976.21	1213.50	1092.01
1984	1634.46	1368.34	1187.28	1257.80	1227.91
1985	1225.26	1411.14	1391.77	1303.70	1360.87
1986	1397.71	1455.27	1376.80	1351.30	1373.74
1987	1675.81	1500.79	1428.01	1400.60	1423.79
1988	1693.25	1547.73	1565.57	1451.70	1522.16
1989	1351.29	1596.14	1664.67	1504.60	1600.13
1990	2046.52	1646.06	1600.61	1559.50	1589.13
1991	1898.87	1697.54	1814.80	1616.50	1732.23
1992	1840.31	1750.64	1904.70	1675.40	1807.73
1993	1900.90	1805.39	1940.83	1736.60	1854.61
1994	2015.70	1861.86	1984.40	1799.90	1906.49
1995	1992.40	1920.09	2053.78	1865.60	1973.62
1996	2326.60	1980.15	2089.00	1933.70	2022.96
1997	1808.30	2042.08	2235.53	2004.30	2134.67
1998	2506.00	2105.95	2137.39	2077.40	2112.74
1999	2305.60	2171.82	2328.13	2153.20	2251.02
2000	1638.00	2239.75	2381.33	2231.80	2314.77
2001	1953.40	2309.80	2161.43	2313.20	2229.36
2002	2214.80	2382.04	2123.07	2397.60	2245.19

Table 2 (*continued*)

2003	2259.60	2456.54	2192.19	2485.10	2321.52
2004	2510.00	2533.38	2254.69	2575.80	2395.57
2005	2581.21	2612.61	2390.11	2669.80	2511.18
2006	2720.00	2694.33	2508.40	2767.20	2618.82
2007	2454.00	2778.60	2560.38	2831.50	2677.95

Y: realized value of target variable
X_1: Fitting value of time series model
X_2: Fitting value of exponential smoothing model
X_3: Fitting value of GM(1,1) model
X_4: Fitting value of composition forecasting model

Chapter 7
An Application of Enhanced Knowledge
Models to Fuzzy Time Series

Chung-Ming Own

Abstract. Knowledge is usually employed by domain experts to solve domain-specific problems. Huarng was the first to embed knowledge into forecasting fuzzy time series (2001). His model involved simple calculations and offers better prediction results once more supporting information has been supplied. On the other hand, Chen first proposed a high-order fuzzy time series model to overcome the drawback of existing fuzzy first-order forecasting models. Chen's model involved limited computing and came with higher accuracy than some other models. For this reason, the study is focused on these two types of models. The first model proposed here, which is referred to as a weighted model, aims to overcome the deficiency of the Huarng's model. Second, we propose another fuzzy time series model, called knowledge based high-order time series model, to deal with forecasting problems. This model aims to overcome the deficiency of the Chen's model, which depends strongly on highest-order fuzzy time series to eliminate ambiguities at forecasting and requires a vast memory for data storage. Experimental study of enrollment of University Alabama and the forecasting of a future's index show that the proposed models reflect fluctuations in fuzzy time series and provide forecast results that are more accurate than the ones obtained when using the to two referenced models.

Keywords: Fuzzy time series, knowledge model, domain specific knowledge.

1   Introduction

The forecasting of time series is crucial in daily life. It is used in forecasting the weather, earthquakes, stock fluctuations, and any phenomenon indexed by variables that change over time. Numerous investigations have solved the

Chung-Ming Own
Department of Computer and Communication Engineering,
St. John's University

W. Pedrycz & S.-M. Chen (Eds.): Time Series Analysis, Model. & Applications, ISRL 47, pp. 139–175.
DOI: 10.1007/978-3-642-33439-9_7   © Springer-Verlag Berlin Heidelberg 2013

associated problems by using the Moving Average, the Integrated Moving Average, and the Autoregressive Integrated Moving Average (Box & Jenkins, 1976; Janacek & Swift, 1993). Song and Chissom (1993) first defined fuzzy time series and modeled fuzzy relationships from historical data (Song & Chissom, 1979). The fuzzy time series is a novel concept that is used to solve forecasting problems that involve historical data with linguistic values. Song and Chissom (1993) used the fuzzy time series model to forecast enrollment at the University of Alabama and provided a step-by-step procedure. However, their method requires a large amount of computation time.

In reference to the time-invariant and time-variant models by Song and Chissom, Sullivan and Woodall proposed the time-invariant Markov model with linguistic labels of probability distribution (Sullivan & Woodall, 1994). Subsequently, Chen proposed a new fuzzy time series model that yielded excellent forecasting results (Chen, 1996). Chen's model simplified the complex computations of the models by Song and Chissom, and forecasted enrollment more accurately than other models. Hwang et al. (1998) proposed a method that focused on relation matrix computing for the variation between enrollments in the current year and those in past years. The Hwang et al. model was more efficient and simpler than most of the other models, although its accuracy was limited. Furthermore, Huarng (2001) solved the forecasting problem integrated the domain-specific knowledge into the fuzzy time series model. Knowledge is typically used by experts to solve domain-specific problems. In the Huarng model, available information was used to assist in the selection of proper fuzzy sets. Knowledge information can be used to solve the forecasting problem easily, and the resulting model outperformed previous models.

Chen (2002) presented a new fuzzy time series model called the high-order fuzzy time series to overcome disadvantages of current fuzzy forecasting models based on the first-order model. The Chen model came with excellent forecasting results. However, the disadvantage of the Chen model is its high dependence on high-order time series preprocessing. Additional methods have been presented to forecast Taiwan Futures Exchanges by using fuzzy forecasting techniques (Huarng & Yu, 2005, 2006a, 2006b, 2008, 2010; Huarng et al., 2007; Chen, 2008; Cheng, 2008). Leu et al. presented the distance-based fuzzy time series model to forecast exchange rates. Tanuwijaya and Chen (2009a, 2009b) also presented a clustering method to forecast enrollments at the University of Alabama. Jilani (2011) proposed a new particle swarm optimization-based multivariate fuzzy time series forecasting method. This model involves five factors with one main factor of interest. This study focused on applying swarm intelligence approaches to forecasting-related problems. Chen et al. (2012) proposed the equal frequency partitioning and fast Fourier transform algorithm to forecast stock prices in Taiwan. The results show the improving forecasting performance, and demonstrated an approach to enhance the efficiency of the fuzzy time-series.

This study proposes two enhanced models. The first model is a weighted model, which is an enhancement of the Huarng model. The proposed weighted model overcomes the disadvantage of the Huarng model, that is, the lack of an efficient measure of the significance of data in a series. Hence, the proposed model involves straightforward computation to defuzzify fuzzy forecasting with the support of knowledge and a weighting measure of the historical fuzzy sets. The second model is a high-order fuzzy time series model, which is an extension of the Chen model. The proposed high-order fuzzy time series model overcomes the disadvantage of the Chen model, which depends strongly on the derivation of highest-order fuzzy time series and requires large memory for data storage. Hence, this model has the advantage of a higher-order model and can apply the knowledge to eliminate the ambiguity in forecasting. An empirical analysis demonstrated that the two proposed fuzzy time series models can capture fluctuations in fuzzy time series and provide superior forecasting results than those coming from other models.

In this study, Section 2 briefly reviews basic concepts of fuzzy time series. In Section 3, we formulate the algorithms of the weighted knowledge and high-order models. Section 4 presents empirical analyses of enrollment and TAIFEX forecasts. Section 5 concludes this study.

2   Fuzzy Time Series

2.1   Basic Concept

Basic concepts related to fuzzy time series are reviewed below. U is the universe of discourse, $U = \{x_1, x_2, \cdots, x_k\}$. A fuzzy set A_i of U is defined as

$$A_i = \mu_{A_i}(x_1) / x_1 + \mu_{A_i}(x_2) / x_2 + \cdots + \mu_{A_i}(x_k) / x_k,$$

where μ_{A_i} is the membership function of A_i, $\mu_{A_i} : U \rightarrow [0, 1]$, and $\mu_{A_i}(x_j)$ represents the grade of membership of x_j in A_i, $\mu_{A_i}(x_j) \in [0, 1]$. The symbols "/" and "+" indicate the "separation" and "union" of elements in the universe of discourse U.

Definition 2.1. Let $Y(t)$ ($t = \ldots, 0, 1, 2, \ldots$), a subset of R ($Y(t) \subseteq R$), be the universe of discourse in which fuzzy sets $u_i(t)$ ($i = 1, 2, \cdots$) are defined. Assume that $F(t)$ consists $\mu_i(t)$ ($i = 1, 2, \ldots$); $F(t)$ is called a fuzzy time series on $Y(t)$.

From Definition 2.1, $F(t)$ can be considered to be a linguistic variable and $u_i(t)$ ($i = 1, 2, \cdots$) can be considered to be the possible linguistic values of $F(t)$. The main difference between fuzzy time series and conventional time series is that the observations in the former are fuzzy sets and those of the latter are real numbers.

Definition 2.2. Suppose that $F(t)$ is determined only by $F(t-1)$; then, there exists a fuzzy relationship $R(t-1,t)$ between $F(t)$ and $F(t-1)$, such that

$$F(t) = F(t-1) \times R(t-1,t),$$

where $\times$ is the composition operator. This relationship can also be represented as $F(t-1) \rightarrow F(t)$.

Definition 2.3. Let $F(t-1) = A_i$ and $F(t) = A_r$; a fuzzy relationship can be defined as $A_i \to A_r$. On the left-hand side of the fuzzy relationship, A_i, is called the current state of the relationship; A_r, on the right-hand side of the fuzzy relationship is called the next state of the relationship.

Definition 2.4. fuzzy relationships with the same current state can be further grouped in a combined fuzzy relationship called the grouped fuzzy relationship.

For example, some fuzzy relationships exist:

$$A_i \to A_{r_1},$$
$$A_i \to A_{r_2},$$
$$\cdots$$

These fuzzy relationships can be grouped together with the same current state and so

$$A_i \to A_{r_1}, A_{r_2}, \cdots$$

can be grouped together into a grouped fuzzy relationship.

Definition 2.5. According to Definition 2.3, if $F(t)$ is caused by more fuzzy sets, $F(t-n)$, $F(t-n+1)$, $\cdots$, and $F(t-1)$, then the fuzzy relationship can be represented as

$$A_{r_1}, A_{r_2}, \cdots, A_{r_n} \to A_j,$$

where $F(t-n) = A_{r_1}$, $F(t-n+1) = A_{r_2}$, $\cdots$, and $F(t-1) = A_{r_n}$. This relationship is called the nth-order fuzzy time series forecasting model. $A_{r_1}, A_{r_2}, \cdots,$ and A_{r_n} are called as the current states of the time series, and A_j is called as the next state of the time series.

Accordingly, the above equation means "If the time series in the year $t-1$, $t-2$, $\cdots$, and $t-n$ are $A_{r_1}, A_{r_2}, \cdots,$ and A_{r_n}, respectively, then that in the year t is A_j".

Definition 2.6. For any t, if $R(t-1,t)$ is independent of t, then $F(t)$ is called a time-invariant fuzzy time series. In contrast, if $R(t-1,t)$ is estimated from most recent observations, then $F(t)$ is called a time-variant fuzzy time series.

In this study, the models are all based on a time-invariant fuzzy time series.

2.2 Configuration of the Fuzzy Time Series

A pure fuzzy system generally comprises four parts: the fuzzifier, the fuzzy rule base, the fuzzy inference engine, and the defuzzifier. The fuzzifier is the input, which transforms a real-valued variable into a fuzzy set. The defuzzifier is the output, which transforms a fuzzy set into a real-valued variable. The fuzzy rule base represents the collection of fuzzy IF-THEN rules from human experts or domain knowledge. The fuzzy inference engine combines these fuzzy IF-THEN rules into a map from fuzzy sets in the input space, U, to fuzzy sets in the output space, V, according to the principles of fuzzy logic (Wang, 1997).

In (Song and Chissom 1993, 1994), Song and Chissom proposed both the time-variant and time-invariant models to forecast enrollments of the University of Alabama. Song and Chissom predicted fuzzy time series using historical data, and the model $F(t) = F(t-1) \times R(t-1,t)$. The procedures of time-variant models can be outlined as follows.

Step 1. Specify the universe of discourse U in which fuzzy sets will be defined;
Step 2. Partition the universe of discourse U into the even length intervals;
Step 3. Define the fuzzy sets on U;
Step 4. Fuzzify the input data x_{t-1} to $F(t-1)$;
Step 5. Forecast by the model $F(t) = F(t-1) \times R(t-1,t)$, and use the past w years data as a relationship;
Step 6. Defuzzify the output.

The main difference between the Song and Chissom's time-invariant and time-variant models is that the relationship $R(t,t-1)$ of the former must be established by all the historical data, whereas that of the latter must be determined only by some of the historical data. Figure 1 is shown the configuration of the fuzzy system to emphasize the distinguishing features of Song and Chissom's models. The fuzzifier is used to map the input to the fuzzy set $F(t)$ (corresponding to Steps 1-4). The fuzzy rule base is established based on all possible relationships. The fuzzy inference engine is used to compute by the model $F(t) = F(t-1) \times R(t-1,t)$ (corresponding to Step 5). The defuzzifier is used to transform the resulting fuzzy sets into a real-valued variable y (corresponding to Step 6).

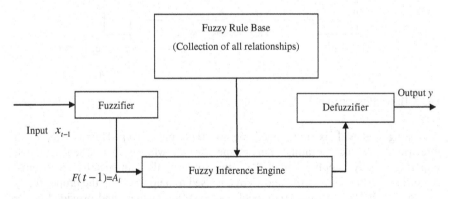

Fig. 1 The configuration of Song and Chissom's Model

The derivation of Song and Chissom's model was very tedious, and the matrix composition required a large amount of computation time. Chen proposed a model that involved straightforward knowledge reasoning to simplify the calculations in

Song and Chissom's model (Song and Chissom 1993). Chen's model not only applied simplified arithmetic operations rather than complicated max-min composition operations, but also provided more accurately forecasts than the other models. The procedures of Chen's model can be outlined as follows.

Step 1. Specify the universe of discourse U in which fuzzy sets will be defined;
Step 2. Partition the universe of discourse U into the even length intervals;
Step 3. Define the fuzzy sets on U, and fuzzify the data;
Step 4. Establish the fuzzy relationships into a group;
Step 5. Forecast;
Step 6. Defuzzify the output by using the arithmetic average.

Figure 2 shows the configuration of Chen's system. The fuzzifier is the same as in Song and Chissom's model (corresponding to Steps 1-4). In the fuzzy rule base, the important feature of Chen's model is that the fuzzy relationships are selected and summarized here (corresponding to Step 5). These computations are simple and straightforward. The defuzzifier takes an arithmetic average operation to derive the result (corresponding to Step 6). For more detail, refer to the (Song and Chissom 1993).

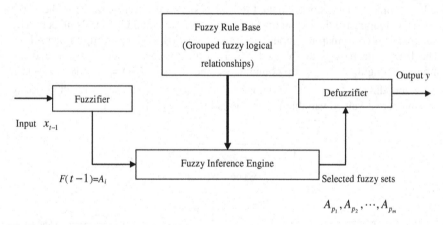

Fig. 2 The configuration of the Chen's model

Huarng's model is introduced below (Huarng 2001). Huarng improved forecasting by incorporating domain-specific knowledge into Chen's model. Experts usually apply knowledge in solving domain-specific problems. Accordingly, domain-specific knowledge is used to help to obtain the proper fuzzy sets during the forecasting. His model was easy to calculate and provided better forecasts as more supporting information was used. The procedures of Huarng's model can be outlined as follows.

Step 1. Specify the universe of discourse U in which fuzzy sets will be defined;
Step 2. Partition the universe of discourse U into the even length intervals;
Step 3. Define the fuzzy sets on U, and fuzzify the data;

Step 4. Establish the fuzzy relationships into a group;
Step 5. Forecast with knowledge assistance;
Step 6. Defuzzify the output by using the arithmetic average.

The inference engine uses two rule bases; one is the same as Chen's model; grouped fuzzy relationships (corresponding to Step 5). The other is the base of domain-specific knowledge. All the other parts are the same as Chen's model. Figure 3 is depicted the configuration of Huarng's model.

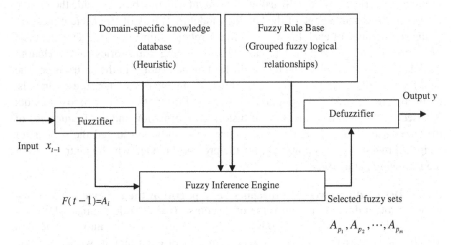

Fig. 3 The configuration of the Huarng's model

Chen (2002) proposed the high-order fuzzy time series model to improve the forecasting accuracy of his model in 1996. This new model can overcome the deficiency of the first-order fuzzy time series, which is inefficiently to eliminate the ambiguity in the forecasting. The procedure of Chen's model is outlined as follows.

Step 1. Specify the universe of discourse U in which fuzzy sets will be defined, and partition the universe of discourse U into the even length intervals;
Step 2. Define the fuzzy sets on U;
Step 3. Fuzzify the input data;
Step 4. Establish the high-order fuzzy relationship groups;
Step 5. Forecast by selecting the appropriate nth-order fuzzy relationship;
Step 6. Defuzzify the output with the elements in the nth-order fuzzy relationship.

Because this model is highly depended on the establishment of high-order fuzzy relationship, the time complexity is $O(p)$, where p denotes the number of grouped fuzzy relationship.

3   The Proposed Model

3.1   *Weighted Knowledge Model*

This model improves the forecasts based on the Huarng model. First, the Huarng model was selected because it is easy to calculate. The Huarng model has the advantage of the straightforward model by Chen. Second, the Huarng model yielded superior forecasts compared to other models. Third, the Huarng model used problem-specific knowledge by using an extra information base to guide the search. However, the disadvantage of the Huarng model is its lack of an efficient measure of the significance of each fuzzy set in fuzzy relationships; that is, every fuzzy set in a grouped fuzzy relationship has the resembling trajectories in the Huarng model. This is reflected by the use of the arithmetic average in the defuzzifier. The significance of fuzzy sets can be stressed by various measures; that is, defuzzification varies according to the observed information. The proposed model is based on the weighted measure of historical information and the frequencies of the fuzzy sets to adjust their ratios. Hence, this study considered the support of weighted measures and knowledge for the proposed model, which is introduced in the following paragraphs.

Step 1. Define the universe of discourse and partition the intervals. According to the problem domain, the universe of discourse U can be determined. Then, let the universe of discourse be partitioned into intervals $u_1 = [a_1, a_2]$, $u_2 = [a_1, a_2]$, $\cdots$, , $u_n = [a_n, a_{n+1}]$ of even length, where u_i is the ith divided interval. The midpoints of these intervals are $m_1, m_2, \cdots, m_n$, respectively.

Step 2. Define the fuzzy sets and fuzzify the data. Subsequently, let $A_1, A_2, \cdots, A_n$ be fuzzy sets, all of which are labeled by possible linguistic values. For example, linguistic values can be applied as fuzzy sets; A_1=(not many), A_2= (not too many), A_3=(many), A_4=(many many), A_5=(very many), A_6=(too many), A_7= (too many many). Hence, A_i is defined on as

$$A_i = \mu_{A_i}(u_1)/u_1 + \mu_{A_i}(u_2)/u_2 + \cdots + \mu_{A_i}(u_n)/u_n, \qquad (1)$$

where u_i is the interval expressed as an element of the fuzzy set, $i = 1, 2, \cdots, n$. $\mu_{A_i}(u_j)$ states the degree to which u_j belongs to A_i, and $\mu_{A_i}(u_j) \in [0,1]$, Then, the historical data are fuzzified by the intervals and expressed in the forms of linguistic values. Note that Eq. (1) uses interval u_j as an equation element.

Step 3. Establish and group fuzzy relationships. According to the definition of relationship in Definition 2.3, the fuzzy relationship can be determined. For instance,

$$A_j \rightarrow A_r,$$
$$A_j \rightarrow A_s,$$
$$A_j \rightarrow A_t,$$
$$\cdots,$$
$$A_m \rightarrow A_r,$$
$$A_m \rightarrow A_q,$$
$$\cdots$$

According to Definition 2.4, the fuzzy relationships can be grouped by the same origin:

$$A_j \rightarrow A_r, A_s, A_t,$$
$$A_m \rightarrow A_r, A_q,$$
$$\cdots$$

Step 4. Measure the frequency of fuzzy sets shown in the fuzzy relationships. Subsequently, according to the fuzzy relationships obtained in Step 3, calculates the frequency of each fuzzy set shown in the fuzzy relationships. For expressive simplicity, the frequency for fuzzy set A_i is denoted as f_i.

Example 3.1: Suppose that the fuzzy relationships calculated from the data set are obtained below:

$$A_r \rightarrow A_{r_1},$$
$$A_s \rightarrow A_{r_3},$$
$$A_r \rightarrow A_{r_1},$$
$$A_r \rightarrow A_{r_1},$$
$$A_r \rightarrow A_{r_2}.$$

Hence, fuzzy set A_{r_1} shown in relationship $A_r \rightarrow A_{r_1}$ occurs three times, denoted as $f_{r_1} = 3$. Fuzzy set A_{r_2} shown in relationship $A_r \rightarrow A_{r_2}$ occurs once, denoted as $f_{r_2} = 1$. Fuzzy set A_{r_3} shown in relationship $A_r \rightarrow A_{r_3}$ occurs once, denoted as $f_{r_3} = 1$.

Step 5. Introduce knowledge and establish selection strategy. In this proposed model, knowledge is used to guide the selection of proper fuzzy sets. Concerning knowledge in this study, changes in time series are used as a variable. According to the changes, the trend in selection strategy specifies the difference between times to an increase, a decrease or no change, and the symbolization of trend, α is set to 1, -1 and 0, respectively. The trend for an increase is used as a trigger to select whose fuzzy sets they have higher ranking in the grouped fuzzy relationship. On the contrary, the trend for a decrease is used to select whose fuzzy sets they have lower ranking. The trend for no change means to select the current state of the grouped fuzzy relationship. The selection strategy is introduced as follows.

Consider all the fuzzy sets A_1, A_2, $\cdots$, A_n are ordered in accordance. That is, A_1, A_2, $\cdots$, A_n are fuzzy sets on intervals $[a_1, a_2]$, $[a_2, a_3]$, $\cdots$, $[a_n, a_{n+1}]$, respectively, where $a_1 < a_2 < \cdots < a_{n+1}$. Suppose that $F(t-1) = A_i$ and the grouped fuzzy relationship of A_i is $A_i \to A_{i_1}, A_{i_2}, \cdots, A_{i_k}, \cdots, A_{i_\ell}$, where $i_1 < i_2 < \cdots < i_k < \cdots < i_\ell$.

Definition 3.1. Accordingly, all the fuzzy sets are partitioned into two parts; high and low parts. High part includes the fuzzy sets in high ranking; on the contrary, low part includes the fuzzy sets in low ranking. If $i_{k-1} < i < i_k$, then fuzzy sets in the low part are $\{A_{i_1}, A_{i_2}, \cdots, A_{i_{k-1}}\}$, and fuzzy sets in the high part are $\{A_{i_k}, A_{i_{k+1}}, \cdots, A_{i_\ell}\}$. Otherwise, If $i = i_k$, then fuzzy sets in the low part are $\{A_{i_1}, A_{i_2}, \cdots, A_{i_k}\}$, and fuzzy sets in the high part are $\{A_{i_k}, A_{i_{k+1}}, \cdots, A_{i_\ell}\}$. Note that, if the low/high part is empty set, then the low/high part needs to include their current stat of the relationship for instead, that is A_i.

Hence, the fuzzy sets in the high part mean to be selected in higher ranking, and fuzzy sets in the low part mean to be selected in lower ranking. The selection strategy of fuzzy sets is: If the trend of time series leads to an increase, the fuzzy sets in the high part are all selected, else if the trend of time series leads to a decrease, the fuzzy sets in the low part are all selected, else if the trend of time series leads to no change, the existing state of affairs would be preferred, the origin fuzzy set A_i is selected.

Step 6. Establish the weighted function. Suppose that the grouped fuzzy relationship of A_i is $A_i \to A_{i_1}, A_{i_2}, ..., A_{i_j}, ..., A_{i_k}, ..., A_{i_m}$, where $j < k < \ell < m$. For the purpose to stress the significance of fuzzy sets, the weighting of each fuzzy set is applied as the factor during the forecasting.

Definition 3.2. The weighting of the fuzzy set A_{i_j} in the low part is computed as the probability of frequency defined by

$$l_{i_j} = \frac{f_{i_j}}{f_{i_1} + f_{i_2} + \cdots + f_{i_Q}}, \tag{2}$$

where $Q = k - 1$, if $i_{k-1} < i < i_k$, and $Q = k$, if $i = i_k$. f_{i_j} is denoted as the frequency of fuzzy set A_{i_j}. On the other hand, the weighting of the fuzzy set A_{i_ℓ} in the high part is computed as the probability of frequency defined by

$$l_{i_j} = \frac{f_{i_\ell}}{f_{i_k} + \cdots + f_{i_m}}. \tag{3}$$

Hence, for each grouped fuzzy relationships, the weighted function is established as follows.

$$h_i(\alpha)\,|_{\alpha=-1,0,1}=$$
$$\frac{\alpha(\alpha-1)}{2}\sum_{j=i_1,\cdots,i_Q}l_j m_j +\frac{2(1+\alpha)(1-\alpha)}{2}m_i+\frac{\alpha(\alpha+1)}{2}\sum_{j=i_k,\cdots,i_m}q_j m_j \qquad (4)$$

where $Q = k-1$, if $i_{k-1} < i < i_k$, and $Q = k$, if $i = i_k$. Note that $h_i(\alpha)$ is derived corresponding to the original fuzzy set A_i of each grouped fuzzy relationships, $i = 1,\ \cdots, n$. α is the variable to select the proper fuzzy sets derived from Step 5. m_j is the midpoint of the interval u_j where the maximum membership value of fuzzy set A_j occurs.

Example 3.2: Consider the same problem as in Example 3.1, in which five fuzzy relationships include. The grouped fuzzy relationships are

$$A_r \rightarrow A_{r_1}, A_{r_2},\ (\text{where}\ f_{r_1} = 3,\ f_{r_2} = 1)$$
$$A_s \rightarrow A_{r_3}.\ (\text{where}\ f_{r_3} = 1)$$

Hence, the weighted functions are

$$h_r(\alpha)\,|_{\alpha=-1,0,1}=\frac{\alpha(\alpha-1)}{2}m_r +\frac{2(1+\alpha)(1-\alpha)}{2}m_r+\frac{\alpha(\alpha+1)}{2}(\frac{3}{4}m_{r_1}+\frac{1}{4}m_{r_2}),$$

$$h_m(\alpha)\,|_{\alpha=-1,0,1}=\frac{\alpha(\alpha-1)}{2}m_{r_3} +\frac{2(1+\alpha)(1-\alpha)}{2}m_s+\frac{\alpha(\alpha+1)}{2}m_s,$$

where the ranking of fuzzy sets are $r < r_1 < r_2$ and $r_3 < s$.

Step 7: **Calculate the forecasted outputs.** Subsequently, suppose that input x_{t-1} in time $t-1$ is fuzzified to $F(t-1) = A_i$, the calculations are carried out as follows. Because the fuzzy set is A_i, then the corresponding weighted function $h_i(\alpha)$ is selected, $i = 1,\ \cdots, n$. In this study, the difference between time $t-1$ and t leads to an increase, a decrease or no change, denoted as parameter $\alpha = 1$, -1 and 0, respectively. Accordingly, the output is derived as $h_i(\alpha)\,|_{\alpha=-1,0,1}$.

According to the configuration of fuzzy systems, the weighted function involves the parts of fuzzy inference engine and defuzzifier. Figure 4 presents the configuration of the weighted model.

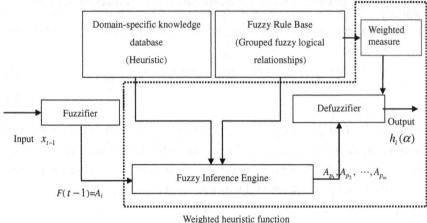

Fig. 4 The configuration of the weighted model

3.2   High-Order Model

This study aimed to overcome the deficiency of the Chen model (2002), which is strongly dependent on the derivation of highest-order fuzzy time series and requires a large amount of memory. In other words, according to Chen's definition, if an ambiguity occurs in the ith-order fuzzy relationship groups, the model seeks a higher order, such as $(i+1)$th-order fuzzy relationship, to perform the forecast. The highest-order fuzzy relationship must be computed before the model can conduct forecasts. Thus, the model requires a large amount of memory to derive the fuzzy relationships from the lowest order to the highest order.

Knowledge was applied to the high-order fuzzy time series model to eliminate the computation "bottleneck." In the domain of expert systems, knowledge is typically considered guides that can be used by domain experts to solve domain-specific problems (Russell & Norvig, 1995). Based on Huarng's assumption (2001), knowledge is used to guide the search for suitable fuzzy sets appropriate for forecasting indices. Therefore, this study enhanced Chen's model by integrating knowledge with high-order fuzzy time series to eliminate ambiguities in forecasting. Thus, the proposed model can be restricted to lower-order fuzzy time series to achieve acceptable forecast accuracy and required memory.

Step 1: **Define the universe of discourse and partition the intervals.** According to the problem domain, the universe of discourse U can be determined. Then, let the universe of discourse be partitioned into intervals $u_1 = [a_1, a_2]$, $u_2 = [a_1, a_2]$, $\cdots$, $u_n = [a_n, a_{n+1}]$ of even length, where u_i is the ith divided interval. The midpoints of these intervals are symbolized as m_1, m_2, $\cdots, m_n$, respectively.

Step 2. Define the fuzzy sets and fuzzify the data. Subsequently, let $A_1, A_2, \cdots, A_n$ be fuzzy sets, all of which are labeled by possible linguistic values. For example, linguistic values can be applied as fuzzy sets; A_1=(not many), A_2= (not too many), A_3=(many), A_4=(many many), A_5=(very many), A_6=(too many), A_7= (too many many). Hence, A_i is defined on as

$$A_i = \mu_{A_i}(u_1)/u_1 + \mu_{A_i}(u_2)/u_2 + \cdots + \mu_{A_i}(u_n)/u_n, \tag{5}$$

where u_i is the interval expressed as an element of the fuzzy set, $i = 1, 2, \cdots, n$. $\mu_{A_i}(u_j)$ states the degree to which u_j belongs to A_i, and $\mu_{A_i}(u_j) \in [0,1]$.

Then, the historical data are fuzzified by the intervals and expressed in the forms of linguistic values. Note that Eq. (5) uses the interval u_j as an element.

Step 3. Establish and group the nth-order fuzzy relationships. The nth-order fuzzy relationships are established based on the fuzzified historical time series. Besides, if there are ambiguities, these fuzzy relationships are grouped together according to Definition 2.4.

Step 4. Introduce knowledge and establish the knowledge function. Changes in time series (the trend) are used as the knowledge to specify the difference between times to an increase, a decrease or no change. For instance, the input in the year $t-1$ is x_{t-1} and year t is x_t. Then the trend leads to an increase, if $x_t - x_{t-1} > 0$, denoted as $\alpha = 1$ for the simplification. The trend leads to a decrease, if $x_t - x_{t-1} < 0$, denoted as $\alpha = 0$. The trend leads to no change, if $x_t - x_{t-1} = 0$, denoted as $\alpha = -1$.

Consider all the fuzzy sets $A_1, A_2, \cdots, A_n$ are well ordered. That is, $A_1, A_2, \cdots, A_n$ are fuzzy sets on intervals $[a_1, a_2], [a_2, a_3], \cdots, [a_n, a_{n+1}]$, respectively, where $a_1 < a_2 < \cdots < a_{n+1}$. Suppose that there is a certain ambiguity in the ith-order fuzzy relationship, and they are grouped as
$A_{r_1}, A_{r_2}, \cdots, A_{r_i} \to A_{j_1}, A_{j_2}, \cdots, A_{j_k}, \cdots, A_{j_\ell}$,
where $j_1 < j_2 < \cdots < j_k < \cdots < j_\ell$.

According to fuzzy set A_{r_i} at the right most side of the current states, all the fuzzy sets in the next states of the grouped fuzzy relationship are partitioned into two parts; high and low parts. If $j_{k-1} < r_i < j_k$, then fuzzy sets in the low part are $\{A_{j_1}, A_{j_2}, \cdots, A_{j_{k-1}}\}$, and fuzzy sets in the high part are $\{A_{j_k}, A_{j_{k+1}}, \cdots, A_{j_\ell}\}$. Otherwise, if $r_i = j_k$, then fuzzy sets in the low part are $\{A_{j_1}, A_{j_2}, \cdots, A_{j_k}\}$, and fuzzy sets in the high part are $\{A_{j_k}, A_{j_{k+1}}, \cdots, A_{j_\ell}\}$.

Hence, the fuzzy sets in the high part mean to be selected in higher ranking, and fuzzy sets in the low part mean to be selected in lower ranking. The selection strategy is: If the trend of time series leads to an increase, the fuzzy sets in the high part are all selected, else if the trend of time series leads to a decrease, the fuzzy sets in the low part are all selected, else if the trend of time series leads to no change, the existing current states are selected. According to the selection strategy, in this proposed model, the knowledge function accepts the relevant trend α, the fuzzy set of the right most current states and grouped fuzzy relationships as parameters. The knowledge function is established as follows:

$$h(\alpha; A_{r_i}; A_{j_1}, A_{j_2}, \cdots, A_{j_\ell}) =$$

$$\begin{cases} \dfrac{(1-\alpha)}{k-1} \displaystyle\sum_{g=j_1, \cdots, j_{k-1}} m_g + \dfrac{\alpha}{(\ell-k+1)} \displaystyle\sum_{g=j_k, \cdots, j_\ell} m_g, & \text{if } \alpha = 0 \text{ or } 1, \text{and } j_{k-1} < r_i < j_k, \\[2ex] \dfrac{(1-\alpha)}{k} \displaystyle\sum_{g=j_1, \cdots, j_k} m_g + \dfrac{\alpha}{(\ell-k+1)} \displaystyle\sum_{g=j_k, \cdots, j_\ell} m_g, & \text{if } \alpha = 0 \text{ or } 1, \text{and } r_i = j_k, \\[2ex] \dfrac{1}{i} \displaystyle\sum_{g=r_1, r_2, \cdots, r_i} m_g, & \text{if } \alpha = -1. \end{cases} \qquad (6)$$

Note that α is the variable. m_{j_k} is the midpoint of the interval μ_{j_k} where the maximum membership value of fuzzy set A_{j_k} occurs.

Step 5. **Calculate the forecasted outputs.** The calculations are implemented as follows.

(1) If the ith-order fuzzified history time series for time t are $A_{r_1}, A_{r_2}, \cdots, A_{r_i}$, where $i \geq 2$, and there is the following fuzzy relationship in the ith grouped order fuzzy relationships shown as follows:

$$A_{r_1}, A_{r_2}, \cdots, A_{r_i} \to A_j.$$

The forecasted fuzzy set at time t is A_j, and the forecasting result is m_j, it is the midpoint of the interval u_j where the maximum membership value of fuzzy set A_j occurs.

(2) If the ith-order fuzzified history time series for time t are $A_{r_1}, A_{r_2}, \cdots, A_{r_i}$, where $i \geq 2$, and there is the following fuzzy relationship in the ith grouped order fuzzy relationships shown as follows:

$$A_{r_1}, A_{r_2}, \cdots, A_{r_i} \to A_{j_1}, A_{j_2}, \cdots, A_{j_k}, \cdots, A_{j_\ell},$$

where $j_1 < j_2 < \cdots < j_k < \cdots < j_\ell$. Then, the function is applied to eliminate the ambiguity and obtain the forecasting result, $h(\alpha; A_{r_i}; A_{j_1}, A_{j_2}, \cdots, A_{j_\ell})$, where the difference between time t-1 and t leads to an increase, a decrease or no change, denoted as parameter $\alpha = 1$, 0 and -1, respectively. Accordingly, the output is derived as $h(\alpha; A_{r_i}; A_{j_1}, A_{j_2}, \cdots, A_{j_\ell})|_{\alpha=1, 0, -1}$.

4   Forecasting Experiment

4.1   Forecasting Enrollment with Weighted Knowledge Model

The proposed weighted knowledge model was applied for effective forecasting of university enrollments. Enrollments from 1971 to 1992 at the University of Alabama were already forecast in a series of experiments. The forecasting of enrollments using the weighted model is detailed in the following paragraphs.

Step 1. As in Table 1, the historical data on enrollments of the University of Alabama yields U=[13000, 20000]. The universe of the discourse is divided into seven equally long intervals u_1, u_2, ..., u_7 with length 1000,   where u_1=[13000, 14000], u_2=[14000, 15000], u_3=[15000, 16000], u_4=[16000, 17000], u_5=[17000, 18000], u_6=[18000, 19000], u_7=[19000, 20000].

Step 2. The enrollments of the University of Alabama can be represented as seven fuzzy sets A_i (i=1, 2, ..., 7). The linguistic values are A_1=(not many), A_2=(not too many), A_3=(many), A_4=(many many), A_5=(very many), A_6=(too many) , A_7=(too many many). Each A_i (i=1, 2, ..., 7) is defined as follows.

$$A_1=1/u_1 + 0.5/u_2 + 0/u_3 + 0/u_4 + 0/u_5 + 0/u_6 + 0/u_7,$$

$$A_2=0.5/u_1 + 1/u_2 + 0.5/u_3 + 0/u_4 + 0/u_5 + 0/u_6 + 0/u_7,$$

$$A_3=0/u_1 + 0.5/u_2 + 1/u_3 + 0.5/u_4 + 0/u_5 + 0/u_6 + 0/u_7,$$

$$A_4=0/u_1 + 0/u_2 + 0.5/u_3 + 1/u_4 + 0.5/u_5 + 0/u_6 + 0/u_7,$$

$$A_5=0/u_1 + 0/u_2 + 0/u_3 + 0.5/u_4 + 1/u_5 + 0.5/u_6 + 0/u_7,$$

$$A_6=0/u_1 + 0/u_2 + 0 /u_3 + 0/u_4 + 0.5/u_5 + 1/u_6 + 0.5/u_7,$$

$$A_7=0/u_1 + 0/u_2 + 0 /u_3 + 0/u_4 + 0/u_5 + 0.5/u_6 + 1/u_7.$$

Table 1 lists the corresponding fuzzy enrollment A_i.

Step 3. The fuzzy relationships are established and grouped. Table 2 lists the fuzzy relationships derived from Table 1. Table 3 lists the grouped fuzzy relationships.

Step 4. Subsequently, the frequency of the fuzzy set in each fuzzy relationship is calculated and recorded in the appendix of Table A-1.

Step 5. The existence knowledge regarding the trend of increase or decrease in university enrollment is referred from the Huarng in [1]. This trend of increase or decrease is used as a guide in selecting the proper fuzzy sets for forecasting enrollment. The increase, unchanged, decrease of trends are symbolized as $\alpha = 1$, -1 or 0, respectively.

Table 1 Enrollment data sets

Years	Enrollments	Fuzzy Set	Years	Enrollments	Fuzzy Set
1971	13055	A_1	1972	13563	A_1
1973	13867	A_1	1974	14696	A_2
1975	15460	A_3	1976	15311	A_3
1977	15603	A_3	1978	15861	A_3
1979	16807	A_4	1980	16919	A_4
1981	16388	A_4	1982	15433	A_3
1983	15497	A_3	1984	15145	A_3
1985	15163	A_3	1986	15984	A_3
1987	16859	A_4	1988	18150	A_6
1989	18970	A_6	1990	19328	A_7
1991	19337	A_7	1992	18876	A_6
1993	18909	A_6	1994	18707	A_6
1995	18561	A_6	1996	17572	A_5
1997	17877	A_5	1998	17929	A_5
1999	18267	A_6	2000	18859	A_6
2001	18735	A_6	2002	19181	A_7
2003	19828	A_7	2004	20512	A_8

Table 2 Enrollment of fuzzy relationships

$$A_1 \rightarrow A_1, \quad A_1 \rightarrow A_2$$
$$A_2 \rightarrow A_3$$
$$A_3 \rightarrow A_3, \quad A_3 \rightarrow A_4$$
$$A_4 \rightarrow A_3, \quad A_4 \rightarrow A_4, \quad A_4 \rightarrow A_6$$
$$A_6 \rightarrow A_6, \quad A_6 \rightarrow A_7$$
$$A_7 \rightarrow A_6, \quad A_7 \rightarrow A_7$$

Table 3 Grouped fuzzy relationships

$$A_1 \rightarrow A_1, A_2$$
$$A_2 \rightarrow A_3$$
$$A_3 \rightarrow A_3, A_4$$
$$A_4 \rightarrow A_3, A_4, A_6$$
$$A_6 \rightarrow A_6, A_7$$
$$A_7 \rightarrow A_6, A_7$$

Hence, according to Definition 3.1, the selection strategy of fuzzy sets is: if the annual trend in university enrollment leads to an increase, then the fuzzy sets in the high part are all selected. Conversely, if the annual trend in university enrollment leads to a decrease, then the fuzzy sets in the low part are all selected, or the annual trend in university enrollment leads to no change, then the original fuzzy sets is selected.

Step 6. According to the grouped fuzzy relationships in Table 3, the corresponding weighted knowledge functions can be established and are listed as follows:

$$h_1(\alpha)\,|_{\alpha=-1,0,1} = \frac{\alpha(\alpha-1)}{2}m_1 + \frac{2(1+\alpha)(1-\alpha)}{2}m_1 + \frac{\alpha(\alpha+1)}{2}(\frac{2}{3}m_1 + \frac{1}{3}m_2),$$

$$h_2(\alpha)\,|_{\alpha=-1,0,1} = \frac{\alpha(\alpha-1)}{2}m_2 + \frac{2(1+\alpha)(1-\alpha)}{2}m_2 + \frac{\alpha(\alpha+1)}{2}m_3,$$

$$h_3(\alpha)\,|_{\alpha=-1,0,1} = \frac{\alpha(\alpha-1)}{2}m_3 + \frac{2(1-\alpha)(1+\alpha)}{2}m_3 + \frac{\alpha(\alpha+1)}{2}(\frac{7}{9}m_3 + \frac{2}{9}m_4),$$

$$h_4(\alpha)\,|_{\alpha=-1,0,1} = \frac{\alpha(\alpha-1)}{2}(\frac{1}{3}m_3 + \frac{2}{3}m_4) + \frac{2(1-\alpha)(1+\alpha)}{2}m_4 + \frac{\alpha(\alpha+1)}{2}(\frac{2}{3}m_4 + \frac{1}{3}m_6)$$

$$h_6(\alpha)\,|_{\alpha=-1,0,1} = \frac{\alpha(\alpha-1)}{2}m_6 + \frac{2(1-\alpha)(1+\alpha)}{2}m_6 + \frac{\alpha(\alpha+1)}{2}(\frac{1}{2}m_6 + \frac{1}{2}m_7),$$

$$h_7(\alpha)\,|_{\alpha=-1,0,1} = \frac{\alpha(\alpha-1)}{2}(\frac{1}{2}m_6 + \frac{1}{2}m_7) + \frac{2(1-\alpha)(1+\alpha)}{2}m_7 + \frac{\alpha(\alpha+1)}{2}m_7,$$

where m_k is the midpoint of the interval u_k, and $m_1 = 13500$, $m_2 = 14500$, $m_3 = 15500$, $m_4 = 16500$, $m_5 = 17500$, $m_6 = 18500$ and $m_7 = 19500$.

Step 7: Subsequently, suppose that input x_{t-1} in the year $t-1$ is fuzzified to $F(t-1) = A_i$, the corresponding weighted function $h_i(\alpha)$ is selected.

Accordingly, the output is derived as $h_i(\alpha)\,|_{\alpha=-1,0,1}$. The following examples are used to demonstrate the procedure of selecting the corresponding weighted function and using the knowledge to derive the forecasts.

[years 1972, 1973, 1974]: The enrollment in 1971 was 13055(A_1), in 1972 was 13563(A_1) and in 1973 was 13867(A_1). While forecasting 1972, the grouped fuzzy relationship of A_1 is $A_1 \rightarrow A_1, A_2$, so the weighted function $h_1(\alpha)$ is selected. Suppose that the knowledge points to an increase for the enrollment forecasts in 1972. Hence, α is set to 1. The forecast in 1972 is

$$h_1(\alpha)\,|_{\alpha=1} = \frac{2}{3}m_1 + \frac{1}{3}m_2 = 13833.$$

That is, the enrollment forecast for the year 1972 is 13833. However, the actual enrollment in 1972 was 13522. Therefore, the forecasting error is 1.99%. The main goal in this paper is to minimize the forecasting error. Meanwhile, the trend in 1973 and 1974 leads the knowledge to an increase. Hence, the forecasts for 1973 and 1974 are both 13833.

[year 1975]: The enrollment in 1974 was 14696 (A_2). The weighted function $h_2(\alpha)$ is determined). Meanwhile, suppose that the knowledge points to an increase for the enrollment forecast in 1975, so $\alpha=1$. Therefore, the weighted function is

$$h_2(\alpha)\,|_{\alpha=1} = m_3 = 15500.$$

That is, the forecast for 1975 is 15500.

[year 1976]: The enrollment of 1975 was 15460(A_3). The weighted function $h_3(\alpha)$ is selected. Meanwhile, suppose that the knowledge points to a decrease for the enrollment forecast in 1976, so $\alpha=-1$. Therefore, the weighted function is

$$h_3(\alpha)\,|_{\alpha=-1} = m_3 = 15500.$$

That is, the forecast for 1976 is 15500.

[years 1977, 1978, 1979]: The enrollment of 1976 was 15311(A_3), 1977 was 15603(A_3), and 1978 was 15861(A_3). The grouped fuzzy relationship of A_3 is $A_3 \rightarrow A_3, A_4$, so the weighted function $h_3(\alpha)$ is selected for forecasting year 1977. Meanwhile, suppose that the knowledge points to an increase for the enrollment forecast in 1977, so $\alpha=1$. The forecast for 1977 is

$$h_3(\alpha)\,|_{\alpha=1} = \frac{7}{9}m_3 + \frac{1}{9}m_4 = 15722.$$

Meanwhile, the enrollment trends in 1978 and 1979 both lead the knowledge to an increase. Hence, the forecasts for 1978 and 1979 are both 15722.

[year 1980]: The enrollment of 1979 was 16807(A_4). The weighted function $h_4(\alpha)$ is selected. Meanwhile, suppose that the knowledge points to an increase for the enrollment forecast in 1980, so $\alpha=1$. Hence, the weighted function is

$$h_4(\alpha)\,|_{\alpha=1} = \frac{2}{3}m_4 + \frac{1}{3}m_6 = 17167.$$

That is, the forecast for the year 1980 is 17167.

Table 4 shows all of the remaining forecasts, and compares various studies of fuzzy time series used for forecasting enrollments. Suppose that knowledge is available. Empirical analysis yields average forecasting errors of 3.22% and 4.38% by Song and Chissom's two models, respectively (1993, 1994), 3.11% by Chen's model (2002), and 2.45% by Huarng's model (2001). This proposed weighted model, however, has an error of 2.24%. In enrollment forecasting, the proposed model outperforms the others.

Table 4 Comparison of enrollment forecasting

Years	Enrollment	Song-I [5]	Song-II [11]	Chen [7]	Huarng [1]	This proposed model
1971	13055					
1972	13563	14000		14000	14000	13833
1973	13867	14000		14000	14000	13833
1974	14696	14000		14000	14000	13833
1975	15460	15500	14700	15500	15500	15500
1976	15311	16000	14800	16000	15500	15500
1977	15603	16000	15400	16000	16000	15722
1978	15861	16000	15500	16000	16000	15722
1979	16807	16000	15500	16000	16000	15722
1980	16919	16813	16800	16833	17500	17167
1981	16388	16813	16200	16833	16000	16167
1982	15433	16789	16400	16833	16000	16167
1983	15497	16000	16800	16000	16000	15500
1984	15145	16000	16400	16000	15500	15500
1985	15163	16000	15500	16000	16000	15500
1986	15984	16000	15500	16000	16000	15722
1987	16859	16000	15500	16000	16000	15722

Table 4 (*continued*)

1988	18150	16813	16800	16833	17500	17167
1989	18970	19000	19300	19000	19000	19000
1990	19328	19000	17800	19000	19000	19000
1991	19337	19000	19300	19000	19500	19500
1992	18876	No forecasting	19600	19000	19000	19000
		3.22%	4.38%	3.11%	2.45%	2.24%

4.2 Robust Forecasting with the Memorizing Capability

In the empirical case, the enrollments for weighted measure and performance forecasting were derived from the same years, and prior knowledge was constructed by analyzing the obtained information. This is referred to as the memorizing capability. Consequently, another robust capability was considered. To evaluate robustness, the weighted measure and performance forecasting must originate from different sources. Therefore, the enrollments of fuzzy relationships were grouped and analyzed from 1971 to 1992 at the University of Alabama, and robustness was tested according to the enrollments from 1993 to 2004. The authors used current knowledge on the annual increase or decrease in university enrollment. This trend of increase, decrease, or no change was used as a guide to select the proper fuzzy sets for forecasting enrollment. For example, forecasting the enrollment in the year t was dependent on the difference between years $t-2$ and $t-1$. The positive difference led to an increase, and $\alpha = 1$. Conversely, the negative difference led to a decrease, and $\alpha = -1$. A difference of less than 100 led to no change, and $\alpha = 0$.

Enrollment forecasting using the weighted model proceeded as described. Table 5 shows various studies on fuzzy time series used for forecasting enrollment. Empirical analysis yielded average forecasting errors of 2.99% when using the Chen model, 2.61% when using the Huarng model, and 2.31% when using the proposed model. Therefore, the proposed model outperformed the other models in robust forecasting.

Table 5 Comparison of enrollment forecasting with robustness (1993 ~ 2004)

	Chen's Model [7]	Huarng's model [1]	This proposed model
Average error	2.99%	2.61%	2.31%

4.3   Forecasting TAIFEX with Weighted Model

Forecasting of the Taiwan Futures Exchange (TAIFEX) was used to demonstrate the advantages of the proposed model (Huarng, 2001). The Taiwan Stock Exchange Capitalization-Weighted Stock Index (TAIEX) was used as knowledge to evaluate the trend over a number of days. In other words, any two consecutive days in the TAIEX reflect gains or losses in the stock market. The TAIFEX and the TAIEX are highly related; therefore, differences between consecutive days in the TAIEX were used as knowledge to forecast the TAIFEX.

In forecasting the TAIFEX, the data range from August 3 to September 30 1988. Forecasting proceeds as follows.

Step 1. From the historical data in Table 6, U=[6100, 7700] is derived. Then, the universe of the discourse is divided into 16 equally long intervals u_1, u_2, ..., u_{16} of length 100, where u_1=[6100, 6200], u_2=[6200, 6300], u_3=[6300, 6400], u_4=[6400, 6500], u_5=[6500, 6600], u_6=[6600, 6700], u_7=[6700, 6800], u_8=[6800, 6900], u_9=[6900, 7000], u_{10}=[7000, 7100], u_{11}=[7100, 7200], u_{12}=[7200, 7300], u_{13}=[7300, 7400], u_{14}=[7400, 7500], u_{15}=[7500, 7600], u_{16}=[7600, 7700].

Step 2. In this case, the linguistic variable "TAIFEX" which can be represented as 16 fuzzy sets; A_i (i=1, 2, ···, 16). The linguistic values are A_1=(lowest), A_2=(very very very low), A_3=(very very low), A_4=(very low), A_5=(low), A_6=(quite low) , A_7=(low medium), A_8=(medium), A_9=(quite medium), A_{10}=(medium high), A_{11}=(quite high), A_{12}=(high), A_{13}=(very high), A_{14}=(very very high), A_{15}=(very very very high), A_{16}=(highest). Each A_i (i=1, 2, ···, 16) is defined in the Table 7.

Table 6 The fuzzy sets of "TAIFEX"

A_1= $1/u_1$ + $0.5/u_2$ + $0/u_3$ + $0/u_4$ + $0/u_5$ + $0/u_6$ + $0/u_7$ + $0/u_8$ + $0/u_9$ + $0/u_{10}$ + $0/u_{11}$

+ $0/u_{12}$ + $0/u_{13}$ + $0/u_{14}$ + $0/u_{15}$ + $0/u_{16}$

A_2= $0.5/u_1$ + $1/u_2$ + $0.5/u_3$ + $0/u_4$ + $0/u_5$ + $0/u_6$ + $0/u_7$ + $0/u_8$ + $0/u_9$ + $0/u_{10}$ +

$0/u_{11}$ + $0/u_{12}$ + $0/u_{13}$ + $0/u_{14}$ + $0/u_{15}$ + $0/u_{16}$

A_3= $0/u_1$ + $0.5/u_2$ + $1/u_3$ + $0.5/u_4$ + $0/u_5$ + $0/u_6$ + $0/u_7$ + $0/u_8$ + $0/u_9$ + $0/u_{10}$ +

$0/u_{11}$ + $0/u_{12}$ + $0/u_{13}$ + $0/u_{14}$ + $0/u_{15}$ + $0/u_{16}$

A_4= $0/u_1$ + $0/u_2$ + $0.5/u_3$ + $1/u_4$ + $0.5/u_5$ + $0/u_6$ + $0/u_7$ + $0/u_8$ + $0/u_9$ + $0/u_{10}$ +

$0/u_{11}$ + $0/u_{12}$ + $0/u_{13}$ + $0/u_{14}$ + $0/u_{15}$ + $0/u_{16}$

Table 6 (*continued*)

$A_{14}= 0/u_1 + 0/u_2 + 0/u_3 + 0/u_4 + 0/u_5 + 0/u_6 + 0/u_7 + 0/u_8 + 0/u_9 + 0/u_{10} + 0/u_{11} +$
$0.5/u_{12} + 1/u_{13} + 0.5/u_{14} + 0/u_{15} + 0/u_{16}$

$A_{15}= 0/u_1 + 0/u_2 + 0/u_3 + 0/u_4 + 0/u_5 + 0/u_6 + 0/u_7 + 0/u_8 + 0/u_9 + 0/u_{10} + 0/u_{11} +$
$0/u_{12} + 0.5/u_{13} + 1/u_{14} + 0.5/u_{15} + 0/u_{16}$

$A_{16}= 0/u_1 + 0/u_2 + 0/u_3 + 0/u_4 + 0/u_5 + 0/u_6 + 0/u_7 + 0/u_8 + 0/u_9 + 0/u_{10} + 0/u_{11} +$
$0/u_{12} + 0/u_{13} + 0.5/u_{14} + 1/u_{15} + 0.5/u_{16}$

The data set of TAIFEX and corresponding fuzzy sets are shown in the appendix of Table A-2.

Step 3. The fuzzy relationships are established and grouped in Table 7.

Step 4. The frequencies of fuzzy sets in each fuzzy relationship are calculated, and shown in Table 8.

Step 5. Daily changes in the TAIEX are used as knowledge to select the proper fuzzy sets for forecasting (listed in Table 8). Hence, the trend specifies increase, decrease or no change. Then the variable in the weighted function is represented as $\alpha = 1$ for an increase, $\alpha = -1$ for a decrease, and $\alpha = 0$ for no change.

Table 7 Grouped TAIFEX fuzzy relationship

$A_1 \rightarrow$	$A_2 \rightarrow$
$A_3 \rightarrow$	$A_4 \rightarrow A_2 A_4 A_6$
$A_5 \rightarrow A_4$	$A_6 \rightarrow A_7$
$A_7 \rightarrow A_5, A_7, A_8, A_9$	$A_8 \rightarrow A_6, A_7, A_8, A_9, A_{10}$
$A_9 \rightarrow A_7, A_8, A_9$	$A_{10} \rightarrow A_8$
$A_{11} \rightarrow$	$A_{12} \rightarrow A_7, A_8, A_9$
$A_{13} \rightarrow A_{12}, A_{13}$	$A_{14} \rightarrow A_{14}, A_{15}$
$A_{15} \rightarrow A_{13}, A_{14}, A_{15}$	$A_{16} \rightarrow$

Table 8 Frequency of fuzz sets in TAIFEX relationship

$A_4 \rightarrow A_6$	1		$A_4 \rightarrow A_4$	1
$A_4 \rightarrow A_2$	1		$A_5 \rightarrow A_4$	1
$A_6 \rightarrow A_7$	2		$A_7 \rightarrow A_9$	1
$A_7 \rightarrow A_8$	2		$A_7 \rightarrow A_7$	4
$A_7 \rightarrow A_5$	1		$A_8 \rightarrow A_{10}$	1
$A_8 \rightarrow A_9$	1		$A_8 \rightarrow A_8$	4
$A_8 \rightarrow A_7$	2		$A_8 \rightarrow A_6$	1
$A_9 \rightarrow A_9$	2		$A_9 \rightarrow A_8$	2
$A_9 \rightarrow A_7$	1		$A_{10} \rightarrow A_8$	1
$A_{12} \rightarrow A_{13}$	1		$A_{12} \rightarrow A_{12}$	4
$A_{12} \rightarrow A_9$	1		$A_{13} \rightarrow A_{13}$	3
$A_{13} \rightarrow A_{12}$	2		$A_{14} \rightarrow A_{15}$	1
$A_{14} \rightarrow A_{14}$	1		$A_{15} \rightarrow A_{14}$	1
$A_{15} \rightarrow A_{13}$	1		$A_{15} \rightarrow A_{15}$	1

Hence, according to Definition 3.1, the selection strategy of fuzzy sets is: if the trend in TAIEX leads to an increase, then the fuzzy sets in the high part are all selected. Conversely, if the trend in TAIEX leads to a decrease, then the fuzzy sets in the low part are all selected. Otherwise, if the trend in TAIEX leads to no change, then the origin fuzzy set is selected.

Step 6. According to the grouped fuzzy relationships in Table 8, the corresponding weighted functions can be established and are listed as follows.

$$h_4(\alpha)\big|_{\alpha=-1,0,1} = \frac{\alpha(\alpha-1)}{2}\frac{1}{2}(-m_2 + \frac{1}{2}m_4) + \frac{2(1-\alpha)(1+\alpha)}{2}m_4 + \frac{\alpha(\alpha+1)}{2}\frac{1}{2}(-m_4 + \frac{1}{2}\alpha m_6),$$

$$h_5(\alpha)\big|_{\alpha=-1,0,1} = \frac{\alpha(\alpha-1)}{2}m_4 + \frac{2(1-\alpha)(1+\alpha)}{2}m_5 + \frac{\alpha(\alpha+1)}{2}m_5,$$

$$h_6(\alpha)\big|_{\alpha=-1,0,1} = \frac{\alpha(\alpha-1)}{2}m_6 + \frac{2(1-\alpha)(1+\alpha)}{2}m_6 + \frac{\alpha(\alpha+1)}{2}m_7,$$

$h_7(\alpha)\,|_{\alpha=-1,0,1}=$

$$\frac{\alpha(\alpha-1)}{2}\frac{1}{5}(-m_5+\frac{4}{5}m_7)+\frac{2(1-\alpha)(1+\alpha)}{2}m_7+\frac{\alpha(\alpha+1)}{2}\frac{4}{7}(-m_7+\frac{2}{7}m_8+\frac{1}{7}m_9),$$

$h_8(\alpha)\,|_{\alpha=-1,0,1}$

$$=\frac{\alpha(\alpha-1)}{2}\frac{1}{7}(-m_6+\frac{2}{7}m_7+\frac{4}{7}m_8)+\frac{2(1-\alpha)(1+\alpha)}{2}m_8+\frac{\alpha(\alpha+1)}{2}\frac{4}{6}(-m_8+\frac{1}{6}m_9+\frac{1}{6}m_{10}),$$

$$h_9(\alpha)\,|_{\alpha=-1,0,1}=\frac{\alpha(\alpha-1)}{2}\frac{1}{5}(-m_7+\frac{2}{5}m_8+\frac{2}{5}m_9)+\frac{2(1-\alpha)(1+\alpha)}{2}m_9+\frac{\alpha(\alpha+1)}{2}m_9,$$

$$h_{10}(\alpha)\,|_{\alpha=-1,0,1}=\frac{\alpha(\alpha-1)}{2}m_8+\frac{2(1-\alpha)(1+\alpha)}{2}m_{10}+\frac{\alpha(\alpha+1)}{2}m_{10},$$

$h_{12}(\alpha)\,|_{\alpha=-1,0,1}=$

$$\frac{\alpha(\alpha-1)}{2}\frac{1}{5}(-m_9+\frac{4}{5}m_{12})+\frac{2(1-\alpha)(1+\alpha)}{2}m_{12}+\frac{\alpha(\alpha+1)}{2}\frac{4}{5}(-m_{12}+\frac{1}{5}m_{13}),$$

$$h_{13}(\alpha)\,|_{\alpha=-1,0,1}=\frac{\alpha(\alpha-1)}{2}\frac{2}{5}(-m_{12}+\frac{3}{5}m_{13})+\frac{2(1-\alpha)(1+\alpha)}{2}m_{13}+\frac{\alpha(\alpha+1)}{2}m_{13},$$

$$h_{14}(\alpha)\,|_{\alpha=-1,0,1}=\frac{\alpha(\alpha-1)}{2}m_{14}+\frac{2(1-\alpha)(1+\alpha)}{2}m_{14}+\frac{\alpha(\alpha+1)}{2}\frac{1}{2}(-m_{14}+\frac{1}{2}m_{15})$$

$$h_{15}(\alpha)\,|_{\alpha=-1,0,1}=\frac{\alpha(\alpha-1)}{2}\frac{1}{3}(-m_{13}+\frac{1}{3}m_{14}+\frac{1}{3}m_{15})+\frac{2(1-\alpha)(1+\alpha)}{2}m_{15}+\frac{\alpha(\alpha+1)}{2}m_{15}.$$

That is, m_k is the midpoint of the interval u_k, and $m_1=6150$, $m_2=6250$, $m_3=6350$, $m_4=6450$, $m_5=6550$, $m_6=6650$, $m_7=6750$, $m_8=6850$, $m_9=6950$, $m_{10}=7050$, $m_{11}=7150$, $m_{12}=7250$, $m_{13}=7350$, $m_{14}=7450$, $m_{15}=7550$, $m_{16}=7650$, respectively.

Step 7. Subsequently, suppose that input x_{t-1} in date $t-1$ is fuzzified to $F(t-1)=A_i$, the corresponding weighted function $h_i(\alpha)$ is selected. Accordingly, the output is derived as $h_i(\alpha)\,|_{\alpha=-1,0,1}$. The following examples are used to demonstrate the procedure of selecting the corresponding weighted function and using the knowledge to derive the forecasts.

[1998/8/4]: The fuzzy set of 1998/8/3 is A_{15} (TAIFEX was 7552). The proper weighted function $h_{15}(\alpha)$ is selected. The TAIEX was 7599 on 1998/8/3 and 7593 on 1998/8/4. The difference between these two days is -6, the trend leads the knowledge to a decrease, so $\alpha=-1$. Hence, the forecast for 1998/8/4 is

$$h_{15}(\alpha)\,|_{\alpha=-1} = \frac{1}{3}m_{13} + \frac{1}{3}m_{14} + \frac{1}{3}m_{15} = 7450\,.$$

[1998/8/6]: The fuzzy set of 1998/8/5 is A_{14} (TAIFEX was 7486). The proper weighted function $h_{14}(\alpha)$ is selected. The TAIEX were 7500 on 1998/8/3 and 7472 on 1998/8/6. The difference between these two days is -28, the trend leads the knowledge to a decrease, so $\alpha=-1$. Hence, the forecast for 1998/8/6 is

$$h_{14}(\alpha)\,|_{\alpha=-1} = m_{14} = 7450\,.$$

[1998/8/7]: The fuzzy set of 1998/8/6 is A_{14} (TAIFEX was 7462). The proper weighted function $h_{14}(\alpha)$ is selected. The TAIEX were 7472 on 1998/8/6 and 7530 on 1998/8/7. The difference between these two days is 58, the trend leads the knowledge to an increase, so $\alpha=1$. Hence, the forecast for 1998/8/7 is

$$h_{14}(\alpha)\,|_{\alpha=1} = \frac{1}{2}m_{14} + \frac{1}{2}m_{15} = 7500\,.$$

[1998/8/10]: The fuzzy set of 1998/8/7 is A_{15} (TAIFEX was 7530). The proper weighted function $h_{15}(\alpha)$ is selected. The TAIEX were 7530 on 1998/8/7 and 77372 on 1998/8/10. The difference between these two days is -158, the trend leads knowledge to a decrease, so $\alpha=-1$. The forecast for 1998/8/10 is

$$h_{A_5}(\alpha)\,|_{\alpha=-1} = \frac{1}{3}m_{13} + \frac{1}{3}m_{14} + \frac{1}{3}m_{15} = 7450\,.$$

Table A-3 in the appendix shows all of the remaining forecasts.

Table 9 compares various studies of fuzzy time series used to forecast TAIFEX. From left to right, the columns in Table 10 present the forecasts by Chen (2002), by Huarng's knowledge models (2001), by this proposed weighted model. The average forecast errors are 1.05%, 1.06%, 0.94%, respectively. Clearly, the proposed model outperforms Chen's model and Huarng's model.

Table 9 Comparison of TAIFEX forecasts

Data Set (1998)	Index	Chen [7]	Huarng [1]	This Proposed Model
8/3	7552			
8/4	7560	7450	7450	7450
8/5	7487	7450	7450	7450
8/6	7462	7500	7450	7450
8/7	7515	7500	7500	7500

Table 9 (*continued*)

8/10	7365	7450	7450	7450
8/11	7360	7300	7350	7350
8/12	7330	7300	7300	7310
8/13	7291	7300	7350	7350
8/14	7320	7183.33	7100	7190
8/15	7300	7300	7350	7350
8/17	7219	7300	7300	7310
8/18	7220	7183.33	7100	7190
8/19	7285	7183.33	7300	7270
8/20	7274	7183.33	7100	7190
8/21	7225	7183.33	7100	7190
8/24	6955	7183.33	7100	7190
8/25	6949	6850	6850	6870
8/26	6790	6850	6850	6870
8/27	6835	6775	6650	6710
8/28	6695	6850	6750	6792.857
8/29	6728	6750	6750	6750
8/31	6566	6775	6650	6710
9/1	6409	6450	6450	6450
9/2	6430	6450	6550	6550
9/3	6200	6450	6350	6350
9/4	6403.2	6450	6250	6250
9/5	6697.5	6450	6550	6550
9/7	6722.3	6750	6750	6750
9/8	6859.4	6775	6850	6807.143
9/9	6769.6	6850	6750	6792.857
9/10	6709.75	6775	6650	6710
9/11	6726.5	6775	6850	6807.143

Table 9 (*continued*)

9/14	6774.55	6775	6850	6807.143
9/15	6762	6775	6650	6710
9/16	6952.75	6775	6850	6807.143
9/17	6906	6850	6950	6950
9/18	6842	6850	6850	6870
9/19	7039	6850	6950	6900
9/21	6861	6850	6850	6850
9/22	6926	6850	6950	6900
9/23	6852	6850	6850	6870
9/24	6890	6850	6950	6900
9/25	6871	6850	6850	6838.889
9/28	6840	6850	6750	6792.857
9/29	6806	6850	6750	6792.857
9/30	6787	6850	6750	6792.857
	1.05%	1.06%		0.94%

4.4 Forecasting TAIFEX with High-Order Model

According to the previous definition of TAIFEX, the author proposed the empirical analysis of knowledge second order model as follows. The data range from August 3 to September 30, 1988.

Step 1. From the historical data in Table 8, $U=[6100, 7700]$ is derived. Then, the universe of the discourse is divided into 16 equally long intervals $u_1, u_2, ..., u_{16}$ of length 100, where $u_1=[6100, 6200]$, $u_2=[6200, 6300]$, $u_3=[6300, 6400]$, $u_4=[6400, 6500]$, $u_5=[6500, 6600]$, $u_6=[6600, 6700]$, $u_7=[6700, 6800]$, $u_8=[6800, 6900]$, $u_9=[6900, 7000]$, $u_{10}=[7000, 7100]$, $u_{11}=[7100, 7200]$, $u_{12}=[7200, 7300]$, $u_{13}=[7300, 7400]$, $u_{14}=[7400, 7500]$, $u_{15}=[7500, 7600]$, $u_{16}=[7600, 7700]$.

Step 2. In this case, the linguistic variable "TAIFEX" which can be represented as 16 fuzzy sets; A_i ($i=1, 2, ...,16$). The linguistic values are A_1=(lowest), A_2=(very very very low), A_3=(very very low), A_4=(very low), A_5=(low), A_6=(quite low) , A_7=(low medium), A_8=(medium), A_9=(quite medium), A_{10}=(medium high),

A_{11}=(quite high), A_{12}=(high), A_{13}=(very high), A_{14}=(very very high), A_{15}=(very very very high), A_{16}=(highest). Each A_i (i=1, 2, …, 16) is defined as follows.

A_1= $1/u_1$ + $0.5/u_2$ + $0/u_3$ + $0/u_4$ + $0/u_5$ + $0/u_6$ + $0/u_7$ + $0/u_8$ + $0/u_9$ + $0/u_{10}$ + $0/u_{11}$ + $0/u_{12}$ + $0/u_{13}$ + $0/u_{14}$ + $0/u_{15}$ + $0/u_{16}$,

A_2= $0.5/u_1$ + $1/u_2$ + $0.5/u_3$ + $0/u_4$ + $0/u_5$ + $0/u_6$ + $0/u_7$ + $0/u_8$ + $0/u_9$ + $0/u_{10}$ + $0/u_{11}$ + $0/u_{12}$ + $0/u_{13}$ + $0/u_{14}$ + $0/u_{15}$ + $0/u_{16}$,

A_3= $0/u_1$ + $0.5/u_2$ + $1/u_3$ + $0.5/u_4$ + $0/u_5$ + $0/u_6$ + $0/u_7$ + $0/u_8$ + $0/u_9$ + $0/u_{10}$ + $0/u_{11}$ + $0/u_{12}$ + $0/u_{13}$ + $0/u_{14}$ + $0/u_{15}$ + $0/u_{16}$,

A_{14}= $0/u_1$ + $0/u_2$ + $0/u_3$ + $0/u_4$ + $0/u_5$ + $0/u_6$ + $0/u_7$ + $0/u_8$ + $0/u_9$ + $0/u_{10}$ + $0/u_{11}$ + $0.5/u_{12}$ + $1/u_{13}$ + $0.5/u_{14}$ + $0/u_{15}$ + $0/u_{16}$,

A_{15}= $0/u_1$ + $0/u_2$ + $0/u_3$ + $0/u_4$ + $0/u_5$ + $0/u_6$ + $0/u_7$ + $0/u_8$ + $0/u_9$ + $0/u_{10}$ + $0/u_{11}$ + $0/u_{12}$ + $0.5/u_{13}$ + $1/u_{14}$ + $0.5/u_{15}$ + $0/u_{16}$,

A_{16}= $0/u_1$ + $0/u_2$ + $0/u_3$ + $0/u_4$ + $0/u_5$ + $0/u_6$ + $0/u_7$ + $0/u_8$ + $0/u_9$ + $0/u_{10}$ + $0/u_{11}$ + $0/u_{12}$ + $0/u_{13}$ + $0.5/u_{14}$ + $1/u_{15}$ + $0.5/u_{16}$.

The data set of TAIFEX and corresponding fuzzy sets are shown in Table 8.

Step 3. The second order fuzzy relationships are established and grouped in Table 10.

Step 4. Daily changes in the TAIEX are used as the knowledge to select the proper fuzzy sets for forecasting (listed in the appendix of Table A-4). Hence, the trend specifies increase, decrease or no change. Then the variable in the weighted function is represented as α=1 for an increase, α=0 for a decrease, and α=-1 for no change.

Hence, if there are ambiguities in the fuzzy relationships, then the selection strategy is: if the trend in TAIEX leads to an increase, then the fuzzy sets in the high part are all selected. Otherwise, if the trend in TAIEX leads to a decrease, then the fuzzy sets in the low part are all selected. Otherwise, if the trend in TAIEX leads to no change, then the origin fuzzy set is selected.

Table 10 Second order grouped TAIFEX fuzzy relationship

$A_{15}, A_{15} \rightarrow A_{14}$
$A_{15}, A_{14} \rightarrow A_{14}$
$A_{15}, A_{13} \rightarrow A_{13}$
$A_{14}, A_{15} \rightarrow A_{13}$
$A_{14}, A_{14} \rightarrow A_{15}$
$A_{13}, A_{13} \rightarrow A_{12}, A_{13}$
$A_{13}, A_{12} \rightarrow A_{12}, A_{13}$
$A_{12}, A_{13} \rightarrow A_{13}$
$A_{12}, A_{12} \rightarrow A_9, A_{12}$
$A_{12}, A_9 \rightarrow A_9$
$A_{10}, A_8 \rightarrow A_9$
$A_4, A_4 \rightarrow A_2$
$A_4, A_2 \rightarrow A_4$
$A_2, A_4 \rightarrow A_6$

Accordingly, let the grouped fuzzy relationship for forecasting $TAIFEX_t$ (TAIFEX at time t) be $A_{r_1}, A_{r_2} \rightarrow A_{j_1}, A_{j_2}, \cdots$. The knowledge function is set as $h(\alpha; A_{r_2}; A_{j_1}, A_{j_2}, \cdots)|_{\alpha=1, 0, -1}$.

Step 5. Subsequently, the second-order forecasting process of TAIFEX $F(t)$ is carried out by the fuzzified input of $F(t$-2$)$ and $F(t$-1$)$. Some examples below are used to illustrate the forecasting process.

[1998/8/5]: The TAIFEX in 1998/8/3 and 1998/8/4 were 7552 (A_{15}) and 7560 (A_{15}). According to the list of second order fuzzy relationship in Table 13, the current states " A_{15}, A_{15} " mapping to the suitable fuzzy relationship is $A_{15}, A_{15} \rightarrow A_{14}$. Because the maximum membership value of the fuzzy set A_{14} occurs at the interval u_{14}, then the midpoint of the interval u_{14} is 7450. Thus, the forecasted TAIFEX of 1998/8/5 is equal to 7450.

[1998/8/6]: The TAIFEX in 1998/8/4 and 1998/8/5 were 7560 (A_{15}) and 7487 (A_{14}). According to the list of second order fuzzy relationship in Table 13, the current states " A_{15}, A_{14} " mapping to the suitable fuzzy relationship is $A_{15}, A_{14} \rightarrow A_{14}$. Because the maximum membership value of the fuzzy set A_{14} occurs at the interval u_{14}, then the midpoint of the interval u_{14} is 7450. Thus, the forecasted TAIFEX of 1998/8/5 is equal to 7450

[1998/8/12]: The TAIFEX in 1998/8/10 and 1998/8/11 were 7365 (A_{13}) and 7360 (A_{13}). According to the list of second order fuzzy relationship in Table 13, the current states " A_{13}, A_{13} " mapping to the suitable fuzzy relationships is $A_{13}, A_{13} \rightarrow A_{12}, A_{13}$. It means that there is an ambiguity. The TAIEX were 7384 on 1998/8/11 and 7352 on 1998/8/12, respectively. The difference between these TAIEX was -32, the trend is positive and $\alpha = -1$. Hence, the knowledge function is $h(\alpha; A_{13}; A_{12}, A_{13})|_{\alpha=-1} = (m_{12} + m_{13}) / 2 = 7300$, where m_{13} is the midpoint of the interval u_{13}.

Table 15 compares various studies of fuzzy time series used to forecast TAIFEX. Mean square errors (*MSE*s) are taken as forecasting errors:

$$MSE = \frac{\sum_{i=1}^{n} (acturall\_TAIFEX - forecasted\_TAIFEX)^2}{n},$$

where i represents the year. From left to right, the columns in Table 15 present the forecasts by Chen's model (Chen 1996), by Huarng's two-variable model and by his three-variable knowledge model (Huarng 2001). The *MSE*s are 9668.94, 7856.5 and 5437.38, respectively.

Table 15 A comparison of the MSE of previous models

	Chen's model (1996)	Huarng's two variable model	Huarng's three variable model
MSE	9668.94	7856.50	5437.58

Table 16 compares the *MSE*s of Chen's restricted model and the proposed knowledge high-order fuzzy time series model. That is to say, Chen's later model (2002) is restricted in lower-order fuzzy time series and lacking the ability to handle the ambiguity well. The averaging operation in Chen's model (1996) is applied to assist to eliminate the ambiguity in Chen's later model (2002) to compare the performance obtained using fuzzy time series of different orders.

Table 16 A comparison of the MSE of Chen's model and this proposed model by using different orders fuzzy time series

Second-order		Third-order		Fourth-order		Fifth-order	
Chen	Knowledge	Chen	Knowledge	Chen	Knowledge	Chen	Knowledge
5900.64	4109.09	3209.98	3052.19	1999.09	1830.2	864.64	864.64

From the left to right, the columns in Table 16 present the *MSE* of Chen second-order model, the knowledge second-order model, Chen third-order model, the knowledge third-order model, Chen fourth-order model, the knowledge fourth-order model, Chen fifth-order model and the knowledge fifth-order model, respectively. The *MSE*s are 5900.64, 4109.09, 3209.98, 3052.19, 1999.09, 1830.2, 864.64 and 864.64, respectively. Obviously, the forecasting accuracy is better than that of Chen's model of the same order. Therefore, the knowledge high-order fuzzy time series model represents an improvement over the Chen's model.

5   Conclusions

Most fuzzy time series models are independent of a specific domain. Among these models, the Chen model uses the simple and straightforward method to find the best forecasting results. In the field of expert systems, experts typically consider knowledge to solve domain-specific problems. Hence, Huarng enhanced the Chen model by integrating knowledge. The weighted model overcomes the disadvantage of the Huarng model, that is, a lack of an efficient measure of the significance of the knowledge. The first proposed model was based on the weighted measure of the fuzzy sets, which differs from the arithmetic average in the traditional defuzzifier. The significance of the derived fuzzy sets was considered in the defuzzification phase. The knowledge model is proposed to forecast time series based on the high-order fuzzy time series and domain-specific knowledge. The proposed model overcomes the deficiency of the Chen model, which is strongly dependent on the highest-order fuzzy time series and requires a large amount of memory.

The results showed that the weighted models can reflect fluctuations in fuzzy time series and provide superior overall forecasting results compared to previous models. The forecasts of university enrollment and the futures index show that domain-specific knowledge can be used with ease to assist forecasting. The efficient measure of the significance of fuzzy relationships provides additional information for improving forecasts. Empirical analysis showed that the proposed high-order model yielded more accurate forecasts than the Chen model when using the same orders. Therefore, the knowledge high-order fuzzy time series model offers the advantages of high-order time series forecasting and the elimination of ambiguity; that is, the forecasting model can be restricted to the acceptable-order fuzzy time series to reduce the amount of memory and computation time required.

References

Box, G.E.P., Jenkins, G.M.: Time series analysis: forecasting and control. Prentice-Hall (1976)

Chen, M.-Y., Fan, M.-H., Chen, C.-C.: Forecasting stock price based on fuzzy time-series with equal-frequency partitioning and fast fourier transform algorithm. In: Computing, Communication and Application Conference, pp. 238–243 (2012)

Chen, S.M.: Forecasting enrollments based on fuzzy time series. Fuzzy Sets and Systems 81, 311–319 (1996)

Chen, S.M.: Forecasting enrollments based on high-order fuzzy time series. Cybernetics and Systems 33, 1–16 (2002)

Cheng, C.H., Chen, T.L., Teoh, H.J., Chiang, C.H.: Fuzzy time-series based on adaptive model for TAIEX forecasting. Expert Systems with Applications 34, 1126–1132 (2008)

Huarng, K.: Knowledge models of fuzzy time series for forecasting. Fuzzy Sets and Systems 123, 369–386 (2001)

Huarng, K.H., Yu, K.H.: A type 2 fuzzy time series model for stock index forecasting. Physica A: Statistical Mechanics and its Applications 353, 445–462 (2005)

Huarng, K.H., Yu, K.H.: Ratio-based lengths of intervals to improve fuzzy time series forecasting. IEEE Transactions on Systems, Man, and Cybernetics – Part B: Cybernetics 32, 328–340 (2006a)

Huarng, K.H., Yu, K.H.: The application of neural networks to forecast fuzzy time series. Physica A: Statistical Mechanics and its Applications 363, 481–491 (2006b)

Huarng, K.H., Yu, K.H., Hsu, Y.W.: A multivariate knowledge model for fuzzy time-series forecasting. IEEE Transactions on Systems, Man, and Cybernetics – Part B: Cybernetics 37, 836–846 (2007)

Hwang, J.R., Chen, S.M., Lee, C.H.: Handling forecasting problems using fuzzy time series. Fuzzy Sets and Systems 100, 217–228 (1998)

Janacek, G., Swift, L.: Time series forecasting, simulation, applications. Ellis Harwood (1993)

Jilani, T.A., Burney, S.M.A., Amjad, U., Siddiqui, T.A.: A particle swarm intelligence based fuzzy time series forecasting model. International Journal of Computer Applications 38, 47–52 (2011)

Leu, Y., Lee, C.P., Jou, Y.Z.: A distance-based fuzzy time series model for exchange rates forecasting. Expert Systems with Applications 36, 8107–8114 (2009)

Song, Q., Chissom, B.S.: Fuzzy time series and its models. Fuzzy Sets and Systems 54, 269–377 (1979)

Song, Q., Chissom, B.S.: Forecasting enrollments with fuzzy time series - part I, Fuzzy Sets and Systems. Fuzzy Sets and Systems 54, 1–9 (1993)

Song, Q., Chissom, B.S.: Forecasting enrollments with fuzzy time series- part II. Fuzzy Sets and Systems 62, 1–8 (1994)

Sullivan, J., Woodall, W.H.: A comparison of fuzzy forecasting and markov modeling. Fuzzy Sets and Systems 64, 279–293 (1994)

Tanuwijaya, K., Chen, S.M.: A new method to forecast enrollments using fuzzy time series and clustering techniques. In: Proceedings of the 2009 International Conference on Machine Learning and Cybernetics, Baoding, Hebei, China (2009a)

Wang, L.X.: A course in fuzzy systems and control. Prentice-Hall, Inc. (1997)

Yu, H.K., Huarng, K.H.: A bivariate fuzzy time series model to forecast the TAIEX. Expert Systems with Applications 34, 2945–2952 (2008)

Yu, T.H.K., Hurang, K.H.: Corrigendum to a bivariate fuzzy time series model to forecast the TAIFEX. Expert Systems with Applications 34, 2945–2952 (2010)

APPENDIX

Table A-1 Frequency of relationships

Fuzzy relationship	Frequency	Fuzzy relationship	Frequency
$A_1 \rightarrow A_1$	2	$A_1 \rightarrow A_2$	1
$A_2 \rightarrow A_3$	1	$A_3 \rightarrow A_3$	7
$A_3 \rightarrow A_4$	2	$A_4 \rightarrow A_3$	1
$A_4 \rightarrow A_4$	2	$A_4 \rightarrow A_6$	1
$A_6 \rightarrow A_6$	1	$A_6 \rightarrow A_7$	1
$A_7 \rightarrow A_6$	1	$A_7 \rightarrow A_7$	1

Table A-2 TAIFEX data set

Date (1998)	Index	Fuzzy Set	Date (1998)	Index	Fuzzy Set
8/3	7552	A_{15}	8/4	7560	A_{15}
8/5	7487	A_{14}	8/6	7462	A_{14}
8/7	7515	A_{15}	8/10	7365	A_{13}
8/11	7360	A_{13}	8/12	7330	A_{13}
8/13	7291	A_{12}	8/14	7320	A_{13}
8/15	7300	A_{13}	8/17	7219	A_{12}
8/18	7220	A_{12}	8/19	7285	A_{12}
8/20	7274	A_{12}	8/21	7225	A_{12}
8/24	6955	A_9	8/25	6949	A_9
8/26	6790	A_7	8/27	6835	A_8
8/28	6695	A_6	8/29	6728	A_7
8/31	6566	A_5	9/1	6409	A_4
9/2	6430	A_4	9/3	6200	A_2
9/4	6403.2	A_4	9/5	6697.5	A_6
9/7	6722.3	A_7	9/8	6859.4	A_8
9/9	6769.6	A_7	9/10	6709.75	A_7

Table A-2 (*continued*)

9/11	6726.5	A_7		9/14	6774.55	A_7
9/15	6762	A_7		9/16	6952.75	A_9
9/17	6906	A_9		9/18	6842	A_8
9/19	7039	A_{10}		9/21	6861	A_8
9/22	6926	A_9		9/23	6852	A_8
9/24	6890	A_8		9/25	6871	A_8
9/28	6840	A_8		9/29	6806	A_8
9/30	6787	A_7				

Table A-3 TAIEX data set

Date (1998)	Index	Difference	Date (1998)	Index	Difference
8/3	7599		8/4	7593	-6
8/5	7500	-93	8/6	7472	-28
8/7	7530	58	8/10	7372	-158
8/11	7384	12	8/12	7352	-32
8/13	7363	11	8/14	7348	-15
8/15	7372	24	8/17	7274	-98
8/18	7182	-92	8/19	7293	111
8/20	7271	-22	8/21	7213	-58
8/24	6958	-255	8/25	6908	-50
8/26	6814	-94	8/27	6813	-1
8/28	6724	-89	8/29	6736	12
8/31	6550	-186	9/1	6335	-215
9/2	6472	137	9/3	6251	-221
9/4	6463	212	9/5	6756	293
9/7	6801	45	9/8	6942	141
9/9	6895	-47	9/10	6804	-91
9/11	6842	38	9/14	6860	18

Table A-3 (*continued*)

9/15	6858	-2	9/16	6973	115
9/17	7001	28	9/18	6962	-39
9/19	7150	188	9/21	7029	-121
9/22	7034	5	9/23	6962	-72
9/24	6980	18	9/25	6980	0
9/28	6911	-69	9/29	6885	-26
9/30	6834	-51			

Table A-4 TAIEX data set

Date (1998)	Index	Difference
8/3	7599	
8/4	7593	-6
8/5	7500	-93
8/6	7472	-28
8/7	7530	58
8/10	7372	-158
8/11	7384	12
8/12	7352	-32
8/13	7363	11
8/14	7348	-15
8/15	7372	24
8/17	7274	-98
8/18	7182	-92
8/19	7293	111
8/20	7271	-22
8/21	7213	-58
8/24	6958	-255
8/25	6908	-50

Table A-4 (*continued*)

8/26	6814	-94
8/27	6813	-1
8/28	6724	-89
8/29	6736	12
8/31	6550	-186
9/1	6335	-215
9/2	6472	137
9/3	6251	-221
9/4	6463	212
9/5	6756	293
9/7	6801	45
9/8	6942	141
9/9	6895	-47
9/10	6804	-91
9/11	6842	38
9/14	6860	18
9/15	6858	-2
9/16	6973	115
9/17	7001	28
9/18	6962	-39
9/19	7150	188
9/21	7029	-121
9/22	7034	5
9/23	6962	-72
9/24	6980	18
9/25	6980	0
9/28	6911	-69
9/29	6885	-26
9/30	6834	-51

Chapter 8
A Wavelet Transform Approach
to Chaotic Short-Term Forecasting

Yoshiyuki Matsumoto and Junzo Watada

Abstract. Chaos theory is widely employed to forecast near-term future values of a time series using data that appear irregular. The chaotic short-term forecasting method is based on Takens' embedding theorem, which enables us to reconstruct an attractor in a multi-dimensional space using data that appear random but rather are deterministic and geometric in nature. It is difficult to forecast future values of such data based on chaos theory if the information that the data provide cannot be reconstructed through wavelet transformation in a sufficiently low-dimensional space. This paper proposes a method to embed data in a small-dimensional space. This method enables us to abstract the chaotic portion from the focal data and increase forecasting precision.

Chaotic methods are employed to forecast near-term future values of uncertain phenomena. The method makes it possible to restructure an attractor of given time-series data set in a multidimensional space using Takens' embedding theory. However, many types of economic time-series data are not sufficiently chaotic. In other words, it is difficult to forecast the future trend of such economic data even based on chaos theory. In this paper, time-series data are divided into wave components using a wavelet transform. Some divided components of time-series data exhibit much more chaotic behavior in the sense of correlation dimension than the original time-series data. The highly chaotic nature of the divided components enables us to precisely forecast the value or the movement of the time-series data in the near future. The up-and-down movement of the TOPICS value is shown to be well predicted by this method, with 70% accuracy.

Keywords: Chaos theory, Short-term forecasting, Wavelet transform.

Yoshiyuki Matsumoto
Faculty of Economics, Shimonoseki City University, 2-1-1,
Daigaku-cho, Shimonoseki, Yamaguchi 751-8510, Japan
e-mail: matsumoto@shimonoseki-cu.ac.jp

Junzo Watada
Graduate School of Information, Production and Systems, Waseda University,
2-7 Hibikino, Wakamatsu-ku, Kitakyushu, Fukuoka 808-0135, Japan
e-mail: junzow@osb.att.ne.jp

W. Pedrycz & S.-M. Chen (Eds.): Time Series Analysis, Model. & Applications, ISRL 47, pp. 177–197.
DOI: 10.1007/978-3-642-33439-9_8   © Springer-Verlag Berlin Heidelberg 2013

1 Introduction

Highly accurate and reliable forecasting systems are essential in modern society. In management and economic activities, forecasting systems play a pivotal role. For example, the decision making required for the development of a production schedule and marketing plan is based on forecasting of sales volume. An inaccurate forecast will have a serious effect on the production schedule and on marketing. When the forecasted demand for a certain product is underestimated, the opportunities for product sales are diminished, and the corporation will experience a loss in sales revenue. On the contrary, when the demand for a product is overestimated, the inventory quantity of the product will increase and the inventory-carrying cost for the corporation will increase. Both of these scenarios have the potential to reduce corporate profits. In a stock market and currency exchange, the forecast will also influence the corporate balance. Corporate inventors such as life insurance companies, other types of insurance companies, and commercial banks commit to capital investments with, for example, the sale or purchase of large amount of stocks and securities to obtain profits. Companies such as these conduct forecasting using economic time-series data and distribute their risks across various investments. Therefore, highly accurate forecasting systems are greatly needed.

Autoregression

Time-series data vary with time. Various methods have been studied to forecast based on time-series data. G.U. Yule, a British statistician, proposed the time-series model to express the varying number of solar spots. This model could forecast the number of solar spots well. The most widely used model for time-series data is the autoregression model, defined as follows:

$$y(t) = \mu' + \phi_1 y(t-1) + \phi_2 y(t-2) + \cdots + \phi_p y(t-p) + u_t \tag{1}$$

where μ', ϕ_1, ϕ_2,$\cdots$,ϕ_p denote parameters and u_t denotes an error variable called disturbance term.

Equation (1) consists of a weighted summation from time 1 to time p of terms $y(t-1)$ to $y(t-p)$ and the disturbance term u_t. This is called a p-th order autocorrelation model (AR(p)).

Another widely used model is the moving average (MA) model. This model is defined as follows:

$$y(t) = \mu + u_t - \theta_1 \mu_{t-1} - \theta_2 \mu_{t-2} - \cdots - \theta_q \mu_{t-q} \tag{2}$$

where μ, θ_1, θ_2,$\cdots$, and θ_q are parameters and u_t denotes an error variable called the disturbance term. In Equation (2), $y(t)$ consists of a weighted summation of the present time to q previous time, 1, $-\theta_1$, $-\theta_2$,$\cdots$,and $-\theta_q$. This model is called a q-th order moving average model (MA(q)).

A more general model is the autoregressive moving average (ARMA) model, which combines the AR and MA models.

$$y(t) = \mu' + u_t + \phi_1 y(t-1) + \phi_2 y(t-2) + \cdots + \phi_p y(t-p)$$
$$+\mu_t - \theta_1 \mu_{t-1} - \theta_2 \mu_{t-2} - \cdots - \theta_q \mu_{t-q} \qquad (3)$$

where μ', ϕ_1, $\phi_2, \cdots, \phi_p$, u_t, θ_1, $\theta_2, \cdots$, and θ_q are parameters. Equation (3) is obtained by combining Equations (1) and (2). Because the autoregressive portion of Equation (3) has p dimensions and the moving average (MA) portion Equation (3) has q dimensions, Equation (3) is called an ARMA(p,q) model.

When finite differences and seasonal variation are included so that the ARMA model can be applied to non-stationary and seasonal data, it is called the "Box-Jenkins" model, the best-known model used in time-series forecasting.

The autoregressive moving average model exhibits random walk behavior. For example, the simplest model, AR(1), is called a random walk model. As Equation (4) shows, the present value is obtained by adding a random value to the value of the previous term. This means that if we know the previous value, the present value can be evaluated. The random term is white noise, a probabilistic variable. The random term has a mean of 0 and some deviations σ.

$$y(t) = y(t-1) + u_t \qquad (4)$$

Time-series data in a random-walk state are influenced only by probabilistic rules, so it is impossible to forecast future movements. Random-walk states or approximately random-walk states are often encountered in the analysis of economic and management time-series data.

Chaotic System

A different forecasting method is employed to forecast the future behavior based on time-series data. In general, economic and management times-series data often exhibit complicated behaviors. When we employ linear equations such as auto-regression models, it is necessary to include a disturbance term to explain the complicated and irregular behaviors of time-series data. The disturbance term is a random variable, as mentioned above, and is not predictable. Tt is well-known, however, that a nonlinear equation in a chaotic system can express complicated phenomena without such disturbance terms.

For instance, the following equation is called a logistic map.

$$x(t+1) = 4x(t) \times (1 - x(t)). \qquad (5)$$

Even though this equation is simple and lacks additional variables, its behavior is quite complicated. This behavior is called chaotic. A chaotic system can express complicated behavior in a different manner than the stochastic process. The most important feature of behavior that can be expressed by a chaotic system is that this behavior is deterministic.

In other words, if times-series data behave in a complicated fashion and have chaotic characteristics, we can find the deterministic regularity. It is possible for us to forecast future behavior based on time-series data. In the real world, many such examples are found. For instance, the movement of the atmosphere including climate variation, is understood as a chaotic phenomenon. N. Lorenz [12] strictly analyzed the movement of the atmosphere using a differential equation and showed that a simple differential equation without any random term can express complicated movement. This means that, given the initial values of pressure, temperature, wind speed and so on, the movement of the atmosphere can be deterministically forecasted depending on physical laws. In the same way, if some economic or management time-series data can be expressed using a chaotic model, it is possible to deterministically forecast such economic or management values. Recently, various types of economic and management time-series data have been studied using chaos theory view to verify their chaotic characteristics. Fluctuations in gold and silver prices, the money supply, and GNP, among other economic indicators, have been as possibly having low-dimensional chaotic features [2] [7] [9]. However, this view has been rejected [1] [5] [16]. For example, most of the literature on the subject supports the position of that GNP des not exhibit chaotic behavior. This issue remains controversial and as yet unresolved. Although the authors of this paper do not consider economic and management time-series data to be low-dimensional chaotic in nature, it is possible that it is high-dimensional chaotic in nature. That is, it is difficult to forecast such economic and management time-series data based on a chaotic forecast model, but if we can successfully improve such chaotic forecasting models, it might be possible to forecast such values within some margin of error. The objective of this paper is to present a wavelet transformation method for improving the forecasting accuracy of chaotic models.

Wavelet Transform

The chaotic short-term forecasting method [8] based on time-series data enables us to predict future values that we could not predict before for some types of data of a chaotic nature. Nevertheless, it may still be difficult to correctly forecast values even for the near future for types of data that are not highly chaotic. Although such data are relatively less chaotic, it is possible to extract the partially chaotic portion from the data [13]. In this research, a wavelet transform [6] is employed to extract chaotic portions from original time-series data. We can identify the more highly chaotic component of the original data by measuring these correlated dimensions. Once we can successfully extract the highly chaotic portion from the original data, we can improve the forecasting precision. The correlation dimension [4] [17] of the transformed components should be smaller than that of the original data if the divided components are more highly chaotic than the original data.

In this research, time-series data are divided into parts using a wavelet transform. It will be shown that the divided orthogonal elements of time-series data can be employed to forecast more precisely than is possible with the original time-series data. Forecasted data are rebuilt into the original form using an inverse wavelet transform.

The remainder of this chapter consists of the following sections. Section 2 explains the basic concept of a chaotic model. The wavelet transform is explained in Section 3. Section 4 explains correlation dimension-based wavelet component selection for improving forecasting accuracy. The results are summerized in Section 5.

2 Chaos Theory

The chaotic nature of phenomena is believed to have been identified in the 19th century by H. Poincare [15]. Poincare proved that the motion equation for a system of three stars cannot be analytically solved if those three stars have correlation with each other. In explaining this motion, he proposed that it be characterized as a complicated nonperiodic motion, that is, a type of chaotic behavior. A paper written by N. Lorenz [12] in 1963 was the starting point in the study of chaotic phenomena. He clarified that a deterministic system can produce nonperiodic motion and that the error can become increasingly large with time if a small change is given as an initial condition. The word "chaos" was first used in reference to this type of system in a paper written by T. Y. Li and J. A. Yorke [11]. Since then, many studies of chaotic systems have been conducted. In this paper, we propose an application of the chaotic model to time-series forecasting. In this section, we explain the basic concepts of the chaotic model, including the correlation dimension that is used in chaotic forecasting and chaotic decision making.

2.1 Chaotic Feature

The term "chaos" has its origins in Greek [$\chi\alpha o\varsigma$, keias]. The common understanding of chaos is a greatly disturbed state, different from an ordered state. Although the scientific meaning of the term "chaos" in science is a disturbed state, the scientific meaning does not necessarily suggest a greatly disturbed state but rather a phenomenon disturbed to a moderate degree, which changes irregularly over time in some manner.

In other words, the scientific meaning of the term "chaos" refers to an irregularity of a changing phenomenon that is controlled by relatively simple rules. A typical example of the chaos system is a logistic map. Logistic mapping can be defined by a simple relationship. The resulting state, however, may seem to exhibit quite random movement on a graph. One characteristic of a chaotic system is sensitivity to the initial state.

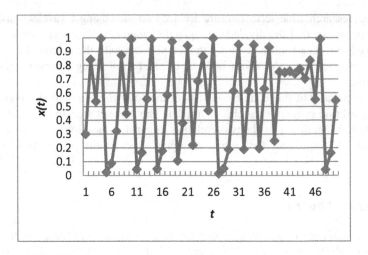

Fig. 1 Logistic map

When an initial state of a chaotic system is changed slightly, the mapping of the system's behavior by a chaotic model exhibits almost the same trajectory in the initial state but shows a very different trajectory beyond some short time. This phenomenon is referred to as a sensitivity to the initial state. Because of sensitivity to the initial state, it is not appropriate to employ the chaos method in forecasting a value in the distant future. That is, it is only possible to predict the state in the near future that is sufficiently influenced by the present state.

The logistic map written in equation (5) follows a simple rule, but the result graph appears to be quite a complicated movement. Equation (5) has only one variable, $x(t)$, but behaves in a complicated manner. Figure 1 illustrates equation (5) for an initial value of 0.3 .

Various functions exist that show irregular movements, such as the following.

$$
\begin{aligned}
x(t+1) &= 2x(t) \quad (0 \le x(t) \le 0.5)\\
&= 2 - 2x(t) \ (0.5 < x(t) \le 1)
\end{aligned} \tag{6}
$$

$$
\begin{aligned}
x(t+1) &= 2x(t) \quad (0 \le x(t) \le 0.5)\\
&= 2x(t) - 1 \ (0.5 < x(t) \le 1)
\end{aligned} \tag{7}
$$

Equation (6) is a tent map, and equation (7) is called a Bernoulli shift. These functions exhibit irregular changes, as does a logistic map.

These functions appear complicated, but mapping them reveals simple behavior. For example, $x(t)$ is mapped to $x(t+1)$ as follows.

$$
x(t) = f(x(t)) \tag{8}
$$

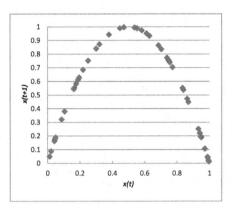

Fig. 2 Logistic map

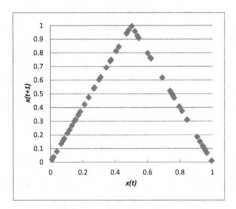

Fig. 3 Tent map

A logistic map is shown as follows.

$$x(t) = f(x(t)) = 4x(t) \times (1 - x(t)). \tag{9}$$

This map is called a one-dimensional map if the value $x(t+1)$ is determined only by the value of $x(t)$. Fig. 1 becomes Fig. 2 if the vertical axis $x(t+1)$ and the horizontal axis $x(t)$ are used. In the same way, a tent map and a Bernoulli shift are illustrated in Fig. 3 and Fig. 4, respectively. Even if, at first glance, a plot of time-series data appears complicated, some simple regularity can often be found by applying a map.

In Figs. 2 to 4, all of the curves have two values of $x(t)$ for the same value of $x(t+1)$. This means the vurves have steep gradients at that point. That is, the two values originated from a similar value and quickly separated. For example, let the difference between two initial values be Δx. In the tent map and the Bernoulli shift, the gradient of a graph is 2 at place. This simplifies the discussion. In this case,

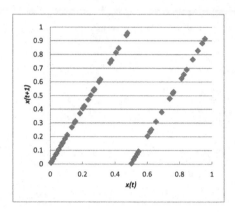

Fig. 4 Bernoulli shift

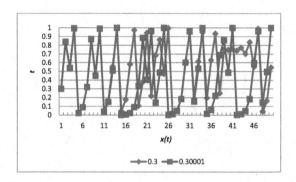

Fig. 5 The comparison of two logistic function using two near initial values

one mapping produces result 2 times larger than Δx. When the difference of two initial values is 10^{-5}, 16 times of maps will produce $\Delta x \times 2^{16} = 10^{-5} \times 2^{16} = 0.66$. When the difference exceeds 0.5, it is highly possible that both of the values will occur on both sides of $x(t) = 0.5$. In this case, the difference will not be double, but the difference in behavior will be evident. This phenomenon is called initial value sensitivity. To illustrate this phenomenon, we show two logistic graphs started with two initial values 0.3 and 0.30001 in Fig. 5, respectively. The two graphs exhibit similar behaviors up to t of 12. Beyond that point, the values are separated. When t is greater than 17, the two behaviors are quite different. When the difference is greater than 0.5, the line is folded, and the behaviors become different. "Expanding" and "folding" are chaotic features.

2.2 Chaotic Forecasting

The objectives of employing the chaos method in forecasting are the following: 1) to find a deterministic structure in a given set of time-series data and 2) to predict a value using this structure for a certain point in time that is in the near future from the present state. Fig. 5 shows the comparison of two logistic functions using two similar initial values that are sufficiently influenced . The chaos method enables us to forecast with high precision the short-term future using time-series data that exhibit highly unpredictable and nonperiodic changes.

This forecasting approach is based on Takens' embedding theory, which states that it is possible to restructure the trajectory of a dynamic system in a high-dimensional space using only the information (that is, time-series data) of one component dimension (variable).

Using time-series data $x(t)$, let us define vector $\mathbf{z}(t)$ as follows:

$$\mathbf{z}(t) = (x(t), x(t-\tau), x(t-2\tau), \cdots ,$$
$$x(t-(n-1)\tau)) \tag{10}$$

Where τ denotes an arbitrary constant time interval. The vector $\mathbf{z}(t)$ shows one point in an n-dimensional space(Data Space). Therefore, changing t generates a trajectory in the n dimensional data space. When n is sufficiently large, this trajectory exhibits a smooth change in the high-dimensional dynamic system. That is, if the dynamic system has some attractor, an attractor transformed from the original one should appear on the data space. In other words, the original attractor of the dynamic system can be embedded in the n dimensional topological space. The number n is called an embedded dimension. Denoting the dimension of the original dynamic system by m, it can be proven that this n dimension is sufficiently large if n holds the following:

$$n = 2m + 1 \tag{11}$$

Equation (11) is a sufficient condition for the embedded dimension. It is required to employ data with more than 3 to 4 samples over time in short-term forecasting.

Next, let us describe the deterministic structure using a restructured trajectory. There are several methods. Figure 6 illustrates short-term forecasting using the chaos method that is embedding discrete time-series data with the equal time interval $\tau = 15$ in embedded dimension $n = 3$.

Observed discrete time-series samples can be mapped into a topological space of 3 embedded dimensions, as shown in Figure 7. As a result, the mapped vector is denoted as follows:

$$\mathbf{z}(i) = (x(i), x(i-1), x(i-2)) \tag{12}$$

Let $\mathbf{z}(i)$ denote a vector of 3 dimensions in which observed data, including the most recent time are mapped on a topological space. Figure 8 illustrates the relationship of the data that are mapped around the neighborhood of $\mathbf{z}(i)$ in the 3-dimensional. These data in the neighborhood of $\mathbf{z}(i)$ are the data observed in the past. The trajectory of $\mathbf{z}(i+1)$ at one step in the future is shown in Figure 8. These

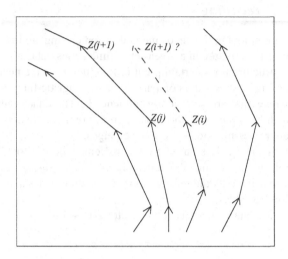

Fig. 6 Forecasting method based on chaotic model

relationships enable us to forecast behavior $\mathbf{z}(i+1)$ in the near future. The future trajectory $x(i+1)$ of the given time-series data $\{x(t),x(t-1),\ldots\}$ can be calculated the following stepts: 1) deciding the nearest point $\mathbf{z}(j)$ included in the neighborhood with diameter ε from $\mathbf{z}(i)$, 2) calculating the distance (I_{t+1}) between $\mathbf{z}(i+1)$ and $\mathbf{z}(j+1)$ using the Jacobian matrix A_j of the nearest point $\mathbf{z}(j)$ and the distance (I_t) between $\mathbf{z}(i)$ and $\mathbf{z}(j)$, and 3) deciding the trajectory $x(i+1)$ in one step in the future of the original time-series data.

2.3 Measurement of the Correlation Dimension

Measurement of the correlation dimension is usually employed to evaluate whether a given set of time-series data is chaotic.

The method of correlation dimension is employed in the assessment of whether time-series data has a chaotic structure by checking whether the time-series data are distributed in a lower-dimensional space than m dimensions, assuming that the data are embedded in an m-dimensional space.

First, let us embed the time-series data into an m-dimensional space.

Then, we draw a circle with radius r at the center of the points that each embedded vector has. We count how many points are included within the drawn circle and denote this number by C. When the radius is large, the number should be included .

Therefore, as C is an increasing function of r, let us denote it as $C(r)$.

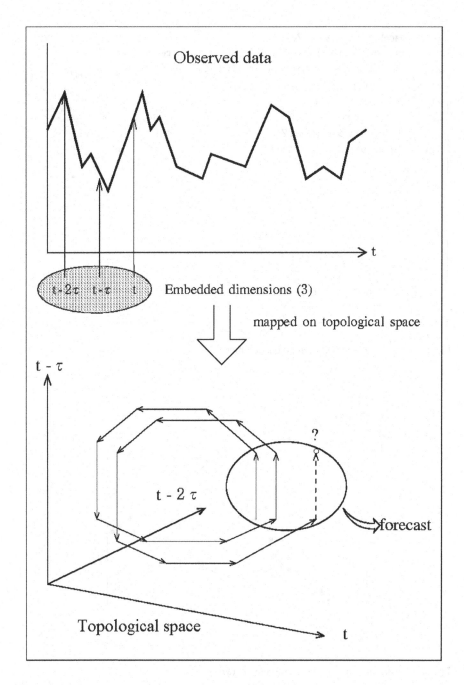

Fig. 7 Forecasting method based on chaotic model

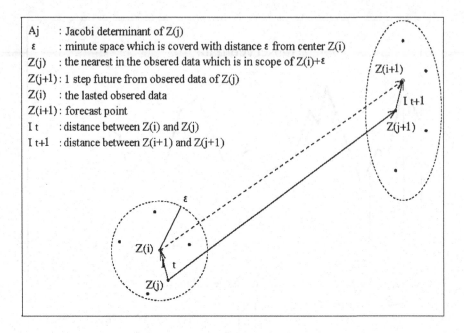

Aj   : Jacobi determinant of Z(j)
ε   : minute space which is coverd with distance ε from center Z(i)
Z(j)   : the nearest in the obsered data which is in scope of Z(i)+ε
Z(j+1) : 1 step future from obsered data of Z(j)
Z(i)   : the lasted obsered data
Z(i+1) : forecast point
I t   : distance between Z(i) and Z(j)
I t+1   : distance between Z(i+1) and Z(j+1)

Fig. 8 Forecasting method based on chaotic model

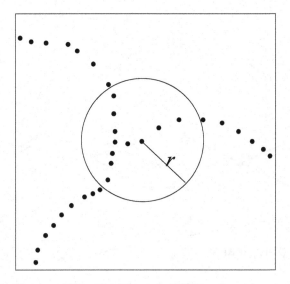

Fig. 9 Measurement of correlation dimension (a)

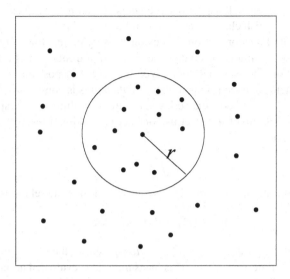

Fig. 10 Measurement of correlation dimension (b)

If the plotted points are distributed evenly in the m-dimensional space, the number of points included within the circle should increase in proportion to the area of the circle as r increases.

$$C(r) = ar^m \tag{13}$$

In contrast, if the structure has any regularity, $C(r)$ should increase in proportion to a value less than the m-powered value:

$$C(r) = br^{(m-x)} \tag{14}$$

The value $(m-x)$ is called the correlation dimension.

In the case of random data, regularity cannot be detected in the space even when the embedded dimension is increased. Therefore, the correlation dimension should increase as the embedded dimension increases.

When time-series data has a deterministic structure in the embedded space, the correlation dimension cannot increase and should be matured at the same value, even if the embedded dimension increases.

It is necessary to check whether time-series data follow a chaotic deterministic rule when chaos theory is applied to the time-series data to forecast a future value. The correlation dimension is used to check whether a set of time-series data exhibits chaotic behavior [10]. The method checks whether same shape can be found as in the original n-dimensional time-series data when the time-series data are embedded in an m-dimensional space, where m is less than n. That is, the embedded space is checked to determine whether m is less than n. When m is less than or equal to 3, this can be determined by visual inspection. It is not possible, however, to detect chaotic behavior by visual inspection when m is greater than 4. Therefore, we need a method to distinguish chaotic behavior in time-series data.

Measurement of the correlation dimension is a way to quantitatively evaluate the embedded shape of time-series data. The correlation dimension is a type of non-integer dimension and is widely employed because it is possible to measure the correlation dimension even if the number of data points is comparatively small.

It is necessary to embed time-series data into an n-dimensional space to measure the correlation dimension. Let us consider embedding time-series data into 2 correlation dimensions.

$$C(r) \approx ar^d \tag{15}$$

Taking the logarithm of Equation (15), we have the following relationship:

$$log(C(r)) \approx dlog(r) + log(\alpha) \tag{16}$$

The gradient of the plots of $log(C(r))$ and $log(r)$ shows the embedded correlation dimension of the time-series data. This measurement is evaluated at only one point. The correlation dimension of real time-series data should be checked to observe the behavior of the correlation dimension average among all of the sample points obtained.

Let us show the formulation as follows:

$$\begin{array}{c} \{x(t)\}, t = 1,2,\cdots,T \\ Z(t) = (x(t), x(t+\tau), x(t+2\tau), \cdots, x(t+(n-1)\tau)) \end{array} \tag{17}$$

$$\begin{array}{c} t = 1,2,\cdots,M \qquad (M = T - (n-1)\tau) \\ C_n(r) = \dfrac{\# \, of \, \{(Z(i),Z(j))|r > |Z(i) - Z(j)|\}}{M^2} \end{array} \tag{18}$$

Assume that the number of time-series data points $x(t)$ is T. Define vector $Z(t)$ using the time-series data $x(t)$ and embed $Z(t)$ into an n-dimensional space. Let τ denote a time interval. Let us measure the embedded correlation dimension. The Euclidian distance between two points $Z(i)$ and $Z(j)$ is denoted by $|Z(i) - Z(j)|$. # of $\{\cdots\}$ denotes the number of elements included in set $\{\cdots\}$. Equation (18) yields the number of the paired distances that are smaller than r divided by the number of whole pairs M^2.

3 Wavelet Transformation

Fast Fourier transform is widely employed to transform a signal into components of different frequencies. A sine function is employed as a base function. The sine function is an infinite smooth function. Therefore, the information obtained by the fast Fourier transform does not include the local information such as the locations and magnitudes of the frequencies the original signals.

On the contrary, the wavelet transform employs a compact portion of a wavelet as a base function. Therefore, it is a time and frequency analysis, in that it enables us to determine the signal using time and frequency.

The mother wavelet transform $(W_\psi f)(b,a)$ of function $f(x)$ can be defined as follows:

$$(W_\psi f)(b,a) = \int_{-\infty}^{\infty} \frac{1}{\sqrt{|\alpha|}} \overline{\Psi\left(\frac{x-b}{a}\right)} f(x)dx \qquad (19)$$

where a is the scale of the wavelet and b is the translate. $\overline{\Psi(x)}$ is a conjunction of a complex number. It is also possible to recover the original signal $f(x)$ using a wavelet transform. That is, we can accomplish the inverse wavelet transform as follows:

$$f(x) = \frac{1}{C_\Psi} \int \int_R (W_\psi f)(b,a) \frac{1}{\sqrt{|\alpha|}\Psi} \left(\frac{x-b}{a}\right) \frac{\partial a \partial b}{a^2} \qquad (20)$$

The wavelet transform is a useful method for identifying know the characteristics of the signal but not an efficient method because the signal has a minimum unit and the wavelet method expresses much-duplicated information. This issue can be resolved by discretizing a dimensional axis. Let us denote a dimension as $(b,1/a) = (2^{-j}k,2^j)$. The discrete wavelet transform can be rewritten as

$$d_k^{(j)} = s^j \int_{-\infty}^{\infty} \overline{\Psi(2^j x - k)} f(x)dx \qquad (21)$$

The inverse wavelet transform is

$$f(x) \sim \sum_j \sum_k d_k^{(j)} \Psi(2^j x - k) \qquad (22)$$

Let us denote the summation of the right term as

$$g_j(x) = \sum_k d_k^{(j)} \Psi(2^j x - k) \qquad (23)$$

Then, let us define $f_j(x)$ as

$$f_j(x) = g_{j-1}(x) + g_{j-2}(x) + \cdots \qquad (24)$$

where an integer j is called a level. If we can denote $f(x)$ as $f_0(x)$, then

$$f_0(x) = g_{-1}(x) + g_{-2}(x) + \cdots \qquad (25)$$

This equation illustrates that the function $f_0(x)$ is transformed into wavelet components $g_{-1}(x)$, $g_{-2}(x)$, $\cdots$. The left side must be transformed uniquely into the right-side, and the left side should be realized by composition from the right-side components as well. This can be accomplished using a mother wavelet Ψ as a base function. The function $f_j(x)$ can be rewritten using a recursive forms

$$f_j(x) = g_{j-1}(x) + f_{j-1}(x) \qquad (26)$$

192 Y. Matsumoto and J. Watada

This equation indicates that the original signal $f_j(x)$ can be transformed into wavelet components $g_{j-1}(x)$ and $f_{j-1}(x)$. This equation enables us to decompose the original equation into the wavelet components step by step. This process is called multi-resolution signal decomposition.

4 Correlation Dimension of Transformed Wavelet Components

Let us transform the time-series data into frequency components by wavelet multiresolution analysis. The Spline4 shown in Fig. 11 is employed as a mother wavelet function and the transformation is conducted until level 4. The time-series data analyzed are Tokyo stock average index (TOPIX) data. The data consists of 2,048 sample points from January 1991. Figure 12 illustrates the results obtained from the multiresolusion analysis. The first graph in Figure 12 is the original one. The smaller value j shows the lower-frequency component.

The results of the wavelet transformation illustrate that each frequency component of the TOPIX data are trasformed smoothly.

We measured the correlation dimension of each component decomposed by the wavelet transform. The results are shown in Figure 13. The original TOPIX data indicate that the correlation dimension is matured at approximately 7. However, the wavelet component $j = -1$ is matured at approximately 6. The correlation dimension of the wavelet component whose j is less than or equal to -2 is matured approximately 4 to 5.

These results illustrate that the transformed components are more chaotic than the original TOPIX time-series data. The measurement of correlation dimensions results in component time-series data that are more chaotic than the original timeseries data. Let us forecast the short-term future using the original data and the decomposed wavelet-transformed component data and compare the results. The data are the same TOPIX data employed above. In this discussion, the data were normalized to a mean of 0 with a variance of 1. The embedded dimensions are examined from dimensions 3 to 9. We measured the forecast errors associated with them. In

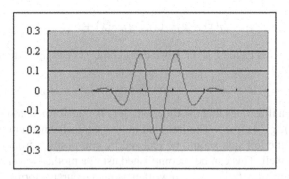

Fig. 11 Spline4: Mother wavelet

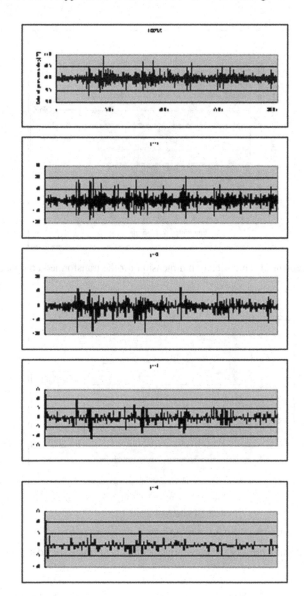

Fig. 12 Divided Time-series Data

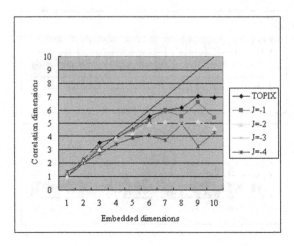

Fig. 13 Measurement of the correlation dimension for the transformed component data and the original data

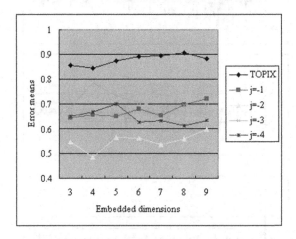

Fig. 14 The prediction error of the transformed component time-series data

the forecast, we employed 100 data points from among the total of 2, 048 points used as described previously. Figure 14 shows the results for the original and the wavelet-transformed component time-series data. The vertical axis denotes error means and the horizontal axis shows the embedded dimension.

The transformed component data exhibit drastically lower forecasting errors than the original data. This result indicates that the component data are much more chaotic than the original TOPIX data.

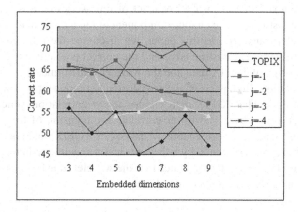

Fig. 15 Correct rates of up-and-down prediction

We can show that the wavelet-transformed component time-series data are more chaotic than the original data by both the measurement of the correlation dimension and the forecasting error.

Let us examine the forecasting of the up-and-down movements of the price of a stock. The goal of this process is to forecast the movement of the price, i.e., the direction of change from today's price to tomorrow's price. This forecast is conducted only for the movement of the price, not the price itself.

Let us check the prediction of up-and-down movements of stock prices based on the forecasted results. The up-and-down prediction was performed for the movement of the price of the stocks on following day using the TOPIX data. The prediction is correct if the actual and predicted directions of the movement are the same. The percentage of correct predictions is shown for the 100 trials. In Figure 15, the vertical axis is the correct prediction rate and the horizontal axis is the embedded dimension.

A much better prediction rate is obtained for the component data obtained using the wavelet transform than for the original time-series data. The transformed component of the original data better predicts the chaotic movement. In the case of $j = -4$, the correct prediction rate is better than 70%, which is considered very high.

5 Conclusions

The objective of this paper is to demonstrate that short-term forecasting can be accomplished using a wavelet transform applied to chaotic data. Original time-series data are divided into components through a wavelet transform. The chaotic short-term forecasting method is applied to the transformed component decomposed by the wavelet transform.

It should be noted that even if a given set of time-series data is relatively less chaotic, we can distinguish the chaotic component of the data from the original data using the wavelet transform. In this paper, we discussed the measurement of the correlation dimension, forecasting error and the correct prediction rate achieved in comparison of original TOPIX data and the wavelet-transformed component of the data. The correlation dimension of the transformed component is lower than that of the original data. The TOPIX data yielded lower correlation dimensions than individual stock price data. The lower correlation dimension of the TOPIX data means that these data are more chaotic. Analysis of TOPIX data showed that the divided component data obtained by wavelet transform were predicted correctly at a higher rate (up to 70This demonstrates that the wavelet transform can extract the chaotic component well from the original data.

References

1. Abhyankar, A., Copeland, L.S., Wong, W.: Nonlinear Dynamics in Real-Time Equity Market Indices: Evidence from the United Kingdom. The Economic Journal 105, 864–880 (1995)
2. Barnett, W.A., Chen, P.: The Aggregation-Theoretic Monetary Aggregates are Chaotic and Have Strange Attractors: An Econometric Application of Mathematical Chaos. In: Barnett, W., Berndt, E., White, H. (eds.) Dynamic Econometric Modeling. Cambridge University Press, Cambridge (1988)
3. Box, G.E.P., Jenkins, G.M.: Time Series Analysis: Forecasting and Control. Holden-Day, San-Francisco (1976)
4. Brock, W.A.: Distinguishing Random and Deterministic Systems: A Bridged Version. Journal of Economic Theory 40, 168–195 (1986)
5. Brock, W.A., Sayers, C.L.: Is the Business Cycle Characterized by Deterministic Chaos? Journal of Monetary Economics 22, 71–90 (1988)
6. Chui, C.K.: Introduction to wavelets. Academic Press, New York (1992)
7. Decoster, G., Mitchell, D.: Nonlinear Monetary Dynamics. Journal of Business & Economic Statistics 9(4), 455–461 (1991)
8. Doyne Farmer, J., Sidorowich, J.J.: Exploiting Chaos to Predict the Future and Reduce Noise. In: Lee, Y.C. (ed.) Evolution Learning and Cognition, pp. 277–330. World Scientific, Singapore (1988)
9. Frank, M.Z., Stengo, T.: Measuring the strangeness of gold and silver rates of return. Review of Economic Studies 56, 553–568
10. Grassberger, P., Procaccia, I.: Measuring the Strangeness of Strange Attractors. Physica D 9, 189–208 (1983)
11. Li, T.Y., Yorke, J.A.: Period three implies chaos. American Mathematical Monthly 82, 985–992 (1975)
12. Lorenz, E.N.: Deterministic non-periodic flow. Journal of the Atmospheric Sciences 20(2), 130–141 (1963)
13. Matsumoto, Y., Watada, J.: Short-term Prediction by Chaos Method of Embedding Related Data at Same Time. Journal of Japan Industrial Management Association 49(4), 209–217 (1998) (in Japanese)
14. Mees, A.I.: Dynamical systems and tessellations: deteting determinism in data. International Journal of Bifurcation and Chaos 1(4), 777–794 (1991)

15. Poincare, H.: Sur l'equilbre d'une masse fluide animee' d'un mouvement de rotation. Acta Mathematica (1885)
16. Ramsey, J.B., Sayers,. C.L., Rothman, P.: The Statistical properties of dimension calculations using small data sets: Some economic applications. International Economic Review 31, 991–1020 (1990)
17. Scheinkman, J.A., LeBaron, B.: Nonlinear Dynamics and Stock Returns. Journal of Business 62(3), 311–337 (1989)
18. Serizawa, H.: Phenomenon knowledge of chaos. Tokyo Books (1993) (in Japanese)
19. Takens, F.: Detecting Strange Attractors in Turbulence. In: Rand, D.A., Young, L.S. (eds.) Dynamical Systems and Turbulence. Lecture Notes in Mathematics, vol. 898, pp. 366–381. Springer, Berlin (1981)

Chapter 9
Fuzzy Forecasting with Fractal Analysis for the Time Series of Environmental Pollution

Wang-Kun Chen and Ping Wang

Abstract. Environmental pollution, which is complicated for forecasting, is a phenomenon related to the environmental parameters. There are many studies about the calculations of concentration variation on pollution time series. A new framework of prediction methodology using the concept of fuzzy time series with fractal analysis (FTFA) was introduced. The FTFA uses the concept of turbulence structure with the fractal dimension analysis to estimate the relationship by fuzzy time series. The candidate indexes of each pattern can be selected from the most important factors by fractal dimension analysis with autocorrelation and cross correlation. Based on the given approach, the relationship between the environmental parameters and the pollution concentration can be evaluated. The proposed methodology can also serve as a basis for the future development of environmental time series prediction. For this reason, the management of environmental quality can be upgraded because of the improvement of pollution forecasting.

Keywords: Fuzzy Theory, Fractal Analysis, Environmental Pollution, Time Series.

1 Time Series of Environmental Phenomenon and Its Physical Nature

Environmental pollution is a kind of natural phenomenon which could be explained by the turbulence structure of fluid dynamics. The environmental properties such as pollution concentration, wind velocity, ocean current and thermal diffusion are all the outcome of natural turbulence. The air pollution is a typical phenomenon caused by pollutants emitted into the atmosphere and diffused with the eddy.

Wang-Kun Chen
Department of Environment and Property Management,
Jinwen University of Science and Technology
email: wangkun@just.edu.tw

Ping Wang
Department of Civil Engineering, Chinyun University
email: pwang@cyu.edu.tw

W. Pedrycz & S.-M. Chen (Eds.): Time Series Analysis, Model. & Applications, ISRL 47, pp. 199–213.
DOI: 10.1007/978-3-642-33439-9_9   © Springer-Verlag Berlin Heidelberg 2013

The environmental phenomenon can be described by its physical properties. The presence of "eddies" in the environment leads to the complexity and variation of the outcome of observation. The length scale of "eddy" differs to several orders of magnitude. For example, typhoon is one of those large eddies exists in the atmosphere, and the sea breeze is the smaller eddy caused by the air-sea exchange. Thus it is very difficult to explain the difference by traditional methods. A better insight into the characteristics of turbulence with these "eddies" will be helpful in understanding the nature of these environmental phenomenon.

The modern principle of fluid dynamics and the theory of fractal analysis are suitable tools to investigate the properties of environmental events. The irregularity and the randomness are the most important characteristics of turbulent flow. These characteristics make it impossible to explain the environmental events using deterministic approach, except invoking statistical methods. The environmental phenomenon can be investigated by long term and large area monitoring. However, the complexity of the time series-based observation has made the interpretation more difficult. The models used to describe environmental turbulence should be able to simulate the non-linear and non-stationery properties of time series. Therefore, the recently developed tool, fractal analysis, can be employed to meet the needs.

Environmental pollution is a phenomenon resulting from the presence of turbulence, characterized by non-linear, randomness, irregularity, and chaos. Thus, turbulence is a complex environmental phenomenon that is difficult to predict precisely through mathematical modeling.

2   Interpretation of Pollution Time Series by Fractal Analysis

Since there are so many different eddies with different scales, the concept of fractal analysis become useful to understand the behavior of environmental turbulence. Fractal analysis, which expresses the complexity using the fractal dimension, is a contemporary method to describe the natural phenomenon. It applies the nontraditional mathematics in analyzing the environmental problem and has been used in the analysis of the scale dependence environmental phenomenon such as rainfall (Olsson & Niemczynowicz,1994,1996), air pollutant concentration (Lee,2002, Lee et al, 2003, Lee & Lin, 2008; Lee et al,2006a,), and earthquake (Lee et al,2006b).

Mandelbrot has defined fractal as a special class of subsets of a complete metric space (Mandelbrot,1982). The fractal dimension, D_F, which is deduced from the scaling rule, is the key concept of fractal analysis. The complexity of environmental phenomenon is due to a change with the variation of turbulence eddies in scale. So it is possible to have many types of fractal dimension, D_F, in an environmental system. These fractal dimensions can be explained in terms of measure of the complexity. Comparing with the change with scale in turbulence, so it is necessary to deduce a scaling system to represent the "patterns of complexity"

Here a simple model is proposed for estimation of pollution concentration influenced by environmental turbulence. Generally, the scaling rule or fractal dimension, D_F, can be represented by two terms, N and ε. The term N is the number of pieces and ε is the scale used to get new pieces. The relationship can be written as:

$$N \propto \varepsilon^{-D_F} \tag{1}$$

which can be further formulated in the form of a scaling rule:

$$N = A \varepsilon^{-D_F} \tag{2}$$

where A is a certain constant. By taking the logarithm of both sides of (2), the variable D_F becomes the ratio of the log of "the number of new parts (N)" to the log of "scale (ε)":

$$D_F = \log N / \log \varepsilon. \tag{3}$$

The scaling rule of fractal dimension helps us explain the variation of pollution time series in the fuzzy time series prediction with fractal analysis.

In turbulence, the attribute of correlated variable helps to characterize the phenomenon. The analysis starts from the average of products, which are computed in the following way. (Tennekes and Lumley, 1972)~

$$\begin{aligned}
\overline{\tilde{u}_i \tilde{u}_j} &= \overline{(U_i + u_i)(U_j + u_j)} \\
&= U_i U_j + \overline{u_i u_j} + \overline{U_i u_j} + \overline{U_j u_i} \\
&= U_i U_j + \overline{u_i u_j}
\end{aligned} \tag{4}$$

The terms consisting of a product of a mean value and a fluctuation vanish if they are averaged, because the mean value is a mere coefficient as far as averaging is concerned, and the average of a fluctuation quantity becomes zero.

If $\overline{u_i u_j} \neq 0$, u_i and u_j are said to be a correlated; if $\overline{u_i u_j} = 0$, the two variable are uncorrelated. Figure 1 illustrates the concept of correlated fluctuating variable. The correlation coefficient C_{ij}, is defined by

$$C_{ij} \equiv \frac{\overline{u_i u_j}}{\sqrt{\overline{u_i}^2 \cdot \overline{u_j}^2}}, \tag{5}$$

where C_{ij} is a measure for the degree of correlation between two variable u_i and u_j. If $C_{ij} = \pm 1$, the correlation is said to be perfect, and could be chosen as the best predictor for forecasting.

In the analysis of turbulence dimension, the "standard deviation" of "root mean square (RMS)" amplitude is defined. For a turbulent flow field, a characteristic velocity, or "velocity scale", might be defined as the mean RMS velocity taken across the flow field at that position. In this way velocity scale could be used as a precise definition in dimensional analysis.

If the evolution of fluctuating function (t) is to be described, it is necessary to know that the value of u at different time is related. The question could be answered by considering a joint density for u(t) and u'(t). The time difference, or time lag, in the property time series is defined by $\tau = t' - t$. The correlation $\overline{u(t)u(t')}$ at two different times is called the autocorrelation, and the correlation between u and v is called the cross-correlation.

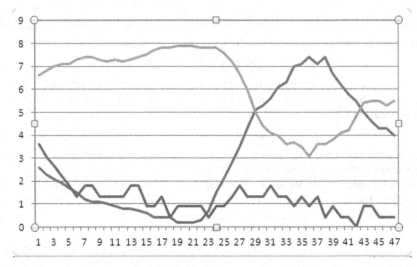

Fig. 1 The example of correlated variable and uncorrelated variable. The green and blue line has a negative correlation; however, the red line is uncorrelated to the two variables.

A tensor R to deal with the correlation between different location x and x+r is given by

$$R_{ij} \equiv \overline{u_i(x,t) \cdot u_i(x+t)}, \tag{6}$$

3   Representation of Environmental Phenomenon by Fuzzy Time Series

After interpretation of environmental phenomenon in terms of turbulence scale and fractal dimension analysis, we return to the problem of predicting the time series value of pollution. Time series is frequently applied to the prediction of environmental events. An example of using time series for pollutant concentration is given by

$$X = \{x_t \quad t = 1,\ldots\ldots, N\} \tag{7}$$

where t is time index and N is the total number of observations. For example, the time series of ozone concentration from a continuous monitoring station, the instantaneous wind velocity at the meteorological observation station is considered as an event for a time series.

The prediction method of fuzzy time series with fractal analysis is a scheme revised from Chen's study of fuzzy time series (Chen, 1996, 2002)(Chen and Hsu, 2004, 2008)(Chen and Hwang, 2000)(Song and Chrisson, 1993, 1994, 2003). The concept of fuzzy time series has been applied to the prediction of pollutant concentration by Chen (Chen, 2011), which produced good results. However, it is still unable to describe the non-linear characteristic of the environmental system.

On the other hand, the database is not extensive enough to generate a complete inference engine to simulate all the possible variation of time series.

In this study, due to an addition of 1440 data sets to the database for generating the inference engine, along with the fractal analysis of turbulence, more insight is given than before into the behavior of pollution concentration in the environment.

The prediction method of fuzzy time series with fractal analysis (FTFA) can be implemented by the following steps: (1) Define the interval. (2) Get the statistical distribution of concentration in each interval. (3) Define each fuzzy set Ai based on the re-divided intervals u_i derived in step 2. (4) Establish fuzzy logical relationship based on the fuzzified concentration. (5) Use the high-order difference to determine the upward or downward trend. (6) Find the appropriate predictors by fractal analysis.

Let $U = \{u_1, u_2, u_3, \ldots\ldots, u_n\}$, where U is the universe of discourse. Fuzzy set A, in the universe of discourse U, is defined as follows:

$$A_i = fA_1(u_1)/u_1 + fA_2(u_2)/u_2 + \cdots + fA_n(u_n)/u_n \quad , \tag{8}$$

where f_A is the membership function of the fuzzy set A, $f_A : U \rightarrow [0,1]$, $f_A(u_i)$ indicates the grade of membership of u_i in the fuzzy set A, $f_A(u_i) \in [0,1]$, and $1 \leq i \leq n$.

Define F(t) as the fuzzy time series of X(t) (t = ..., 0, 1, 2, ...), and X(t) (t = ..., 0, 1, 2, ...) is the universe of discourse in X (t). In order to extract the knowledge from the time series database, assume there exists a fuzzy relationship R (t, t-1) such that

$$F(t) = F(t-1) \cdot R(t, t-1) \tag{9}$$

Where, R(t, t-1) denotes the fuzzy relationship between F(t) and F(t-1). If fuzzy set F(t-1)=A_i, and F(t)=A_j, the fuzzy relationship is called the first order fuzzy time series.

More hidden relationships could be found in the time series database. If F(t) is caused by F(t-1), F(t-2), F(t-3) ......,and F(t-n), then there is a high-order fuzzy time series which can be represented by

$$F(t-n),\ldots\ldots,F(t-2), F(t-1) \rightarrow F(t). \tag{10}$$

The fuzzy interval of pollution time series can be an equal-length interval (ELI) or an un-equal length interval (ULI). ULI is the improved model of ELI by adjusting the length of each interval in the universe of discourse This is called the multi-step fuzzy time series for the forecasting of concentration (Chen, 2008) (Chen, 2010). The proposed method of FTFA is presented as follows:

Step 1: Define the interval

Let U be the universe of discourse, $U = [D_{min} - D_1, D_{max} + D_2]$, where D_{min} and D_{max} denote the minimum and maximum concentration.

Step 2: Form the statistical distribution of concentration in each interval..

Sort the interval based on the number of concentration data falling into each interval in a descending sequence.

Step 3 : Define each fuzzy set Ai based on the re-divided intervals u_i derived in step 2, and fuzzify the historical concentration.

The interval with no data distributed was discarded. The interval with more data was divided into more sub-intervals. The idea behind the determination of interval and sub-interval is to divide the interval containing a higher number of historical concentration data into more sub-intervals to improve the accuracy of predict.

Step 4: Establish fuzzy logical relationship based on the fuzzified concentration.

If the fuzzified concentration of month i and i+1 are A_j and A_k, respectively, then construct the fuzzy logical relationship "$A_j \rightarrow A_k$", where A_j and A_k are called the current state and the next state of the concentration.

If the fuzzified concentration of month i is A_j and the fuzzy logical relationship is shown as: $A_j \rightarrow A_{k1(x1)}, A_{k2(x2)}...A_{kp(xp)}$, then the estimated concentration of month i is calculated as

$$C(i) = \frac{X_1 \times m_{k1} + X_2 \times m_{k2} + \cdots + X_p \times m_{kp}}{X_1 + X_2 + \cdots + X_p} \tag{11}$$

where X_i denotes the number of fuzzy logical relationships "$A_j \quad A_k$" in the fuzzy logical relationship group, $1 \leq i \leq p$, $m_{k1}, m_{k2}, \ldots$ and m_{kp} are the mid point of the intervals $u_{k1}, u_{k2}, \ldots$ and u_{kp} respectively, and the maximum membership values of $A_{k1}, A_{k2}, \ldots$ and A_{kp} occur at intervals $u_{k1}, u_{k2}, \ldots$ and u_{kp}, respectively.

Step 5: Use the high-order difference to determine the upward or downward trend.

The difference of the second order difference between any two neighboring time segments of the historical concentration can be used for forecasting the trend. The second order difference is calculated by the equation: $Y_n = Y_{n-1} - Y_{n-2}$.

The α-cut value determines the fuzzified concentration in the interval. It is quite usual to use the triangle function and chose the value of α-cut equal to 0.5 for estimation. Another important factor is the value of high-order difference; it will dominate the trend of concentration variation.

Step 6: Selecting the appropriate predictor by fractal analysis

There are many factors which may influence the concentration variation of time series. To improve the accuracy of prediction, it is better to incorporate these factors into a more advanced model. The properties of pollution, which are influenced by many factors, can be described in terms of an appropriate function and α-cut value. The triangular function, trapezoidal function, or Gaussian membership function are all possible choices.

The autocorrelation and cross correlation in the knowledge space phase of fractal dimension analysis help us find the best predictor. The autocorrelation coefficient between X_t and $X_{t-\tau}$ is calculated as follows

$$C_{auto}(X_t, X_{t-\tau}) = \frac{E(X_t X_{t-\tau} - E(X_t)^2)}{E\{[X_t - E(X_t)^2]\}} \tag{12}$$

Where, τ is the time lag of the two time segments X_t and $X_{t-\tau}$. The autocorrelation coefficient helps us define the fractal dimension of the environmental system.

The cross correlation help us know the relationship between two properties. It is calculated as

$$C_{cross}(X_t,Y_t) = \frac{E(X_tY_t - E(X_t)^2)}{E\{[X_t - E(X_t)^2]\}}$$ (13)

The fractal dimension is determined by finding the maximum and minimum values of these two correlation coefficients in the time plot.

4   Statistical Pattern Recognition of Environmental Concentration in Space-Time Series

The space-time series could be described by the average values and fluctuating quantities such as U and $\overline{uv}$. It is also important to know how fluctuations are related to the adjacent fluctuations in time or space next to each other. The statistical pattern recognition helps us examine how fluctuations are distributed around an average value in the space-time series. Some statistical properties are introduced for the purpose of pattern recognition of environmental concentration time series such as probability density function and its Fourier transform, the autocorrelation and its Fourier transform, etc.

A steady time series is statistically stable, calculated by the mathematical function. The probability density function $B(\tilde{u})$ is defined by

$$B(\tilde{u})\Delta(\tilde{u}) = \lim_{T \to \infty}\frac{1}{T}\sum(\Delta t),$$ (14)

where $\tilde{u}$ denotes the fluctuation value.

$$B(\tilde{u}) \geq 0, \int_{-\infty}^{\infty} B(\tilde{u})d\tilde{u} = 1.$$ (15)

The mean values of the various powers of $\tilde{u}$ are called moments. The means value is the first moment defined by

$$U = \int_{-\infty}^{\infty} \tilde{u}B(\tilde{u})d\tilde{u},$$ (16)

The variance, or the mean square departure, σ^2, from the mean value U is the second moment, which is defined by

$$\sigma^2 = \overline{u^2}$$

$$= \int_{-\infty}^{\infty} u^2 B(\tilde{u})d\tilde{u},$$ (17)

$$= \int_{-\infty}^{\infty} u^2 B(u)du$$

where σ is the standard deviation of root mean square(rms) amplitude.

The third moment, skewness (S), K, helps us discriminate the symmetric and anti symmetric parts of the time series. It is defined by

$$\overline{u^3} = \int_{-\infty}^{\infty} u^3 B(u)du , \tag{18}$$

and the value of skewness is

$$S = \frac{\overline{u^3}}{\sigma^3} . \tag{19}$$

The fourth moment, kurtosis or flatness factor K is represented as

$$K = \frac{\overline{u^4}}{\sigma^4} \tag{20}$$

$$= -\int_{-\infty}^{\infty} e^{-iku} B(u)du$$

The other important measurement is the Fourier transform of B(u), which is defined as

$$\phi(K) = \int_{-\infty}^{\infty} e^{-iku} B(u)du, B(u) \tag{21}$$

$$= \frac{1}{2\pi} \int_{-\infty}^{\infty} e^{-iku} \phi(K)dK$$

The Fourier transform of B(u) is the characteristic function which is more convenient to see the pattern of flow property and behavior of pollution concentration.

In order to examine the feasibility of this model, the numerical experiment was conducted. The experimental data for time series analysis were acquired from the observed air quality data of Taiwan Environmental Protection Administration. Different pollutant concentration such as carbon monoxide, sulfur dioxide, nitrogen oxide, $PM_{2.5}$, and ozone were used. Data for analyzed were obtained from the year Sep, 2010, to AUG, 2011.

The results of the pattern of space-time series was analyzed by the autocorrelation to know the fractal dimension for different pollutants.

Figure 2 is the autocorrelation coefficient time series plot of ozone concentration time series. The amplitude of autocorrelation coefficient gradually decreases with time lag. There are totally five peaks in five days (120hours) of monitoring results, which means c(t) is correlated with itself every twenty four hours. The maximum value, which is near 1, is close to the first hours. This result reveals that the best predictor for ozone concentration is the time segment prior to the time of prediction.

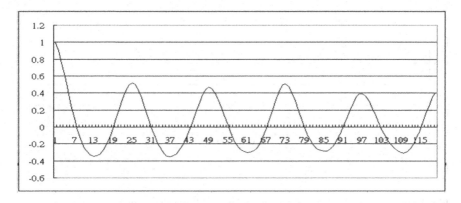

Fig. 2 Autocorrelation coefficient of ozone concentration time series

Figure 3 is the autocorrelation coefficient time series plot of nitrogen dioxide concentration. The results exhibit a wave-like trend of decreasing amplitude. The reason why the trend of nitrogen dioxide is the same as that of ozone is that they are all photochemical pollutants. Their formations are mostly governed by solar radiation. Therefore, all the graphical results show the same daily cycle.

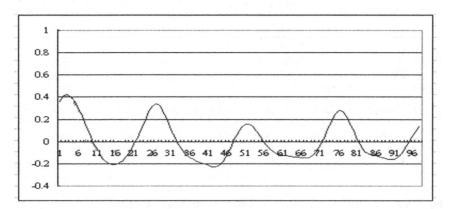

Fig. 3 Autocorrelation coefficients of nitrogen dioxide concentration time series

Figure 4 shows the plot of autocorrelation coefficient time series of carbon monoxide concentration. Two peaks, including a higher peak and a lower one, are found in one day. This represents two possible sources of carbon monoxide emission in the morning and in the evening. The time lag of twenty four hours has the highest value and the one of twelve hours has the second highest value. It is expected that the autocorrelation coefficient will approaches zero as the time lag approaches infinity.

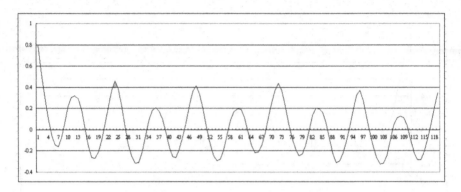

Fig. 4 Autocorrelation coefficients of carbon monoxide concentration time series

Figure 5 includes the autocorrelation coefficient time series plot of PM-10 concentration. The irregular trend fails to indicate significant correlation between u(t) and time lag. An assumption can be made that the correlation is significant only up to 8-hour time lag corresponding to the least coefficient of 0.2.

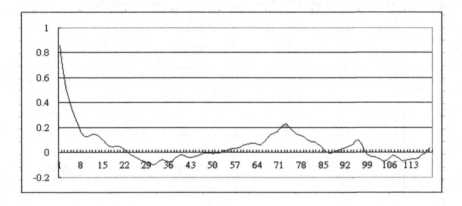

Fig. 5 Autocorrelation coefficients of PM-10 concentration time series

Figure 6 presents the autocorrelation coefficient of Non-methane hydrocarbon concentration. The figure reveals a very sharp decrease in the autocorrelation coefficient, which means that there are many factors which influence the variation of non-methane hydrocarbon concentration. It is more difficult to predict the concentration of non-methane hydrocarbon in time series.

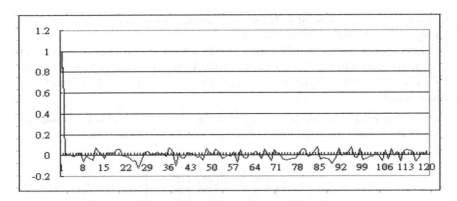

Fig. 6 Autocorrelation coefficients of Non-methane hydrocarbon concentration time series

5   Prediction of Environmental Pollution by Fuzzy Forecasting

Once understood space-time series pattern by turbulence theory and fractal dimension analysis, an attempt is made to apply fuzzy forecasting approach for prediction. The FTFA method, including a variety of schemes with different fuzzy intervals, multi-step fuzzy time series and the high-order fuzzy time series, is used to predict pollution concentration. These methods are used to determine the trend of data by adjusting the length of each interval in the universe of discourse.

The Mean Square Error was calculated as follows:

$$MSE = \frac{\sum_{1}^{m}(C_{obs} - C_{est})^2}{m} \qquad (22)$$

where C_{obs} denotes the actual particulate concentration of time step I, C_{est} denotes the forecasting concentration, and m denotes the historical data.

The FTFA model is compared with other forecasting methods:

(1) The linear regression model: This model uses a linear trend over time to estimate the concentration:

$$C = a X + b \qquad (23)$$

where C is the estimated concentration at specific time X.

(2) The autoregressive model: The autoregressive model uses the previous data to estimate the concentration as follows:

$$C = r_1 C_{t-1} + \varepsilon \qquad (24)$$

where C denotes the regression result at time t, C_{t-1} denotes the concentration at time t-1, r_1 denotes the regression coefficient , and ε denotes the predicting error.

The auto-regression model can be modified by the two time step estimation as follows:

$$C = r_1 C_{t-1} + r_2 C_{t-2} + \varepsilon \qquad (25)$$

where r_1 and C_{t-1} are defined as before, C_{t-2} denotes the concentration at time t-2, and r_2 denotes the regression coefficient.

The concentration predicted by the above FTFA method is listed in Table 1and graphically shown in Fig 7. As shown in table 1, the mean square error from the FTFA model is least, exhibiting its superiority over other models in predicting concentration.

Table 1 Example of the observed concentration and fuzzy concentration C_{obs} : observed concentration;   Int : concentration intervals C_{fuzzy} : fuzzified concentration

time	C_{obs}	Int	C_{fuzzy}
T_1	46.63	[46,47]	47
T_2	51.35	[51,52]	51
T_3	63.33	[63,64]	63
T4	58.77	[58,59]	59

The fuzzy logical relationship is listed in Table 2. For example, the following fifth-order fuzzy logical relationship: A_{17}, A_{17}, A_{16}, A_{16}, $A_{15} \rightarrow A14$, where the fuzzy logical relationship denotes the fuzzified concentration.

Table 2 Fuzzy logical relationship

Number of steps	fuzzy logical relationship
One step	$A_1 \rightarrow A_{13}$
Two steps	$A_{17}, A_{13} \rightarrow A_{20}$
Three steps	$A_{12}, A_{15}, A_{16} \rightarrow A_{19}$
Four steps	$A_{15}, A_{15}, A_{16}, A_{16}, \rightarrow A_{18}$
Five steps	$A_{17}, A_{17}, A_{16}, A_{16}, A_{15}, \rightarrow A_{14}$

Each interval is equally divided into four subintervals, where the points at 0.25 and 0.75 are used as bases to make forward or backward prediction. From the fuzzy logical relationship described above, the forecasted concentration can be determined. The results of several different prediction methods are shown in Table 3.

The value of mean square error for multi-steps fuzzy model (MSF) is the smallest among all the prediction methods, as shown in Table 3. It indicates that the proposed method is better than other models based on intervals with 10, 20 and 30 equal spaces. The prediction results of traditional auto-regressive model and linear egression model are both worse than those of the FTFA model. The reason is that the traditional statistical method and pattern recognition are either parametric or non-parametric models, but the high-order fuzzy time series recognize the pattern in other ways.

There are usually some data with unknown pattern in our forecasting procedure. The pattern recognition of concentration prediction is important in forecasting the particulate concentration, therefore more advanced tool has to be used.

An example is taken to compare the "patterned" and "un-patterned" data in this study. The method has the capability to catch the pattern of the concentration variation in the atmosphere. Also, the mean square error from the forecasted results of FTFA models was lower than that from the linear regression model and autoregressive model. Since the prediction of pollution concentration also involves hourly, daily, concentration, a more sophisticated analysis should take space and time into account in predicting variation of concentration.

Table 3 includes the result of comparison of different forecasting method.

Table 3 Comparison of the results by different prediction method Note: $\hat{C}$: actual concentration A :(10) equal interval fuzzy model B: (20) equal interval fuzzy model C: (30) equal interval fuzzy model D: multi steps fuzzy model E: simple linear model F: AR(1) model G: AR(2) model σ: standard deviation MSE: mean square error

statistical Property	$\hat{C}$	forecasted concentration						
		A	B	C	D	E	F	G
mean	7049	7109	7064	7055	7055	7100	7321	7230
σ	4022	4168	4055	4033	4233	8965	5655	4699
MSE		827	143	38	34	16323	3266	6803

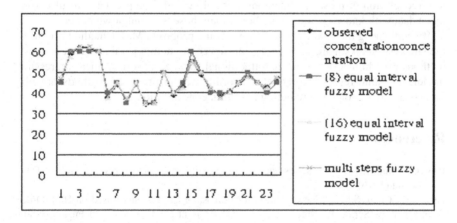

Fig. 7 Forecasted results with different interval

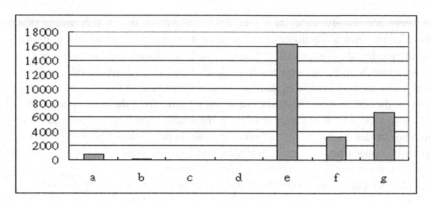

Fig. 8 Comparison of mean square error by different forecasting methods A : (10) equal interval fuzzy model B: (20) equal interval fuzzy model C: (30) equal   interval fuzzy model D: multi steps fuzzy model E: simple linear model F: AR(1) model G: AR(2) model

6   Conclusions

In this paper, an attempt has been made to predict pollution concentration by two methods, i.e. multi-step fuzzy time series (MSFT) and different interval fuzzy time series (DIFT). The MFT method was implemented by adjusting the length of each interval in the universe of discourse and using the "second order difference" of concentration to predict the variation of concentration. The characteristics of pattern recognition of these two methods were discussed. The predicted results from those data with known pattern were better than those with unknown pattern.

By comparing the results, it is shown that the proposed MSFT method produces the smallest mean square error among the seven predicting methods. That is, these methods give higher accuracy than traditional fuzzy time series, linear regression model, and auto-regressive model. Accordingly, the proposed methods are obviously a better choice in predicting pollution concentration.

References

Box, G., Jenkins, G.: Time series analysis: forecasting and control, rev. ed. Holden-Day, Oakland (1976)

Hsu, C.C., Chen, S.M.: A New Mothod for Forcasting Enrollments Based on High-Order Fuzzy Time Series. In: Proceeding of the 2003 Joint Conference on AI Fuzzy System and Grey System, Taipei, Taiwan, Republic of China (2003)

Lee, C.K.: Multifractal characteristics in air pollutant concentration time series. Water Air Soil Poll. 135, 389–409 (2002)

Lee, C.K., Ho, D.S., Yu, C.C., Wang, C.-C.: Fractal analysis of temporal variation of air pollutant concentration by box counting. Environmental Modeling & Software 18(2003), 243–251 (2003)

Lee, C.K., Lin, S.C.: Chaos in Air Pollutant Concentration (APC) Time Series. Aerosol and Air Quality Research 8(4), 381–391 (2008)

Lee, C.K., Yu, C.C., Wang, C.C., Hwang, R.D., Yu, G.K.: Scaling characteristics in ozone concentration time series (OCTS). Chemosphere 62(2006), 934–946 (2006a)

Lee, C.K., Yu, C.C., Wang, C.C., Hwang, R.D., Yu, G.K.: Scaling characteristics in aftershock sequence of earthquake. Physica A 371(2006), 692–702 (2006b)

Olsson, J., Niemczynowicz, J.: On the possible use of fractal theory in rainfall applications. Wat. Sci. Tech (1994)

Olsson, J., Niemczynowicz, J.: Multifractal analysis of daily spatial rainfall distributions. J. Hydrol. 187, 29–43 (1996)

Huarng, J.R., Chen, S.M., Lee, C.H.: Handling Forcasting Problem Using Fuzzy Time Series. Fuzzy Sets and Systems 100(2), 217–229 (1998)

Huang, K.: Heuaistic Models of Fuzzy Time Series for Forcasting. Fuzzy Sets and System 123(3), 369–386 (2001)

Huang, K.: Effective Length of Intervals to Improve Forcasting in Fuzzy Time Series. Fuzzy Sets and Systems 123, 387–394 (2001)

Zadeh, L.A.: Fuzzy sets. Information and Control 8(3), 338–353 (1965)

Zadeh, L.A.: Fuzzy logic and approximate reasoning. Synthese 30, 407–428 (1975)

Mandelbrot, B.B.: The fractal Geometry of Nature. Freeman, New York (1982)

Song, Q., Chissom, B.S.: Fuzzy Time Series and Its Models. Fuzzy Sets and Systems 54(3), 269–277 (1993)

Song, Q., Chissom, B.S.: Forcasting Enrollments with Fuzzy Time Series – Part I. Fuzzy Sets and Systems 54(1), 1–9 (1993)

Song, Q., Chissom, B.S.: Forcasting Enrollments with Fuzzy Time series Part II. Fuzzy Sets and Systems 62(1), 1–8 (1994)

Chen, S.M.: Forcasting Enrollments Based on Fuzzy Time Series. Fuzzy Sets and Systems 81(3), 311–319 (1996)

Chen, S.M.: Forecasting Enrollments Based on High-Order Fuzzy time Series. Cybernetics and System, An International Journal 33(1), 1–16 (2002)

Chen, S.M., Hsu, C.C.: A New Approach for Handling Forcasting Problem Using High-Order Fuzzy Time series. Intelligent Automation and Soft Computing 14(1), 29–43 (2008)

Chen, S.M., Hwang, J.R.: Temperature Predicting Using Fuzzy Time Series. IEEE Transaction on System, Man, and Cybernetics- Part B: Cybernetics 30(2), 263–275 (2000)

Chen, S.M., Hsu, C.C.: A New Method to Forecast Enrollments Using Fuzzy Time Series. International Journal of Applied Science and Engineering 2(3), 234–244 (2004)

Tennekes, H., Lumley, J.L.: A First Course in Turbulence. MIT Press, Cambridge (1972)

Chen, W.K.: Environmental applications of granular computing and intelligent systems. In: Pedrycz, W., Chen, S.-M. (eds.) Granular Computing and Intelligent Systems- Design with Information Granules of High Order and Higher Type. Springer (2011)

Chapter 10
Support Vector Regression with Kernel Mahalanobis Measure for Financial Forecast

James N.K. Liu and Yan-xing Hu

Abstract. For time series forecasting which have data sets coming from an unstable and nonlinear system such as the stock market. Support Vector Regression (SVR) appears to be an efficient tool which has been widely used in recent years. It is also reported to have a higher accuracy and generalization ability than other traditional methods. The SVR method deals with the nonlinear problem by mapping the input feature space into a high dimensional space so that it becomes a linear problem. Kernel function is one of the crucial components in SVR algorithm as it is used to calculate the inner product between vectors in the mapped high dimensional space. The kernel function of Radial Basis Function (RBF), which is based on the Euclidean distance, is the most commonly used kernel function in SVR. However, the SVR algorithm may neglect the effect of correlation among the features when processing the training data in time series forecasting problems due to the limitation of Euclidean distance. In this chapter, a Mabalanobis distance RBF kernel is introduced. It is well known that when we need to calculate similarity between two vectors (samples), the use of Mahalanobis distance can take into account the correlation among the features. Thus, the SVR with Mahalanobis distance kernel function may follow the behavior of the data sets better so that it can give more accurate result. From the comparative investigation, we find that in some circumstances, the Mabalanobis distance RBF kernel based SVR can outperform the Euclidean distance based SVR.

1 Introduction

In the past ten years, Support Vector Regression (SVR) has been widely applied to deal with time series forecasting problems in different domains; especially for those

James N.K. Liu · Yan-xing Hu
Department of Computing,
The Hong Kong Polytechnic University, Hong Kong
e-mail: {csnkliu, csyhu}@comp.polyu.edu.hk

W. Pedrycz & S.-M. Chen (Eds.): Time Series Analysis, Model. & Applications, ISRL 47, pp. 215–227.
DOI: 10.1007/978-3-642-33439-9_10   © Springer-Verlag Berlin Heidelberg 2013

domains characterized as some complex, nonlinear and unstable discipline. For example, it has been reported that the SVR performed well in financial forecasting [1-3] and also in temperature forecasting [4,5]. When applied to these problems, empirical results showed many advantages with the SVR algorithm; for example, higher generalization ability and the capability of avoiding overfitting problems. Due to its well-founded statistical learning theory, the SVR showed a better performances than other methods such as, Artificial Neural Networks (ANN) or Auto-Regressive and Moving Average (ARMA) Model in some previous works[6,7].

In 1995, Vapnik first introduced the Support Vector Machine (SVM) [8]. The SVM was originally applied to deal with classification problems and soon extended to regression problems [9,10]. Compared with other estimation models such as ANN and ARMA, SVR substitutes the traditional Empirical Risk Minimization (ERM) principle with Structure Risk Minimization (SRM) to address the overfitting problem and able to offer higher generalization ability. Accordingly, as we discussed above, when applied in a complex, nonlinear and unstable system, the SVR can demonstrate better performance than other methods.

The kernel function is one of the crucial components in the SVM algorithm. By using the kernel function, we can map all the samples into high dimensional feature space so that the nonlinear problem can be solved as the linear problem. The choice of kernel function may affect the performance of SVM algorithm. However, till now, there is still no guideline how to determine which kernel function can provide the best performance of SVM. Among all the kernel functions, the Radial Basis Function (RBF) kernel $K(x_i, x_j) = \exp(\frac{-\|x_i - x_j\|^2}{2\gamma^2})$ is most frequently used because the number of parameters of RBF kernel function is less than in other kernel functions. Moreover, previous experiments also showed that in most conditions, RBF kernel function could provide better performance than other kernel functions[11]. Many previous experiments that used SVR to deal with time series forecasting problems also considered RBF kernel function.

In this chapter, we introduce a new Mahalanobis distance based RBF kernel function. We focus our investigation on the comparison of SVRs with the two (Mahalanobis distance based and Euclidean distance based) different kinds of RBF kernel functions and their performance on financial time series forecasting problems. It is well known that compared with Euclidean distance, Mahalanobis distance takes into account the correlations among attributes (features) of the data set. Based on Euclidean distance, SVM algorithm neglects the correlations among attributes (features) of the training samples. Some previous researches have noticed this limitation of SVM, and subsequently introduced the Mahalanobis distance into SVM to take into consideration of the effect of correlations among attributes (features) in the training process. Some encouraging results have been obtained when Mahalanobis distance based SVM is applied to deal with classification problems [12,13]. Particularly, Wang and Yeung analysed some conventionally used kernel functions in SVM and employed Mahalanobis distance to modify the kernel function so as to improve the classification accuracy [14]. Nevertheless till now, there is little investigation on the effect of Mahalanobis distance based SVM on regression problems.

Since that in the calculation process of SVR, the Euclidean distance is applied to express the distances among the sample points, the SVR algorithm will inevitably accept the limitation of Euclidean distance: the correlations among the input variables are neglected. However, for a typical time series forecasting problem, we have features selected according to a certain time interval, such as stock index of $(t-1)$ day, stock index $(t-2)$ day, ... , stock index $(t-n)$ day. Obviously, these features are not independent from each other. When using classical SVR to deal with such types of time series forecasting problems, the correlation influence will be neglected. Therefore, it would be beneficial to improve the performance of SVR if we could take into account correlation in the training process. Therefore in this chapter, a Mahalanobis distance based RBF kernel function will be used in the SVR to deal with the financial time series forecasting problems; and we will investigate the performance of the proposed kernel function through a series of experiments.

The chapter is organized as follows: in Section 2 we will briefly introduce some background, and analysis of the Mahalanobis distance based RBF kernel and the Euclidean distance based RBF kernel in Section 3. Section 4 discusses the experiments along with some comparative analysis. The last section gives the conclusion and future work.

2 Background Knowledge

2.1 Support Vector Regression

Based on the structural risk minimization (SRM) principle, SVM method seeks to minimize an upper bound of generalization error instead of the empirical error as in other neural networks. Additionally, SVM models generate the regression function by applying a set of high-dimensional linear functions. The SVR function is formulated as follows:

$$y = w\phi(x) + b \tag{1}$$

where $\phi(x)$ is called the feature, which is nonlinear and mapped from the input space $\Re^n$. y is the target output value we want to estimate. The coefficients w and b are estimated by minimizing:

$$R = \frac{1}{2}\|w\|^2 + \frac{1}{n}C\sum_{i=1}^{n}L_\varepsilon(d_i, y_i) \tag{2}$$

where:

$$L_\varepsilon(d, y) = \begin{cases} |d-y| - \varepsilon, |d-y| \geq \varepsilon \\ 0, \text{otherwise} \end{cases} \tag{3}$$

Eq. (2) is the risk function consisting of the empirical error and a regularization term that is derived from the SRM principle. The term $\frac{1}{n}\sum_{i=1}^{n}L_\varepsilon(d_i, y_i)$ in Eq. (2) is the empirical error (risk) measured by the ε-insensitive loss function (ε-insensitive tube) given by Eq. (3); in the meanwhile, the term $\frac{1}{2}\|w\|^2$ is the regularization term.

The constant $C > 0$ is taken as the regularized constant that determines the trade-off between the empirical error (risk) and the regularization term. Increasing the value of C will add importance to the empirical risk in the risk function. ε is called the tube size of the loss function and it is equivalent to the accuracy approximation placed on the training data points. Both C and ε are user-prescribed parameters.

Then the slack variables ζ and ζ^* which represent the distance from the actual values to the corresponding boundary values of ε-insensitive tube are introduced. With these slack variables, Eq. (3) can be transformed to the following constraint based optimization:

Minimize:

$$R(w,\zeta,\zeta^*) = \frac{1}{2}ww^T + C(\sum_{i=1}^{n}(\zeta+\zeta^*)) \tag{4}$$

Subject to:

$$\begin{aligned} w\phi(x_i)+b_i-d_i &\leq \varepsilon+\zeta_i^* \\ d_i-w\phi(x_i)-b_i &\leq \varepsilon+\zeta_i \\ \zeta_i,\zeta_i^* &\geq 0, i=1,2,\cdots,n \end{aligned} \tag{5}$$

Finally, by introducing the Lagrangian multipliers and maximizing the dual function of Eq. (4), it can be changed to the following form:

$$\begin{aligned} R(\alpha_i-\alpha_i^*) &= \sum_{i=1}^{n} d_i(\alpha_i-\alpha_i^*) - \varepsilon \sum_{i=1}^{n}(\alpha_i-\alpha_i^*) \\ &- \frac{1}{2}\sum_{i=1}^{n}\sum_{j=1}^{n}(\alpha_i-\alpha_i^*) \times (\alpha_j-\alpha_j^*)(\Phi(x_i)\cdot\Phi(x_k)) \end{aligned} \tag{6}$$

with the constraints:

$$\sum_{j=1}^{n}(\alpha_i-\alpha_i^*) = 0, 0 \leq \alpha_i \leq C, 0 \leq \alpha_i^* \leq C, i=1,2,\cdots,n \tag{7}$$

In Eq. (7), α_i and α_i^* are called Lagrangian multipliers which satisfy $\alpha_i \times \alpha_i^* = 0$, the general form of the regression estimation function can be written as:

$$f(x,\alpha_i,\alpha_i^*) = \sum_{i=1}^{l}(\alpha_i-\alpha_i^*)K(x,x_i)+b \tag{8}$$

In this equation, $K(x_i \cdot x)$ is called the kernel function. It is a symmetric function $K(x_i \cdot x) = (\Phi(x_i)\cdot\Phi(x))$ satisfying Mercer's conditions. When the given problem is a nonlinear problem in the primal space, we may map the sample points into a high-dimensional feature space where the linear problem can be performed. Linear, Polynomial, Radial Basis Function (RBF) and sigmoid are four main kernel functions in use. As we discussed above, in most of the time series forecasting problems, the SVR employs RBF kernel function to estimate the nonlinear behavior of the forecasting data set because RBF kernels tend to give good performance under general smoothness assumptions.

2.2 Euclidean Distance Measure verse Mahalanobis Distance Measure

It is well known that the Euclidean distance is the most widely used measure to define the distance between two points in Euclidean space. In Euclidean space, for any two points $x_i = (x_{i1}, x_{i2}, \ldots, x_{in})$ and $x_j = (x_{j1}, x_{j2}, \ldots, x_{jn})$, the Euclidean distance between these two points can be calculated as:

$$d_E(X_i, X_j) = \sqrt{\sum_{k=1}^{n} \left| x_{ik} - x_{jk} \right|^2} \tag{9}$$

Although the Euclidean distance is widely used, it also has an obvious limitation. As discussed earlier, different features of the samples are considered as equal in the calculation of Euclidean distance; also, the correlations among the features are neglected.

One of the methods to address the limitation of Euclidean distance is to use the Mahalanobis distance [15]. Let X be a $l \times n$ input matrix containing l random observations $x_i \in \Re^n$, $i = 1, \ldots, l$. The Mahalanobis distance d_M between any two samples x_i and x_j can be calculated as follows:

$$d_M(x_i, x_j) = \sqrt{(x_i - x_j)^T \sum{}^{-1} (x_i - x_j)} \tag{10}$$

$\sum$ is the covariance matrix which can be calculated as:

$$\sum = \frac{1}{l} \sum_{k=1}^{l} (xk - \mu) \cdot (xk - \mu)^T \tag{11}$$

where μ is a mean vector of all samples.

Originally, the Mahalanobis distance can be defined as a dissimilarity measure between two random vectors of the same distribution with covariance matrix $\sum$.

From the definition of Mahalanobis distance we can see that the Mahalanobis distance is based on correlations between variables where different samples that can be identified and analyzed. It differs from Euclidean distance based on the correlations of the data set and is scale-invariant. Then again, if the covariance matrix is the identity matrix, the Mahalanobis distance will be equal to the Euclidean distance.

Considering that samples are locally correlated, a local distance measure incorporating the samples' correlation might be a better choice as a distance measure. Mahalanobis distance can take into account the covariance among the variables in calculating distances. Accordingly, in some circumstances, it may be a more suitable measure to calculate the distance and evaluate the similarity between two points[13].

3 Mahalanobis Distance RBF Kernel Based SVR

3.1 The Analysis of Kernel Functions in SVR

In SVR, to enable the nonlinear problem to be estimated by a linear function as shown in Eq. (8), we have to map the original input feature space into a high-dimensional feature space. Note that the mapping is $\Phi(x)$ as given in Eq. (6), we have to calculate the inner products of every two vectors in the transformed high-dimensional feature space. Thus, the curse of dimensionality[16] will emerge.

To deal with this problem, we can obtain a kernel function that meets this requirement $K(x_i \cdot x_j) = (\Phi(x_i) \cdot \Phi(x_j))$ and calculate the $K(x_i,x_j)$ instead of calculating the inner products of the vectors in the transformed high-dimensional feature space. In fact, every kernel function meeting the Mercer's Theorem can be used in SVM algorithm[9]. Usually, Linear, Polynomial, Radial Basis Function (RBF) and Sigmoid are the four main kernel functions in use.

Table 1 shows the form of the four kernel functions.

Table 1 The main kernel functions used in SVM

Linear	$K(x_i,x_j) = x_i \cdot x_j$
Polynomial	$K(x_i,x_j) = (c + x_i \cdot x_j)^d$
Radial Basis Function (RBF)	$K(x_i,x_j) = \exp(\frac{-\|x_i - x_j\|^2}{2\gamma^2})$
Sigmoid	$K(x_i,x_j) = \tanh(c(x_i \cdot x_j) + \theta)$

The RBF kernel function is the most commonly used among the four kernel functions in real applications. From what we notice of the linear kernel function, the Polynomial kernel function and the Sigmoid kernel function are all based on the inner products of the vectors. In other words, these kernel functions can be considered as functions with the variable $(x_i \cdot x_j)$. Unlike other three kernel functions, the RBF kernel function is based on the Euclidean distance between two points in the feature space: having examined the format of RBF kernel function we can observe that the variable of the RBF kernel function can be considered as the Euclidean distance between two points denoted as $\|x_i - x_j\|$ [14].

In fact, the RBF kernel function is a measure of the similarity between two vectors in the Euclidean feature spaces. If x_i and x_j are very close in Euclidean distance ($\|x_i - x_j\| \approx 0$), the value of the RBF kernel function will tend to be 1, conversely, if x_i and x_j are quite far apart in Euclidean distance ($\|x_i - x_j\| >> 0$), the value of the RBF kernel function will tend to be 0.

3.2 Substituting Euclidean Distance with Mahalanobis Distance in RBF

The Euclidean distance has the limitation that it neglects the correlations among the features. From the above discussion, we can see that the RBF kernel function can be considered as a Euclidean distance variable-based function. Accordingly, the RBF kernel function inherits the limitation of Euclidean distance.

One possible method of addressing this limitation is the use of Mahalanobis distance instead. We substitute the Euclidean distance with Mahalanobis distance as the variable in RBF kernel function. The Mahalanobis distance based RBF kernel function is:

$$K_M(x_i, x_j) = \exp(\frac{-((x_i - x_j)^T \Sigma^{-1}(x_i - x_j))}{2\gamma^2}) \tag{12}$$

Σ is the covariance matrix which can be calculated as:

$$\Sigma = \frac{1}{l}\sum_{k=1}^{l}(xk - \mu)\cdot(xk - \mu)^T \tag{13}$$

where μ is a mean vector of all samples. The format of linear estimate function in Eq.(8) can now be transformed to:

$$f(x) = \sum_{i=1}^{l}(\alpha_i - \alpha_i^*)K_M(x, x_i) + b \tag{14}$$

By introducing the Mahalanobis distance into the RBF kernel function, we can measure the similarity between two vectors with the Mahalanobis distance rather than Euclidean distance. The Mahalanobis distance based kernel function can take into account the correlations among attributes of the samples in the SVM training processing. When SVR with the proposed kernel function is applied to deal with time series forecasting problems, it should be beneficial to the performance improvement of the SVR forecasting result.

4   Experimental Results and Analysis

In this chapter, our investigation mainly focuses on the performance of Mahalanobis distance RBF kernel function in SVR and its performance in financial time series forecasting. To evaluate the performance of Mahalanobis distance BRF kernel function based SVR in time series forecasting, a series of experiments are conducted. 15 financial data sets about time series forecasting problem are applied in our experiment. Three forecasting methods are used in our experiment for comparison: we use the Mahalanobis distance RBF based SVR, Euclidean distance RBF based SVR and the BP neural network to estimate the target values and analyze the result with comparison.

4.1 Data Collection

As discussed in the above, the SVR is reported to be very suitable in dealing with complex, unstable and nonlinear forecasting problems such as problems in financial forecasting domain. In our experiment, 15 financial datasets from the real world are collected to evaluate the performance of the SVR. We have chosen 6 stocks from China A share market in Shanghai and 7 stocks from China A share market in Shenzhen. We aim to forecast the close prices of the 13 stocks. In addition, the Stock Indexes of the two markets: Shanghai composite index and Shenzhen composite index are also used as data sets in our experiment. This data covers the period from the 15th, September 2006 to the 31st, December, 2009. Thus, each of the data sets contains more than 750 samples.

Table 2 The features for the stock price forecasting in China A share markets

1	Today's lowest price
2	Today's highest price
3	The lowest price of the last trading day
4	The highest price of the last trading day
5	The moving average lowest price of the last 5 trading days
6	The moving average highest price of the last 5 trading days
7	Today's open price
8	The highest price of the last trading day
9	The moving average highest price of the last 5 trading days
10	Today's turnover
11	The turnover of the last trading day
12	The moving average turnover of the last 5 trading days
13	Today's volume
14	The volume of the last trading day
15	The moving average volume of the last 5 trading days

Table 3 The features for the Shanghai/Shenzhen composite index forecasting

1	Today's daily open index
2	The open index of the last trading day
3	The open index of the $(t-2)$ trading day
4	The open index of the $(t-3)$ trading day
5	The open index of the $(t-4)$ trading day
6	The open index of the $(t-5)$ trading day
7	The close index of the last trading day
8	The close index of the $(t-2)$ trading day
9	The close index of the $(t-3)$ trading day
10	The close index of the $(t-4)$ trading day
11	The close index of the $(t-5)$ trading day

For different data types, we select different features for constructing the regression models. Table 2, shows features for the stock price forecasting in China A share market in Shanghai and Shenzhen; Table 3, shows the input features for the forecasting of Shanghai composite index and Shenzhen composite index.

4.2 Data Pre-processing

4.2.1 Shift Windows

In this experiment; to test the learning capability of the algorithms and to follow as well as forecast the trend of the stock price movement, a shift window was designed. For each of the data set, there were 30 samples in one window, approximately 5% of the total samples, and the first 25 of these samples were used as training data and the last 5 samples as testing data. We then shifted forward this window by the shift step of 5 days. For example, the first shift window contains 30 trading days of data from 15th, September, 2006 to 9th, November, 2006, in this shift window, the first 25 samples, which began on 15th, September and finished on 2nd, November, are used as training set, and the data of the following five days, from 3rd, November to 9th, November, are used as testing data. We predict the stock price of these five days, and compare that with the actual price of these five days. Then we shift the window forward and the training set started from 21st, September till 9th, November. The actual stock prices of these 25 samples are used as training set to predict the following 5 day's stock price. Analogically, we can predict all the stock prices of our set by shifting the windows. In every window, the ratio of training samples and testing samples is 5:1.

4.2.2 Normalization of Data

When we use Euclidean distance RBF kernel function-based SVR to do the prediction, the data set should be normalized to avoid features that may contain a greater numeric value range from dominating the features; that have smaller numeric ranges in the process of training and regression. In this experiment, the formula we applied to normalize the data is:

$$v' = \frac{v - \min_\alpha}{\max_\alpha - \min_\alpha},$$ (15)

where v' is the normalized value and v is the original value. After the process, all the values of the features were normalized within the range of [0, 1].

4.3 Evaluation Criteria

The prediction performance can be evaluated by the following statistical metrics[17]:

Normalized Mean Squared Error (NMSE) measures the deviation between the actual values and the predicted values. The smaller the values are, the closer the predicted values to the actual values. The formula of NMSE is:

$$\text{NMSE} = 1/(\delta^2 n) \sum_{i=1}^{n} (a_i - p_i)^2 \qquad (16)$$

where

$$\delta^2 = 1/(n-1) \sum_{i=1}^{n} (a_i - p_i)^2 \qquad (17)$$

Directional symmetry (DS) indicates the correctness of the predicted direction of predicted value in terms of percentages. The formula of DS is:

$$\text{DS} = (100/n) \times \sum_{i=1}^{n} d_i \qquad (18)$$

where

$$d_i = \begin{cases} 1, (a_i - a_{i-1})(p_i - p_{i-1}) \geq 0 \\ 0, \text{otherwise} \end{cases} \qquad (19)$$

4.4 Experimental Results and Discussion

Table 4, shows the results of the experiment. The stocks which have a stock number starting with "6" are from China A share market in Shenzhen, the stocks which have a stock number starting with "0" are from China A share market in Shanghai. The columns denoted as MRBFSVR present the results of Mahalanobis distance RBF based SVR, the columns denoted as ERBFSVR present the results of Euclidean distance RBF based SVR, and BPNN is short for BP neural network. Fig. 1 and Fig. 2 give the comparison results of the three methods.

The results in Fig. 1 and Fig. 2 show that both of the two SVRs, the Mahalanobis distance RBF based SVR and Euclidean distance RBF based SVR; outperform the BP neural network with respect to the criteria of NMSE and DS in most of the 15 data sets. Obviously, from these results, we can observe that the SVM regression method is more suitable for time series forecasting problems in financial forecasting than the BP neural network algorithm.

From Table 4, we cannot conclude that the Mahalanobis distance RBF based SVR is definitely a better algorithm than the Euclidean distance RBF based SVR. We can observe that the criteria of NMSE, the Mahalanobis distance RBF based SVR reduces to a lower NMSE value in 8 of the 15 data sets than the Euclidean distance RBF based SVR; although, for the criteria of DS, the Mahalanobis distance RBF based SVR receives a higher DS value in 10 of the 15 data sets than the Euclidean distance RBF based SVR. These results are not enough to support the conclusion that the Mahalanobis distance RBF based SVR is superior to the Euclidean distance RBF based SVR when applied to time series forecasting.

Table 4 The NMSE and DS values of the three algorithms for the 15 data sets

	NMSE MRBFSVR	NMSE RERBFSVR	NMSE ERBFSVR	DS MRBFSVR	DS RERBFSVR	DS ERBFSVR
Stock600111	1.452	2.031	3.251	0.723	0.721	0.692
stock600839	1.234	1.252	2.187	0.832	0.751	0.747
stock600644	2.122	2.574	3.145	0.765	0.862	0.691
stock600688	3.217	3.202	4.012	0.658	0.723	0.735
Stock601318	1.231	1.439	2.809	0.852	0.635	0.821
Stock600031	3.381	3.226	4.515	0.721	0.696	0.734
Stock000858	1.535	2.024	2.991	0.890	0.791	0.695
Stock000014	1.213	1.201	2.715	0.724	0.713	0.627
Stock000024	2.642	2.412	2.499	0.635	0.731	0.522
Stock000002	1.924	2.213	2.758	0.592	0.591	0.670
Stock000063	1.983	71.327	2.301	0.832	0.751	0.753
Stock000100	2.910	1.045	1.703	0.901	0.912	0.715
Stock000527	1.523	1.213	1.609	0.749	0.812	0.842
Shanghai index	1.237	2.341	2.764	0.831	0.826	0.731
Shenzhen index	1.101	1.923	2.113	0.877	0.841	0.687
Avarage	1.913	1.961	2.758	0.772	0.757	0.710

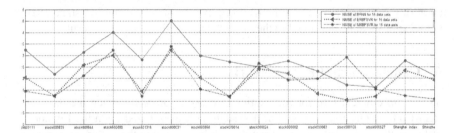

Fig. 1 The comparison of NMSE value of the three algorithms

This phenomenon can be explained by the limitation of Mahalanobis distance. Compared with Euclidean distance, the Mahalanobis distance takes into account the effect of correlation among the features of the training samples; but it also has the limitation that the Mabalanobis distance may enlarge the effect of correlation among the features. Such an enlargement might have generated some negative effect for some data sets.

However, from Table 4, we can see that for 6 of the 15 data sets, the Mahalanobis distance RBF based SVR outperforms the Euclidean distance RBF based SVR in both of the criteria of NMSE and criteria of DS; and for 12 of the 15 data sets, the Mahalanobis distance RBF based SVR gives a better performance based on at least one criterion. With only 3 data sets, the Mahalanobis distance RBF based SVR obtains the worst performance in both of the criteria of NMSE and criteria of DS.

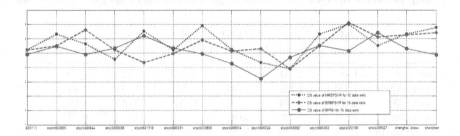

Fig. 2 The comparison of DS value of the three algorithms

What is more, for the average of the results of 15 data sets, the Mahalanobis distance RBF based SVR shows better performance for both of the two criteria of NMSE and criteria of DS. Hence, we can conclude that on the whole the Mahalanobis distance RBF based SVR performs better than the Euclidean distance RBF based SVR. But for a certain new data set, we cannot determine which one can achieve a better performance if there is no prior knowledge.

In summary, we find that under certain circumstances when applied to time series forecasting problems; the Mabalanobis distance RBF kernel can be a better choice than the traditional Euclidean distance RBF kernel function for SVR. In consideration of the correlation among the features, the use of Mabalanobis distance RBF kernel appears to be beneficial to the improvement of the forecasting result.

5   Conclusion and Future Work

SVR is an efficient tool for time series forecasting problems when the data sets stem from an unstable and nonlinear system. However, based upon the Euclidean distance, the SVR neglects the effect of correlation among the features when processing the training data in time series forecasting problems. Since the Mabalanobis distance can address the limitation of Euclidean distance, a Mabalanobis distance RBF kernel is introduced in this chapter. From the comparison investigation, we find that in some circumstances, the Mabalanobis distance RBF kernel based SVR can outperform the Euclidean distance based SVR. Consequently, when people use the SVR to deal with the time series forecasting problems, the proposed Mabalanobis distance RBF kernel based SVR is worthy for consideration.

One of the limitations of our work is that we have not yet provided the means for determining which RBF kernel may attain better results for a certain data set. Some of our current work reveals that there could be some influence due to the selection and distribution of the features in the data set. This will be our further work in future.

References

1. Kim, K.: Financial time series forecasting using support vector machines. Neurocomputing 55(1-2), 307–319 (2003)
2. Khemchandani, R., Chandra, S.: Regularized least squares fuzzy support vector regression for financial time series forecasting. Exp. Sys. with Appl. 36(1), 132–138 (2009)
3. Huang, W., Nakamori, Y., Wang, S.: Forecasting stock market movement direction with support vector machine. Comput Operat. Res. 32, 2513–2522 (2005)
4. Radhika, Y., Shashi, M.: Atmospheric temperature prediction using support vector machines. Int. J. Comput. Theor. and Eng. 1(1), 1793–8201 (2009)
5. Liu, B., Su, H., Huang, W., Chu, J.: Temperature prediction control based on least squares support vector machines. J. Contr. Theor. and Appl. 2(4), 365–370 (2004)
6. Bhasin, M., Raghava, G.: Prediction of CTL epitopes using QM, SVM and ANN techniques. Vaccine 22(23-24), 3195–3204 (2004)
7. Hossain, A., Zaman, F., Nasser, M., Islam, M.: Comparison of GARCH, Neural Network and Support vector machine in financial time series prediction. Pat. Rec. and Mach. Intel., 597–602 (2009)
8. Vapnik, V.N.: The Nature of Statistical Learning Theory. Springer, New York (1995)
9. Smola, A.J., Schlkopf, B.: A tutorial on support vector regression. Stat. and Comput. 14(3), 199–222 (2004)
10. Vapnik, V., Golowich, S.E., Smola, A.J.: Support vector method for function approximation, regression estimation, and signal processing. In: Michael, C., Michael, I., Thomas, P. (eds.) Advances in Neural Information Processing Systems 9, pp. 281–287. MIT Press (1996)
11. Cherkassky, V., Ma, Y.: Practical selection of SVM parameters and noise estimation for SVM regression. Neur. Netw. 17(1), 113–126 (2004)
12. Tsang, I.W., Kwok, J.T.: Learning the Kernel in Mahalanobis One-Class Support Vector Machines. In: The 2006 IEEE International Joint Conference on Neural Network Proceedings, Vancouver, BC, pp. 1169–1175 (2006)
13. Zhang, Y., Xie, F., Huang, D., Ji, M.: Support vector classifier based on fuzzy c-means and Mahalanobis distance. J. of Intel. Inf. Sys. 35, 333–345 (2009)
14. Wang, D., Yeung, D.S., Tsang, E.C.C.: Weighted mahalanobis distance kernels for support vector machines. IEEE Trans. on Neural Networks 18, 1453–1462 (2007)
15. Mahalanobis, P.C.: On the generalized distance in statistics. In: Proceedings National Institute of Science, India, pp. 49–55 (1936)
16. Bellman, R., Kalaba, R.: On adaptive control processes. IEEE Trans. on Auto. Contr. 4(2), 1–9 (1959)
17. Makridakis, S., Wheelwright, S.C., Hyndman, R.J.: Forecasting methods and applications. Wiley-India (2008)

Chapter 11
Neural Networks and Wavelet De-Noising for Stock Trading and Prediction

Lipo Wang and Shekhar Gupta

Abstract. In this chapter, neural networks are used to predict the future stock prices and develop a suitable trading system. Wavelet analysis is used to de-noise the time series and the results are compared with the raw time series prediction without wavelet de-noising. Standard and Poor 500 (S&P 500) is used in experiments. We use a gradual data sub-sampling technique, i.e., training the network mostly with recent data, but without neglecting past data. In addition, effects of NASDAQ 100 are studied on prediction of S&P 500. A daily trading strategy is employed to buy/sell according to the predicted prices and to calculate the directional efficiency and the rate of returns for different periods. There are numerous exchange traded funds (ETF's), which attempt to replicate the performance of S&P 500 by holding the same stocks in the same proportions as the index, and therefore, giving the same percentage returns as S&P 500. Therefore, this study can be used to help invest in any of the various ETFs, which replicates the performance of S&P 500. The experimental results show that neural networks, with appropriate training and input data, can be used to achieve high profits by investing in ETFs based on S&P 500.

1 Introduction

Stock prices are highly dynamic and bear a non-linear relationship with many variables such as time, crude oil prices, exchange rates, interest rates, as well as factors like political and economic climate. Hence stock prices are very hard to model by even the best financial models. Future stock prices can be studied merely by historical prices.

Lipo Wang · Shekhar Gupta
School of Electrical and Electronic Engineering
Nanyang Technological University
Block S1, 50 Nanyang Avenue,
Singapore 639798
e-mail: elpwang@ntu.edu.sg

W. Pedrycz & S.-M. Chen (Eds.): Time Series Analysis, Model. & Applications, ISRL 47, pp. 229–247.
DOI: 10.1007/978-3-642-33439-9_11   © Springer-Verlag Berlin Heidelberg 2013

With the globalization and ease of investment in international and national markets, many people are looking towards stock markets for gaining higher profits. There is a high degree of uncertainty in the stock prices, which makes it difficult for the investors to predict price movements.

Hence the study of prediction of stock prices has become very important for financial analysts as well as the general public, so as to gain high profits and reduce investment risk. With vast investments in the equity market, there has been a huge motivation for a system which can predict future prices. The Efficient Market Hypothesis (EMH) [1] states that no information can be used to predict the stock market in such a way as to earn greater profits from the stock market. There have been studies to show the accountability of the EMH [2], but some later studies have implied otherwise [3]. There are various views that oppose the EMH and indicate the predictability of some stock markets.

Stock prediction methods may be categorized into Fundamental Analysis and Technical Analysis [1]. Specific modeling techniques include multivariate regression [2] and artificial neural networks (ANN) [3,4]. This chapter is concerned only with ANN approaches. Rodrigues [5] used a relatively simple neural network to predict and trade in the Madrid Stock Market Index. Rodrigues [5] used nine lagged inputs to predict the prices and make buy/sell decisions, which gave evidence that ANN is a superior strategy to predict the index prices as compared to various other analyses. Although this model did not perform well in a bullish market.

Due to the ability of ANNs to form a complex model between training inputs and targets values, ANNs give an opportunity to model highly complex and dynamic relation in the stock prices [6,7]. There are many areas where the neural networks have been used, e.g., signal processing, speech recognition, control, and many types of neural networks have been created [5,6,24]. According to Chang et al [8], ANNs are believed to have limitations due to noise in and complexity of stock prices.

Neural network prediction systems can be divided into 2 categories, i.e., using (1) the past prices of the index and (2) fundamental data, such as exchange rates, gold prices, interest rates etc. [22-26]. For example, in the first category, [9] developed a neural network model based on past prices by using three neural networks. They were able to obtain around 16% returns per annum with a weekly prediction model, but the ANNs failed in daily prediction models. In the second category, ANNs in [10] performed much better compared to traditional stock valuation systems using financial models.

This chapter describes some of our attempts to successfully predict S&P 500 index [11], which is then developed into a trading model, in order to achieve relatively high rates of return over a long period of time. This involves the use of ANNs and wavelet de-noising in the input time series.

2   Artificial Neural Networks

Artificial neural networks attempt to mimic the biological counterparts. A neural network is composed of a number of interconnected processing elements (neurons) working in unison to solve specific problems. Learning involves adjustments to the synaptic connections between the neurons, as well as other parameters, such as the biases in neurons [5,6].

In a feed-forward neural network or a multilayer perceptron, there can be 3 layers, i.e., an input layer, an output layer, and a hidden layer. The number of hidden layers, as well as the number of neurons in each layer, can vary according to given requirements. A simple 3-layer neural network is shown below:

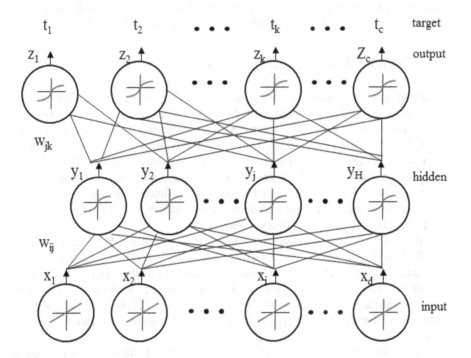

Fig. 1 A simple 3-layer neural network

The architecture of a network consists of the number of layers, the number of neurons in each layer, each layer's transfer function, and the weights by which the layers connect to each other. The best architecture to use depends on the type of problem to be solved by the network [5,6,12].

There can be two training methodologies used in the training of the neural networks, namely, incremental (online) training and batch training. In incremental training, the weights and biases are updated after each input is presented, while in batch training, weights and biases are updated after all of the inputs and targets are presented.

The inputs to a neuron include its bias and the sum of its weighted inputs. The output o_i of neuron i depends on the neuron's inputs and on the transfer function:

$$f(x) = \frac{1}{1 + \exp(-\beta x)} \tag{1}$$

$$o_j = f(\sum_i w_{ji} o_i + \theta_j) \tag{2}$$

where β is the gain, θ_j is the bias of neuron j, and w_{ji} is the connection weight from neuron i to neuron j. There are many training algorithms and different network structures which can be selected according the problem requirements. In our work, we use the Levenberg-Marquardt training algorithm. Sometimes, over-training of the network can lead to poor generalization. To prevent this from happening, a technique called early stopping is used with the help of validation data.

3   Wavelet Analysis and De-noising

Wavelets are mathematical functions used to decompose a given function or continuous-time signal into components of different scales. Usually one can assign a frequency range to each scale component. A wavelet transform is the representation of a function by wavelets. The wavelets are scaled and translated copies (known as "daughter wavelets") of a finite-length or fast-decaying oscillating waveform (known as the "mother wavelet"). Wavelet transforms have advantages over traditional Fourier transforms for representing functions that have discontinuities and sharp peaks, and for accurately deconstructing and reconstructing finite, non-periodic and/or non-stationary signals [13,14].

There are two types of wavelet transform, i.e., Discrete Wavelet Transform (DWT) and Continuous Wavelet Transform (CWT). In CWT, during computation, the analyzing wavelet is shifted smoothly over the full domain of the analyzed function [12]. Calculating wavelet coefficients at every possible scale can be very tedious and data generated can be difficult to analyze. By choosing the scales by the power of two, the analysis can become more accurate and faster. Hence, we use the DWT in this study.

There are numerous types of wavelets available for different types of series being analyzed. The Daubechies (db) wavelet is used in our analysis for the reduction of noise in the time series, which are later fed as an input to the neural network for training.

There have been research efforts on wavelet neural networks, in which wavelets are used as neuron transfer functions. But we will restrict ourselves to studying the effect of de-noising by wavelets and then feeding the reconstructed de-noised signal into the ANN, instead of training the neural network with wavelet coefficients [5,15].

4   Data and Experiments

This study attempts to predict future index prices, solely on the basis of past index prices, along with effects of NASDAQ 100 index on the prices of S&P 500. Another aspect of this chapter is to study the effect of wavelet de-noising on the raw S&P 500 time series. The number of inputs ranges from 10 day lagged values to 40 day lagged values of the S&P 500 index closing prices.

The main source of historical index prices is from the Yahoo Finance. We download 2 sets of data to study the effect and relevance of historical prices. The first set of data involves the closing prices of S&P 500 index from 9 January 1950 to 15 January 2010. The second set of data involves the closing prices of S&P 500 and NASDAQ 100 index from 7 January 1991 to 15 January 2010.

Due to public holidays, there are data missing on various days in the raw time series, which need to be adjusted to account for the missing values. A 5-day lagged average is used to fill in the missing data:

$$Missing\ Value\ (x_t) = \frac{x_{t-1} + x_{t-2} + x_{t-3} + x_{t-4} + x_{t-5}}{5} \tag{3}$$

The raw data are divided into 3 parts, i.e., for training, validation, and testing.

a.   Wavelet De-noising

The wavelet toolbox of MATLAB is used to de-noise the raw S&P 500 time series. The first set of time series, i.e., from 9 January 1950 to 15 January 2010 is fed into the Signal Extension toolbox for the wavelet analysis, in order to make it compatible for stationary wavelet analysis (SWT) and de-noising. The data are then fed into the SWT De-noising 1-D toolbox, where the signal is decomposed by a db wavelet at Level 5. The de-noising tool is used to remove the white noise, with different thresholds for each level of decomposition and reconstruction of the de-noised signal. This signal is then divided into matrices of suitable sizes for training, validation and testing of the neural network.

b.   Neural Network Architecture and Training

The feedforward back propagation neural network is used in this chapter, with a Levenberg-Marquardt training algorithm. The performance criterion used is the mean square error (MSE), with the TANSIG transfer function in MATLAB.

The number of hidden layers varies from 1 to 2, with one input layer and one output layer. The number of neurons in the input layer varies from 10 to 40, with a search interval 5, that is, we try out 10, 15, 20, ..., and 40 input neurons, in order to find the optimal number of input neurons. The output layer consists of only 1 neuron which is the predicted price on the next working day. The number of neurons in the hidden layers varies according to the number of neurons in the input layer.

After creating the neural network, the raw time series is divided into 3 sets of data, i.e., training, validation and testing data. These data sets in the form of p*1 matrices are divided into n*m matrices, where n is the number of inputs in the neural network and m is the number of training sets in batch training. For training, p = 15,040, n = 10, 15, 20, or 40, and m =15,000.

The training parameters are now selected. Since the Levenberg-Marquardt algorithm is a fast learning algorithm, the maximum number of epochs is limited to 100. The minimum gradient is selected to be 1e-6.

The validation data are an n*t matrix, where n = 10, 15, 20 or 40 (same as the training data) and t can vary according to the number of validations required, in our case, from 200 to 400, depending on the size of the training data. The validation is used for early stopping for the network, so as to prevent over fitting of the neural network and maintain its generalization. The testing for the trained network is done over different periods of time, ranging from 1 year to 2 years, i.e., from 250 to 420 prediction points. The input test matrix is of the size n*u, where n = 10, 15, 20 or 40 (same as that of the training and validation matrices, i.e., the number of inputs) and u varies from 250 to 420, i.e., the number of testing sets in the batch. The target test matrix is of the size 1*u.

c.   Neural Network Simulations of Test Data

After the training of the neural network, the test data are used to generate the predicted outputs which are compared with the actual values. The test is done over various periods of time, ranging from 1 to 2 years, and with different market conditions, i.e., before, during, and after recessions.

d.   Trading System

The predicted outputs during testing are exported into a trading system, to obtain the directional efficiency and rate of returns for the specified period of time. The trading system is developed for daily trading. When the predicted price for the next day is less than today's price, a sell decision is made. And when the predicted price for the next day is more than today's price, then a buy decision is made. Based on these trading rules, the rate of returns is calculated over a period of time.

Below is an example of the results of the trading system, which are shown in the cells G4:I6, i.e., directional efficiency, return, and the rate of return for the specified period of time. These results are calculated for different neural networks, with varying inputs, hidden layers, training data and are then compared in the next section.

	A	B	C	D	E	F	G	H	I
1	Actual	Predicted							
2	1385.7	1384.2			1.5				
3	1377.7	1387.2	Buy	Wrong	-9.5				
4	1377.2	1373.3	Sell	Correct	3.9		Efficiency	52.05993	%
5	1404.1	1375.1	Sell	Wrong	29		Return	322.5	
6	1360.7	1401.8	Sell	Correct	-41.1		Rate of Return	35.80443	%
7	1361.8	1365.6	Buy	Correct	-3.8				
8	1358.4	1358.7	Sell	Correct	-0.3				
9	1335.5	1354.2	Sell	Correct	-18.7				
10	1339.9	1331.3	Sell	Wrong	8.6				

Fig. 2 An example output of the trading system

5 Results and Discussions

We carry out various experiments to study the future index prices of S&P 500, i.e.,

a) effects of wavelet de-noising;
b) effects of gradual sub-sampling the past data;
c) effects of NASDAQ 100 on the S&P 500 index.

We will first discuss the effects of the above mentioned factors in detail, followed by selecting the best results based on the returns and risk for a long period of time.
The data set has been divided into 2 sets, i.e.,

a) Data Set 1: January 1950 – January 2010
b) Data Set 2: January 1991 – January 2010

a. Effects of Wavelet De-Noising

We now use the 1st set of data, i.e., the data from January 1950 till January 2010, for training and testing. The training and validation data are selected from January 1950 till April 2008. Then testing is done for 2 time intervals, i.e.,

a) Period 1: May 2008 till January 2010 (1.5 years)
b) Period 2: January 2009 till January 2010 (1 year).

This experiment is done with and without wavelet de-noising and the results with the maximum efficiency are shown as below:

Table 1 Results without wavelet de-noising in training from 1950 – 2008

Network Structure	10-10-1		10-20-1	
Period	1	2	1	2
Rate of Return	6.3%	26.5 %	5.5%	25.2%
Directional Effi-ciency	51.5%	53.4%	51.2%	52.3%

Table 2 Results with wavelet de-noising in training from 1950 – 2008

Network Structure	10-10-1		10-21-1	
Period	1	2	1	2
Rate of Return	5.2%	23.9%	4.4%	21.1%
Directional Effi-ciency	51.1%	52.3%	50.1%	50.4%

From the above results, it can be seen that the effect of wavelet de-noising is not satisfactory, as compared to the training of the neural networks by the raw time series. This may indicate that there exists minimum noise in the raw financial data.

b. Effects of Gradual Data Sub-Sampling

An experiment is carried out with a novel technique of gradual data sub-sampling [22], whereby data of historically distant past are given less significance and recent data are given more significance. This technique is applied on the 1st data set, i.e., from January 1950 till April 2008, during training. Originally there are 15,200 points of training data, which are reduced to 6,700 in the following way:

a)   900 training data are selected from the 1st 5,400 data;
b)   2,000 training data are selected from the next 6,000 data;
c)   all the remaining 3,800 data are selected.

This technique ensures that the historical trends are not ignored and, at the same time, the system is more related to the current market situations. Thus the number of lagged values also varies to show the effect of number of inputs as well as wavelet de-noising. Again, the testing is done for Period 1 and Period 2 stated above. The most efficient results are shown below:

Table 3 Results without wavelet de-noising with gradual data sub-sampling

Network Structure	10-5-1		15-26-1		20-22-1	
Period	1	2	1	2	1	2
Rate of Return	8.9%	34.4%	15.3%	44.7%	3.5%	12.7%
Directional effi-ciency		52.6%		50.0%	53.9%	52.8%	52.2%	51.6%

Table 4 Results with wavelet de-noising with gradual data sub-sampling

Network Structure	10-18-1		15-21-1		20-24-1	
Period	1	2	1	2	1	2
Rate of Return	6.0%	36.5%	1.6%	21.5%	9.5%	5.3%
Directional efficiency		51.5%	53.2%	50.4%	49.6%	50.9%	47.4%

Again from the above results, it can be seen that wavelet de-noising is not effective in predicting the future index prices. After obtaining these results, we decide not to include wavelet de-noising in further experiments in this chapter, since the results are much better with the raw S&P 500 time series. This may mean that the financial data used here are not noisy.

Our results also show an increase in the efficiency with gradual data sub-sampling. A comparison between the original data series and the series with gradual data sub-sampling is shown in Table 5:

Table 5 Effect of gradual data sub-sampling without wavelet de-noising

Network Structure	10-10-1 (Original data)		10-5-1(Gradual Data Sub-sampling)	
Period	1	2	1	2
Rate of Return	6.3 %	26.5 %	8.9%	34.3%
Directional effi-ciency	51.5%	53.4%	52.6%	50.0%

It is evident from Table 5 that the rates of return with gradual data sub-sampling are much better than those with the original time series, for both Periods 1 and 2.

We now discuss results for data set 2, i.e., the data from January 1991 till January 2010, with data and without data sub-sampling. The testing period is kept the same as the previous section, so as to compare all the models under the same market conditions.

In this approach the training and validation of the neural network are done with data sets from January 1991 till April 2008. Again, testing is done for Period 1 and Period 2 stated above. The results are shown in Table 6:

Table 6 Results with original data (no sub-sampling) from 1991 – 2010

Network	Period 1		Period 2	
	Rate of Return	Directional efficiency	Rate of Return	Direc-tional efficiency
40-16-4-1:	11.7%	55.2 %	53.7 %	58.7 %
40-120-80-25-1:	14.8 %	51.9 %	36.3 %	49.6 %
40-20-1:	17.8%	54.8%	30.1%	52.4 %
40-13-1:	19.8 %	54.0 %	32.8 %	53.6 %

We now describe the results for the data set 2, i.e., from January 1991 till January 2010, with gradual data sub-sampling. Originally there are 4,500 data sets for training and validation, which are reduced to 2,900 data sets as follows:

a)   800 data sets are selected from the first 2,400 data;
b)   all the remaining 2,100 data sets are selected.

The above training set is used to train the network with different lagged inputs. Again, the testing is done for Period 1 and Period 2 stated previously. The most efficient results are shown below:

Table 7 Results with gradual data sub-sampling for Data Set 2

Network Structure	40-22-1		30-30-1		10-22-1	
Period	1	2	1	2	1	2
Rate of Return	36.0%	47.2%	24.5%	45.6%	24.5%	30.7%
Directional efficiency	55.7%	55.6%	55.2%	56.3%	53.3%	51.6%

Table 7 shows that the results with 40 lagged input values are the best.

Now we will further reduce the historical data and give even more preference to the more recent data, by reducing the training and validation inputs from 2,900 to 1,800 data sets. The most efficient results are shown in the table below:

Table 8 Results with further sub-sampling for Data Set 2

Network Structure	10-9-1	
Period	1	2
Rate of Return	47.2%	44.1%
Directional efficiency	57.6%	54.8%

The results in the above table are quite good.

We compare the results of training with the original data, sub-sampled data, and further sub-sampled data in the following table:

Table 9 Comparison between various training data sets

Network Structure	40-13-1 (Original data)		40-22-1 (Data Sub-sampling)		10-9-1 (Further Sub-sampling)	
Period	1	2	1	2	1	2
Rate of Return	19.8%	32.8%	36.0%	47.20%	47.2%	44.1%
Directional efficiency	54.0 %	53.6%	55.7%	55.6%	57.6%	54.8%

It can be seen that the effect of gradual data sub-sampling has a great influence on the directional efficiency and the rate of return. With reducing the training and validation data set from 4,500 to 1,800, the rate of returns increased from 19.8% to 47.2% for period 1 and from 32.8% to 44.1% for period 2. Hence, the effect of data sub-sampling is of major importance.

c.   Effects of NASDAQ 100 Index

In this section the training data also include the NASDAQ 100 lagged index values along with the lagged values of S&P 500 index (Data Set 2, i.e., from January 1991 till January 2010). 4500 data points are used for training and validation, and there are 20 inputs to the neural network, i.e., 10 lagged values each for S&P 500 and NASDAQ 100. The testing is again done for Period 1 and Period 2 stated previously. The most efficient results are shown below:

Table 10 Results with effect of NASDAQ 100

Network Structure	20-7-1	
Period	1	2
Rate of Return	34.8%	72.4%
Directional effi-ciency	57.9%	59.6%

The above table shows that NASDAQ 100 data can be used to predict the S&P 500 index prices and these results are comparable to the effect of gradual data sub-sampling.

In the next section, we will compare the best results from each of the above sections and discuss which model would be the best suited for prediction of S&P 500 index, and therefore, making the maximum profit in the long run. Finally, we make comparisons of our results with other published work and discuss the feasibility of our model for investment.

d   Discussions of Various Models for Prediction

Now we will discuss about the best results from the above 3 sections and compare them. Below is the table which consists of the best results from each section.   The comparison is between:

a)   The results obtained after the wavelet de-noising of the original data from January 1950 till January 2010.

b)   The results obtained after the further data sub-sampling for the data set from January 1991 till January 2010.

c)   The results obtained by including the NASDAQ 100 index to study the index movements for S&P 500, for data set from January 1991 till January 2010.

Table 11 Comparisons between Wavelet, Data sub-sampling and NASDAQ 100

Network Structure	10-10-1 (Wavelet)		10-9-1 (Further Data Sub-sampling)		20-7-1 (NASDAQ 100)	
Period	1	2	1	2	1	2
Rate of Return	5.2%	23.9%	47.2%	44.1%	34.8%	72.4 %
Directional Efficiency	51.1%	52.3%	57.6%	54.8%	57.9%	59.6 %

From the above comparisons, we draw the following conclusions:

a) Wavelet de-noising does not help much to predict the future index prices.
b) The technique of gradual data sub-sampling has been proved to be effective in prediction of index values. 10 lagged values as inputs are optimal.
c) NASDAQ 100 index has been proved to be effective for the prediction of S&P 500 index.

Period 1 (May 2008 till January 2010) used in testing includes the time period when the economies were in and coming out of recession, and there was a very slow improvement in the overall index movements. Period 2 (January 2009 till January 2010) includes the time when the economy was recovering and coming out of the recession, and when the increase in index was relatively stronger than that in Period 1.

Since the wavelet de-noising has been proved to be an ineffective technique to train the neural network with de-noised inputs, we will compare the results of data sub-sampling and NASDAQ 100 effects for the above mentioned 2 prediction periods.

The rate of return is as high as 47% for the 1st period and 44% for the 2nd period for the data sub-sampling technique, and around 34% and 72%, respectively, for the 2 periods with NASDAQ 100 index. This shows that the technique of data sub-sampling can provide much more stable results over a long period of time, in different markets conditions.

On the other hand, the technique of including S&P 500 index with NASDAQ 100 also leads to good results, which varies significantly for the 2 periods of study. The returns are around 34% for Period 1 and 72% for Period 2. This shows that this model is very effective to predict prices in good market conditions and leads to high returns, but is not very effective to deliver stable returns over a long period of time.

Table 12 Risk and return comparisons of the 2 models

	Returns	Risk
Model 1: Data Sub-sampling	Stable Returns	Low Risk
Model 2: NASDAQ 100	High Returns	High Risk

Hence an investor who would like to take less risk and want steady returns would consider using the gradual data sub-sampling technique, while an investor who wants enormously high returns can invest using the model with NASDAQ 100 index, if he thinks that the market conditions are relatively good.

Table 13 Best models for investment

Network Structure	10-9-1 (Further Data Sub-sampling)		20-7-1 (NASDAQ 100)	
Period	1	2	1	2
Rate of Return	47.2%	44.1%	34.8%	72.4%
Directional Efficiency	57.6%	54.8%	57.9%	59.6%

Next we show the actual vs. predicted prices for both the models. Both figures indicate that the actual and predicted prices coincide with each other quite well, which also makes it evident that these 2 model are able to predict well and provide good returns.

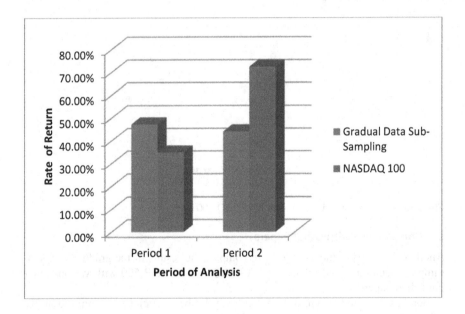

Fig. 3 Comparison of rate of returns for Period 1 and Period 2

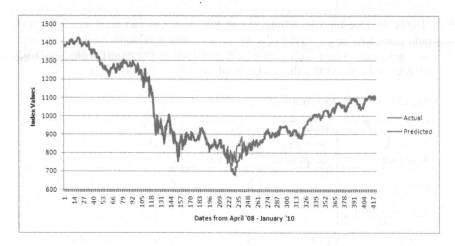

Fig. 4 Actual vs predicted values for gradual data sub-sampling (10-9-1)

Fig. 5 Actual vs. predicted values with NASDAQ 100

e.   Comparisons with Other Papers

Another set of experiments is done to compare the results of the gradual data sub-sampling technique and effects of NASDAQ 100 on S&P 500 with various other published papers.

For this part, the 2 neural networks with the best efficiencies from the above sub-sections are used, i.e.:

a) A network trained with the gradual data sub-sampling technique with 10 lagged values.
b) A network trained with the inclusion of both the NASDAQ 100 index and S&P 500 index, 10 lagged values each.

The results of our research are compared with 3 sources, i.e.,

a) Trading Solutions [16], an online website for a trading software package.
b) A model integrating a piecewise linear representation method and a neural network [18].
c) A system of 3 neural networks (3-ANN) [9], one network each for bullish, bearish and choppy markets.

The period of comparison is from January 2004 till December 2004. The results are as follows:

Table 14 Comparisons of rate of return by various techniques for January - December 2004

Technique	Trading Solutions [16]	IPLR [18]	3 – ANN System [9]	Gradual Data Sub-sampling	Effect of NASDAQ
Rate of Return	11.0%	35.7%	18.4%	25.4%	16.3%

It can be seen that the Trading Solutions, the 3-ANN system and the effect of NASDAQ are not very effective for this period. The rate of return for this period is the highest for the IPLR system, followed by the new technique of data sub-sampling.

We now show the rates of return for our gradual data sub-sampling for a number of periods:

Table 15 Rates of return with gradual data sub-sampling for different periods

Time of Testing	January 2004 – December 2004	April 2008 - January 2010	January 2009 – January 2010
Rate of Return	25.4%	47.2%	44.1%

Although the rate of return is not very high with our technique of gradual data sub-sampling for the period of January – December 2004, but for the period from April 2008 till January 2010, the rate of return is as high as 47.2%. Data is not available for us to comment on the rate of return from the IPLR technique for the period from April 2008 till January 2010. But we can still compare the situation of the markets in the two periods.

Below are the rates of return with the effect of NASDAQ 100:

Table 16 Rate of returns with effect of NASDAQ for different periods

Time Period	January 2004 – December 2004	April 2008 - January 2010	January 2009 – January 2010
Rate of Return	16.3%	34.8%	72.4%

Below are 3 graphs for the trend of S&P 500 index for the 3 periods:

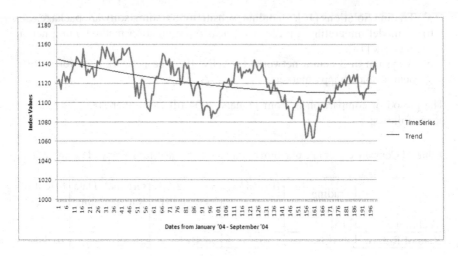

Fig. 6 Trend for the period from January - September 2004

Fig. 7 Trend for the period from April 2008 - January 2010

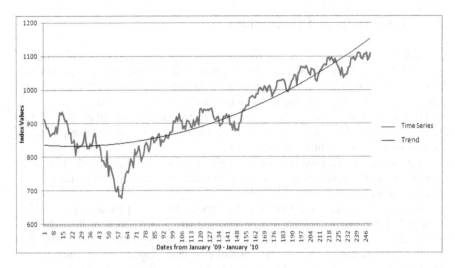

Fig. 8 Trend for the period from January 2009 - January 2010

From the above 3 figures, it can be concluded that:

a) January 2004 – September 2004 was a relatively stable period.
b) April 2008 – January 2010 involved a huge dip in price index (recession) followed by a recovery.
c) January 2009 – January 2010 had an increasing trend.

Table 17 Trading models vs. investor risk profile

Risk Profile \ Model	Data Sub-sampling (Moderate to High and Stable Returns)	NASDAQ 100 (Stable Returns, High Returns in bullish market)	IPLR (High Returns in Stable Markets, Unknown Returns in other market conditions)
Low	✓		
Moderate		✓	
High			✓

From the analysis above, it can be seen that our system with the innovative technique of gradual data sub-sampling is relatively stable and efficient in delivering good rates of return in all market situations and high rates of return in special market situations involving more movement in the index values. Also, the effect of NASDAQ is suitable to deliver high rates of return, i.e., around 72.4%, in good market conditions.

5   Conclusions and Discussions

It is evident from this chapter that feed forward back propagation neural networks are useful in predicting the future index prices and the research also proves that suitable network structure and appropriate data sub-sampling could lead to maximization of rates of return and hence deliver large profits to investors.

Historical data of the index, as well as historical data of other indices like NASDAQ 100, along with the innovative technique of data sub-sampling, are helpful in determining future index prices.

Trial-and-error was used at various stages of our experiments to select the most suitable network architecture with optimum number of hidden layers, lagged values, and neurons.

This trading system could be used by various financial products, such as Exchange Traded Funds (ETFs), which replicate the performance of the S&P 500 index, thereby giving investors an opportunity to invest following the S&P 500 trends. We emphasize that neural networks in this chapter are not used to get the exact future index prices, but is used to determine the directional trend of the index.

Finally, this study shows that the technique of gradual data sub-sampling and the effect of NASDAQ 100 on the prediction of S&P 500 index prove to be very beneficial, leading to high rates of return.

There are rooms for improvement and further research. We have studied here the effect of only NASDAQ 100 on S&P 500 prediction. Various other factors, such as crude oil prices, gold prices, other indices, exchange rates, interest rates, etc., could be used to study their impact on the price movements of S&P 500.

The feed forward back propagation neural network has been used in this chapter. There are various other types of neural networks, e.g., the cascaded back propagation neural networks, radial basis function, etc., which could be used to study the index prices.

The gradual data sub-sampling technique can be studied together with other systems, such as IPLR. A real time trading system could be developed, in which prices are downloaded live from the server. These prices are then fed into the system to predict prices at various intervals in a particular day, which could further be used for intra-day trading or high frequency trading.

References

1. Malkiel, B.G.: A Random Walk Down Wall Street. W. W. Norton & Company, London (1999)
2. Fama, E.F., Fisher, L., Jensen, M., Roll, R.: The Adjustment of Stock Market Prices to New Information. International Economic Review X, 1–21 (1969)
3. Thaler, R.: Does the Stock Market Overreact? Journal of Finance 40(3), 793–805 (1985)
4. Yahoo Finance, http://finance.yahoo.com
5. Wang, L.P., Fu, X.J.: Data Mining with Computational Intelligence. Springer, Berlin (2005)

6. Zurada, J.M.: Introduction to Artificial Neural Systems. Jaico Books, Mumbai (1992)
7. MATLAB Version 7.0.1 Help Manual, The MathWorks Inc.
8. Azoff, E.M.: Neural Network Time Series Forecasting of Financial Markets. John Wiley & Sons, New York (1994)
9. Haddad, A.N., Akansu, R.A.: Multiresolution Signal Decomposition: Transforms, Subbands, Wavelets. Academic Press (1992)
10. Kaiser, G.: A Friendly Guide to Wavelets. Birkhäuser (1994)
11. Tak, B.: A New Method for Forecasting Stock Prices using Artificial neural network and Wavelet Theory. UMI Dissertation Services (1995)
12. Timmermann, M., Pesaran, H.: Forecasting stock returns: An examination of stock market trading in the presence of transaction. Journal of Forecasting 13, 335–367 (1994)
13. Chang, P.C., Wang, Y.W., Yang, W.N.: An investigation of the hybrid forecasting models for stock price variation in Taiwan 21(4), 358–368 (2004)
14. Chi, S.C., Chen, H.P., Cheng, C.H.: A forecasting approach for stock index future using grey theory and neural networks. In: Proc. IEEE Int. Joint. Conf. Neural Netw., vol. 6, pp. 3850–3855 (1999)
15. Femdndez-Rodriguez, F., Gonzdlez-Martel, C., Sosviall-Rivero, S.: On the Profitability of Technical Trading Rules based on Artificial neural networks: Evidence from the Madrid Stock Market. Economics Letter 69(1), 89–94 (2000)
16. Yoo, P.D., Kim, M.H., Jan, T.: Financial Forecasting: Advanced Machine Learning Techniques in Stock Analysis. In: Proceedings of the 9th IEEE International Conference on Multitopic, pp. 1–7 (2005)
17. Siddique, M.N.H., Tokhi, M.O.: Training neural networks: Backpropagation vs Genetic Algorithms. In: Proceedings of the International Joint Conference on Neural Networks, IJCNN, pp. 2673–2678 (2001)
18. Chang, P.-C., Fan, C.-Y., Liu, C.-H.: Integrating a Piecewise Linear Representation Method and a neural network Model for Stock Trading Points Prediction. IEEE Transactions on Systems, Man, and Cybernetics—Part C: Applications and Reviews 39, 80–92 (2009)
19. Fu, C.: Predicting the Stock Market using Multiple Models. In: ICARCV (2006)
20. Chenoweth, T., Obradović, Z.: A Multi-Component Nonlinear Prediction System for the S & P 500 Index. Neurocomputing 10, 275–290 (1996)
21. NeuroDimension, Inc., Trading Solutions (2008),
   http://www.tradingsolutions.com/
22. Gupta, S., Wang, L.P.: Stock Forecasting with Feedforward Neural Networks and Gradual Data Sub-Sampling. Australian Journal of Intelligent Information Processing Systems 11, 14–17 (2010)
23. Zhu, M., Wang, L.P.: Intelligent trading using support vector regression and multilayer perceptrons optimized with genetic algorithms. In: The 2010 International Joint Conference on Neural Networks (IJCNN), pp. 1–5 (2010)
24. Wang, L.P. (ed.): Support Vector Machines: Theory and Application. Springer, Berlin (2005)
25. Rather, A.M.: A prediction based approach for stock returns using autoregressive neural networks. In: 2011 World Congress on Information and Communication Technologies (WICT), pp. 1271–1275 (2011)
26. de Oliveira, F.A., Zarate, L.E., de Azevedo Reis, M., Nobre, C.N.: The use of artificial neural networks in the analysis and prediction of stock prices. In: 2011 IEEE International Conference on Systems, Man, and Cybernetics (SMC), pp. 2151–2155 (2011)

Chapter 12
Channel and Class Dependent Time-Series Embedding Using Partial Mutual Information Improves Sensorimotor Rhythm Based Brain-Computer Interfaces

Damien Coyle

Abstract. Mutual information has been found to be a suitable measure of dependence among variables for input variable selection. For time-series prediction mutual information can quantify the average amount of information contained in the lagged measurements of a time series. Information quantities can be used for selecting the optimal time lag, τ, and embedding dimension, Δ, to optimize prediction accuracy. Times series modeling and prediction through traditional and computational intelligence techniques such as fuzzy and recurrent neural networks (FNNs and RNNs) have been promoted for EEG preprocessing and feature extraction to maximize signal separability to improve the performance of brain-computer interface (BCI) systems. This work shows that spatially disparate EEG channels have different optimal time embedding parameters which change and evolve depending on the class of motor imagery (movement imagination) being processed. To determine the optimal time embedding for each EEG channel (time-series) for each class an approach based on the estimation of partial mutual information (PMI) is employed. The PMI selected embedding parameters are used to embed the time series for each channel and class before self-organizing fuzzy neural network (SOFNN) based predictors are specialization to predict channel and class specific data in a prediction based signal processing framework, referred to as neural-time-series-prediction-preprocessing (NTSPP). The results of eighteen subjects show that subject-, channel- and class-specific optimal time embedding parameter selection using PMI improves the NTSPP framework, increasing time-series separability. The chapter also shows how a range of traditional signal processing tools can be combined with multiple computational intelligence based approaches including the SOFNN and practical swarm optimization (PSO) to develop a more autonomous parameter optimization setup and ultimately a novel and more accurate BCI.

Damien Coyle
Intelligent Systems Research Centre,
University of Ulster, Derry, BT48 7JL, UK

W. Pedrycz & S.-M. Chen (Eds.): Time Series Analysis, Model. & Applications, ISRL 47, pp. 249–278.
DOI: 10.1007/978-3-642-33439-9_12   © Springer-Verlag Berlin Heidelberg 2013

1 Introduction

The human brain contains approximately 10^{11} neurons interconnected through over 100 trillion synapses. Each neuron, containing many different compartments made up of many different chemicals and neurotransmitters, emits tiny electrical pulses every millisecond. The electroencephalogram (EEG), recorded from the scalp surface, is a measure of the aggregate activity of many post-synaptic-potentials (PSPs) of these neurons and includes information from many different brain sources along with background noise from other non-neural signals. EEG is therefore inherently complex and non-stationary, rendering it very difficult to associate a particular EEG time series pattern or dynamic with a specific mental state or thought.

Coupling EEG dynamics to a person's thoughts or intent, expressed in the form of mental imagery, is the objective of non-invasive brain-computer interface (BCI) technology. BCIs enable people to communicate with computers and devices without the need for neuromuscular control or the normal communication pathways and therefore have many potential applications [1]-[3]. BCI has applications in assistive technologies for the physically impaired [4][5], rehabilitation after stroke [7], awareness detection in disorders of consciousness (DoC) [6] and in non-medical applications such as games and entertainment [8]. Voluntarily modulation of sensorimotor rhythms (SMR) forms the basis of non-invasive (EEG-based) motor imagery (MI) BCIs. Planning and execution of hand movement are known to block or desynchronize neuronal activity which is reflected in an EEG bandpower decrease in mu band (8-12Hz). Inhibition of motor behaviour synchronizes neuronal activity [1]. During unilateral hand imagination, the preparatory phase is associated with a contralateral mu and central beta event related desynchronization (ERD) that is preponderant during the whole imagery process [9]-[11]. BCIs utilize a number of self-directed neurophysiological processes including the activation of sensorimotor cortex during motor imagery (MI). However, as outlined, the dynamical and non-stationary patterns in the time series must be dealt with to ensure information can be discriminated and classified precisely so that BCI technology is robust enough to be made available to those who need it most: those who are severely physically impaired due to disease or injury. Maximizing the capacity for computer algorithms to separate noise from source, distinguish between two or more different mental states or one mental task (intentional control (IC) state) from all other possible mental states (no control (NC) state) has been the goal of many BCI focused researchers for the past 20 years. Linear and non-linear approaches to classification have been applied to classifying the EEG signals [12]-[14]. Times series modeling and prediction through traditional and computational intelligence techniques such as fuzzy and recurrent neural networks (FNNs and RNNs) have been promoted for EEG pre-processing and feature extraction to maximize signal separability [15]-[24].

Coyle et al [22][23] have proposed an approach were specific self-organizing FNNs (SOFNN) are trained to specialize in predicting EEG time-series recorded from various electrode channels during different types of motor imagery (left/right movement imagination). The networks become specialized on the dynamics of each time series and the relative difference in the predictions provided by the networks can produce information about the times series' that are being fed to the networks e.g., if two networks are specialized on two particular time series (left or right motor imagery) and unlabeled time series are fed to both networks, the network that produces the lowest prediction error can be indicative of the times series being processed and thus the information can be used to classify (or label) the unlabeled time series. This idea has been extended to include multiple time series, multiple classes and integrated with a range of other signal processing techniques to aid in the discrimination of sensorimotor based activations for BCI. A critical element in the neural time-series-prediction pre-processing (NTSPP) framework [21]-[24] is predictor (network) specialisation. This can be achieved through network optimization techniques and self-organising systems assuming that there are underlying differences in the time series being processed. Coyle et al [21] have shown in preliminary studies that subject specific time-embedding of the time-series can assist in specializing networks to improve BCI performance but that generally an embedding dimension, $\Delta=6$ and a time lag, $\tau=1$, works well for one-step-ahead EEG time series prediction.

The aim of this chapter is to show that spatially disparate EEG channels have different optimal time embedding parameters which change and evolve depending on the motor imagery or mental task being processed. To determine the optimal time embedding for each EEG channel (time-series) a recently proposed method based on the estimation of partial mutual information (PMI) is employed [25][26]. Mutual information has been found to be a suitable measure of dependence among variables for input variable selection and quantifies the average amount of common information contained in Δ measurements of a time series. Information quantities can be used for selecting the optimal time lag, τ, and embedding dimension, Δ, to optimize prediction accuracy. The PMI selected embedding parameters are used to embed the time series for each channel and class before SOFNN specialization is performed in the NTSPP framework. The results of eighteen subjects show that subject-, channel- and class-specific optimal time embedding parameter selection using PMI improves the NTSPP framework, increasing time-series separability and therefore overall BCI performance.

The following section describes the data used in the chapter to validate the proposed approach. Section 3 includes a description of the methods employed where section 3.1 describes the BCI including the NTSPP approach and other stages of signal processing such as spectral filtering, common spatial patterns, feature extraction and classification. Section 3.2 outlines the partial mutual information based input variable selection (PMIS) approach and the implications of applying this in the NTSPP framework for BCI. A description of how the BCI is setup and parameters are optimized is contained in Section 3.3. A discussion of the results and findings is presented in the remaining sections along with suggested future work for improvements to the proposed methodology.

2 Data Acquisition and Datasets

Data from 18 participants partaking in BCI experiments are used in this work. All datasets were obtained from the fourth international BCI competitions, BCI-IV, [27][28], which include datasets 2A and 2B [29]. Table 1 below provides a summary of the data.

Table 1 Summary of datasets used from the International BCI competition IV

Competition	Dataset	Subjects	Labels	Trials	Classes	Channels
BCI-IV	2B	9	S1-9	720	2	3
BCI-IV	2A	9	S10-18	576	4	22

Dataset 2B - This data set consists of EEG data from 9 subjects (S1-S9). Three bipolar recordings (C3, Cz, and C4) were recorded with a sampling frequency of 250 Hz (downsampled to 125Hz in this work). The placement of the three bipolar recordings (large or small distances, more anterior or posterior) were slightly different for each subject (for more details see [29][31]). The electrode position Fz served as EEG ground. The cue-based screening paradigm (cf. Fig 1(a).1) consisted of two classes, namely the motor imagery (MI) of the left hand (class 1) and the right hand (class2). Each subject participated in two screening sessions without feedback recorded on two different days within two weeks. Each session consisted of six runs with ten trials each and two classes of imagery. This resulted in 20 trials per run and 120 trials per session. Data of 120 repetitions of each MI class were available for each person in total. Prior to the first motor imagery training the subject executed and imagined different movements for each body part and selected the one which they could imagine best (e.g., squeezing a ball or pulling a brake). For the three online feedback sessions four runs with smiley feedback were recorded whereby each run consisted of twenty trials for each type of motor imagery (cf. Fig 1(a) for details of the timing paradigm for each trial). Depending on the cue, the subjects were required to move the smiley towards the left or right side by imagining left or right hand movements, respectively. During the feedback period the smiley changed to green when moved in the correct direction, otherwise it became red. The distance of the smiley from the origin was set according to the integrated classification output over the past two seconds (more details can be found in [31]). The classifier output was also mapped to the curvature of the mouth causing the smiley to be happy (corners of the mouth upwards) or sad (corners of the mouth downwards). The subject was instructed to keep the smiley on the correct side for as long as possible and therefore to perform the correct MI as long as possible. A more detailed explanation of the dataset and recording paradigm is available [31]. In addition to the EEG channels, the electrooculogram (EOG) was recorded with three monopolar electrodes and this additional data can be used for EOG artifact removal [32] but was not used in this study.

Dataset 2A - This dataset consists of EEG data from 9 subjects (S10-S18). The cue-based BCI paradigm consisted of four different motor imagery tasks, namely the imagination of movement of the left hand (class 1), right hand (class 2), both feet (class 3), and tongue (class 4) (only left and right hand trials are used in this investigation). Two sessions were recorded on different days for each subject. Each session is comprised of 6 runs separated by short breaks. One run consists of 48 trials (12 for each of the four possible classes), yielding a total of 288 trials per session. The timing scheme of one trial is illustrated in Fig 1(b). The subjects sat in a comfortable armchair in front of a computer screen. No feedback was provided but a cue arrow indicated which motor imagery to perform. The subjects were asked to carry out the motor imagery task according to the cue and timing presented in Fig 1(b). For each subject twenty-two Ag/AgCl electrodes (with inter-electrode distances of 3.5 cm) were used to record the EEG; the montage is shown in Fig 1(c) left. All signals were recorded monopolarly with the left mastoid serving as reference and the right mastoid as ground. The signals were sampled with 250 Hz (downsampled to 125Hz in this work) and bandpass filtered between 0.5 Hz and 100 Hz. EOG channels were also recorded for the subsequent application of artifact processing although this data was not used in this work. A visual inspection of all data sets was carried out by an expert and trials containing artifacts were marked.

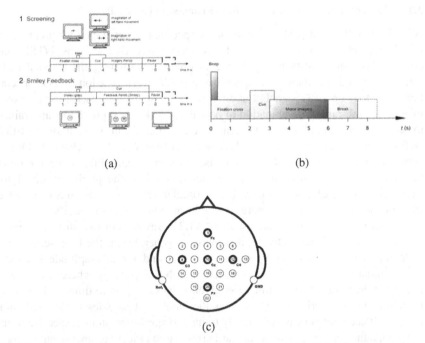

(a)   (b)

(c)

Fig. 1 (a) Timing scheme of the paradigm for recording dataset 2B; 1) the first two sessions provided training data without feedback, and 2) the last three sessions with smiley feedback. (b) Timing scheme of recording for dataset 2A; (c) electrode montage for recording dataset 2A; For dataset 2B electrodes positions were fine-tuned around positions c3, cz and c4 electrodes used to derive bipolar channels for each subject [31].

To summarize, in this work only electrodes positioned anteriorly and posteriorly to positions C3, Cz and C4 are used to derive 3 bipoloar channels. These channels are located over left, right hemisphere and central sensorimotor areas – areas which are predominantly the most active during motor imagery. In dataset 2A only 2 of the available 4 classes are used (left and right hemisphere). As outlined all data was downsampled to 125 Hz in this work. The data splits (training and testing) were the same as those used for the BCI Competition IV [30]. For dataset 2A, one session (2 classes consisting of 72 trials each) are used for training and the remaining session is used for final testing. For dataset 2B the first two sessions are not used, session 3, the first feedback session, is used for training (160 trials) and feedback sessions 4E and 5E are used for final testing. All parameter selection is conducted on the training data using cross validation as described in section 3.3 and the system setup is tested on the final testing sessions.

3 Methods

3.1 BCI Description

3.1.1 Neural-Time-Series-Prediction-Processing (NTSPP)

NTSPP, introduced in [21], is a framework specifically developed for preprocessing EEG signals associated with motor imagery based BCI systems. NTSPP increases class separability by predictive mapping and filtering the original EEG signals to a higher dimensional space using predictive/regression models specialized (trained) on EEG signals for different brain states i.e., each type of motor imagery. A mixture or combination of neural network-based predictors are trained to predict future samples of EEG signals i.e., predict ahead the state of the EEG. Networks are specialized on each class of signal from each EEG channel. Due to network specialization, prediction for one class of signals differ from the other therefore introducing discriminable characteristics into the predicted signal for each class of signal associated with a particular brain state. Features extracted from the predicted signals are more separable and thus easier to classify.

Consider two EEG times-series, x_i, $i \in \{1,2\}$ drawn from two different signal classes c_i, $i \in \{1,2\}$, respectively, assuming, in general, that the time series have different dynamics in terms of spectral content and signal amplitude but have some similarities. Consider also two prediction NNs, f_1 and f_2, where f_1 is trained to predict the values of x_1 at time $t+\pi$ given values of x_1 up to time t (likewise, f_2 is trained on time series x_2), where π is the number of samples in the prediction horizon. If each network is sufficiently trained to specialize on its respective training data, either x_1 or x_2, using a standard error-based objective function and a standard training algorithm, then each network could be considered an ideal predictor for the data type on which it was trained[1] i.e., specialized on a particular data type. If each prediction NN is an ideal predictor then each should predict the time-series on which it was trained perfectly, leaving only error residual equiva lent to white or Gaussian noise with zero mean.

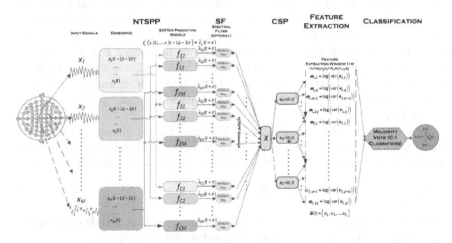

Fig. 2 An illustration of a generic multiclass or multichannel neural-time-series-prediction-preprocessing (NTSPP) framework with spectral filtering, CSP, feature extraction and classification

In such cases the expected value of the mean error residual given predictor f_1 for signal x_1 is $E[x_1-f_1(x_1)]=0$ and the expected power of the error residual, $E[x_1-f_1(x)]^2$, would be low (i.e., in relative terms) whereas, if x_2 is predicted by f_1 then $E[(x_2-f_1(x_2)] \neq 0$ and $E[(x_2-f_1(x_2)]^2$ would be high (i.e., again in relative terms). The opposite would be observed when x_i, $i \in \{1,2\}$, data are predicted by predictor f_2. Based on the above assumptions, a simple set of rules could be used to determine which signal class an unknown signal type, u, belongs too. To classify u one or both of the following rules could be used:-

1. *If* $E[u-f_1(u)] = 0$ *&* $E[u-f_2(u)] \neq 0$ *then* $u \in C_1$, *otherwise* $u \in C_2$.

2. *If* $E[u-f_1(u)]^2 < E[u-f_2(u)]^2$ *then* $u \in C_1$, *otherwise* $u \in C_2$.

These are simple rules and may only work successfully in cases where the predictors are ideal and specialized sufficiently. Due to the complexity of EEG data and its non-stationary characteristics, and the necessity to specify a NN architecture which approximates universally, predictors trained on EEG data will not consistently be ideal however; when trained on EEG with different dynamics e.g., left

---

[1] Multilayered feedforward NNs and adaptive-neuro-fuzzy-inference-systems (ANFIS) are considered universal approximators due to having the capacity to approximate any function to any desired degree of accuracy with as few as one hidden layer that has sufficient neurons [33][34].

and right MI, predictor NNs can introduce desirable characteristics in the predicted outputs which render the predicted signals more separable than the original signals and thus aid in determining which brain state produced the unknown signal. This predictive filtering modulates levels of variance in the predicted signals for data types and most importantly manipulates the variances differently for different classes of data. Instead of using only one signal channel, the hypothesis underlying the NTSPP framework is that, if two or more channels are used for each signal class and advanced feature extraction techniques and classifiers are used instead of the simple rules outlined above, additional advantageous information relevant to the differences introduced by the predictors for each class of signal can be extracted to improve overall feature separability thereby improving BCI performance.

In general, the number of time-series available and the number of classes governs the number of specialized predictor networks employed and the resultant number of predicted time-series from which to extract features, such that

$$P = M \times C \tag{1}$$

where P is the number of networks (=no. of predicted time-series), M is the number of EEG channels and C is the number of classes. For prediction,

$$\hat{x}_{ci}(t + \pi) = f_{ci}\langle x_i(t), ..., x_i(t - (\Delta - 1)\tau)\rangle \tag{2}$$

where t is the current time instant, Δ is the embedding dimension and τ is the time delay, π is the prediction horizon, f_{ci} is the prediction network trained on the ith EEG channel, x_i, $i=1,..,M$, for class c, $c=1,..C$, where C is the number of classes and $\hat{x}_{ci}$ is the predicted time series produced for channel i by the predictor for class c and channel i. An illustration of the NTSPP framework is presented in Fig. 2.

Many different predictive approaches can be used for prediction in the NTSPP framework [21][22][24]. In this work the self-organizing fuzzy neural network (SOFNN) is employed [23][36][37]. The SOFNN is a powerful prediction algorithm capable of self-organizing its architecture, adding and pruning neurons as required. New neurons are added to cluster new data that the existing neurons are unable to cluster while old, redundant neurons are pruned ensuring optimal network size, accuracy and training speed (cf. [23] for details of the SOFNN and recent improvements to the SOFNN learning algorithm and its autonomous hyper-parameter-free application in BCIs).

Earlier work [21] has shown $\Delta=6$ and $\tau=1$ provide good performance in a two class MI-BCI however this chapter shows how NTSPP can be enhanced by selecting channel- and class-specific embedding parameters using partial mutual information selection as described in section 3.2. Firstly, the other signal processing components of the BCI are described.

3.1.2 Common Spatial Patterns (CSP)

CSP maximizes the ratio of class-conditional variances of EEG sources [38][39]. To utilise CSP, Σ_1 and Σ_2 are the pooled estimates of the covariance matrices for two classes, as follows:

$$\Sigma_c = \tfrac{1}{I_c}\sum\nolimits_{i=1}^{I_c} X_i X_i^t \quad (c \in \{1,2\}) \tag{3}$$

where I_c is the number of trials for class c and X_i is the $M{\times}N$ matrices containing the ith windowed segment of trial i; N is the window length and M is the number of EEG channels – when CSP is used in conjunction with NTSPP, $M{=}P$ according to (1). The two covariance matrices, Σ_1 and Σ_2, are simultaneously diagonalized such that the Eigenvalues sum to 1. This is achieved by calculating the generalised eigenvectors W:

$$\Sigma_1 W = (\Sigma_1 + \Sigma_2) WD \tag{4}$$

where the diagonal matrix D contains the Eigenvalues of Σ_1 and the column vectors of W are the filters for the CSP projections. With this projection matrix the decomposition mapping of the windowed trials X is given as

$$E = WX. \tag{5}$$

To generalize CSP to 3 or more classes (multiclass paradigm), spatial filters are produced for each class vs. the remaining classes (one vs. rest approach). If q is the number of filters used then there are $q{\times}C$ surrogate channels from which to extract features. To illustrate how CSP enhances separability among 4 classes the hypothetical relative variance level of the data in each of the 4 classes are shown in Fig. 3.

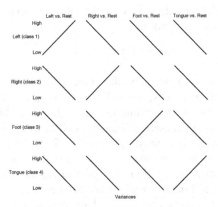

Fig. 3 Hypothetical relative variance level of the CSP transformed surrogate data

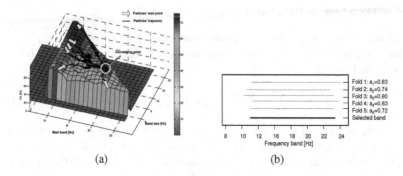

(a)   (b)

Fig. 4 (a) Frequency band selection using PSO. The graphical representation of particles'
motion over progressive generations where classification accuracy(CA) is assessed at teach
generation ; (b) Frequency band selection over 5-folds (where a_i is the probability of correct
classification in fold I and the selected band is the weighted average (weighted by a_i) of the
band selected for each fold).

3.1.3 Spectral Filtering (SF)

Prior to the calculation of the spatial filters, X can be preprocessed with NTSPP
and/or spectrally filtered in specific frequency bands. Optimal frequency bands are
selected autonomously in the offline training stage using particle swarm atomiza-
tion (PSO) [16][40][41] to band pass filter the data before CSP is applied. The
search space is every possible band size in the 8 - 28Hz range as shown in Fig.
4(a). These bands encompass the μ and β bands which are altered during sensorimotor processing [42][43] and can be modulated via motor imagery.

3.2 *Feature Extraction and Classification*

Features are derived from the log-variance of preprocessed/surrogate signals
within a two second sliding window:

$$\bar{\omega} = \log(\mathrm{var}(E)) \tag{6}$$

The dimensionality of $\bar{\omega}$ depends on the number of surrogate signals used from
E. The common practice is to use several (q between 2 and 4) eigenvectors from
both ends of the eigenvector spectrum, i.e., the columns of W. Using NTSPP the
dimensionality of X can increase significantly. CSP, can be used to reduce the dimensionality therefore the benefits of combining NTSPP with CSP are twofold; 1)
increasing separability and 2) maintaining a tractable dimensionality [22].

Linear Discriminant Analysis (LDA) is used to classify the features at the rate
of the sampling interval. Linear classifiers are most commonly used for classifying
motor imagery in BCI applications. Before describing how the parameters associat
ed with these stages of signal processing are optimized in section 3.3, the following section describes the main novelty of this chapter for enhancement of this
framework where the embedding parameters are selected using PMIS.

3.3 Partial Mutual Information

The selection of an optimal embedding dimension and its corresponding time lags is often referred to as the input variable selection (IVS) problem. The IVS problem is defined as the task of appropriately selecting a subset of k variables, from the initial candidate set C which comprise the set of all potential inputs to the model (i.e., candidates) [26]. Mutual information has been found to be a suitable measure of dependence among variables for IVS and quantifies the average amount of common information contained in Δ measurements of a time series. Information quantities can be used for selecting the optimal τ and Δ, to optimize prediction accuracy. Evaluation of mutual information and redundancy-based statistics as functions of τ and Δ can further improve insight into dynamics of a system under study.

In essence, two successive measurements of a random variable have no mutual information (in the case of more than two variables mutual information is commonly replaced by the term redundancy) but data based on an underlying rule may have some association; mutual information is proportional to the strength of that association. Utilising only one or two observations of a time series, x, may not provide enough information about a future value of x to make a reliable prediction. Generally, for periodic, quasi-periodic and even chaotic data redundancy tends to rise as each additional measurement of x (i.e., Δ is increased) is involved in the redundancy calculation, at a fixed lag. Mutual information is an arbitrary measure and makes no assumption of the structures of dependence among variables, be they linear or non-linear. It has also been shown to be robust to noise and data transformations.

Although mutual information is a strong candidate for IVS there are number of issues associated with applying the algorithm such as the ability of the selection algorithm to consider the inter-dependencies among variables (redundancy handling) and the lack of appropriate analytical methods to determine when the optimal set has been selected. One method involves the estimation of marginal redundancy, ς, which quantifies the average amount of information contained in the variables $x_{t+(\Delta-1)\tau}, \ldots x_{t+\tau}$ about the variable x_t and the quantity is the difference between two successive R calculations ($\varsigma = R_{\Delta+1} - R_\Delta$). Depending on the complexity of the data, usually ς increases as Δ is increased. Eventually further increases in Δ provide a lesser increase in ς. Finally, ς becomes approximately constant or begins to decrease. A constant ς indicates that further increases in Δ does not improve the ability of a sequence of measurements to predict the last measurement in the sequence at that value of τ (i.e., there is no advantage in increasing Δ). Another method of estimating the optimum value of Δ can be realised by plotting ς as a function of τ with Δ as a third variable. The relationship between plots of ς versus τ becomes closer as Δ is increased. The optimum Δ is chosen as the smallest Δ for which the plotted relations become relatively close to each other. For more information see [43][44][46][47][48]. In some cases the results from this type of redundancy analysis can be subjective and may not be fully conclusive whilst the calculation can also be time consuming.

Sharma [25] proposed an alternative algorithm and one that overcomes the difficulties in terms of determining the optimal sets of variables with mutual information by using the concept of partial mutual information (PMI). The approach was further assessed and developed by May et al. [26]

3.4 Estimation of Partial Mutual Information

The mutual information calculation stems from Shannon's information theory [49] formulated in (7)

$$I_{Y;X} = \iint p(x, y) \log \frac{p(x, y)}{p(x)p(y)} \tag{7}$$

where x and y are observations of random variables X and Y, respectively i.e., $y \in Y$ and $x \in X$. Considering Y is an output variable for which there is uncertainty around its observation and is dependent upon the random input variable x then the mutual observation of (x,y) reduces this uncertainty, since knowledge of x allows inferences of the values of y and vice versa. Within a practical context the true functional forms of the pdfs in (7) are typically unknown. In such cases the estimates of the densities are used instead. Substitution of the density estimates into a numerical approximation of the integral in (7) gives

$$I_{Y;X} = \sum_{i=1}^{N_s} \sum_{j=1}^{N_s} p(x_i, y_j) \log_2 \frac{p(x_i, y_j)}{p(x_i)p(y_j)} \tag{8}$$

where N_s is the number of bins used for calculating the probability $p(x_i)$ of signal measurement x, occurring in bin x_i and the probability $p(y_j)$ of signal measurement y occurring in bin y_j. $p(x_i, y_j)$ is the joint probability of occurrence of both measurements of the signal. Equation (2) can be generalised to calculate redundancies among variables in a time series, as shown in (3)

$$R(x_t, x_{t+\tau}, ..., x_{t+(\Delta-1)\tau})$$
$$= \sum^{N_r} p(x_t, x_{t+\tau}, ..., x_{t+(\Delta-1)\tau}) \log_2 \frac{p(x_t, x_{t+\tau}, ..., x_{t+(\Delta-1)\tau})}{p(x_t)p(x_{t+\tau})...p(x_{t+(\Delta-1)\tau})} \tag{9}$$

where x_t is the measurement of the signal sampled at time t and N_r is the number of phase space routes (i.e., the number of combinations). Equations (8) and (9) can be derived in the probability form or entropy form (H):-

$$R(.) = H(x_t) + ... + H(x_{t+(\Delta-1)\tau}) - H(x_t, ..., x_{t+(\Delta-1)\tau}) \tag{10}$$

Full derivations can be found in [45][46][47]. Depending on the number of measurements of the signal and the number of bins, the joint probability, $p(x_t, x_{t+\tau}, ..., x_{t+(\Delta-1)\tau})$, can encompass a very large number of sequence probabilities. For example, if $\Delta = 5$ and $N_s = 20$ then the number of sequence probabilities to be estimated is $N_r=N_s^D=3.2 \times 10^6$, increasing exponentially as Δ is increased. Estimating redundancies for $\Delta>5$ can be significantly time consuming.

Mutual information estimation is therefore largely dependent on the technique employed to estimate the marginal and joint pdfs. Non-parametric techniques such as kernel density estimation (KDE) are considered suitably robust and accurate although somewhat computationally intensive compared to alternative approaches such as the histogram approach. Substitution of the density estimates into a numerical approximation of the integral in (7) and (8) gives

$$I_{Y;X} \approx \frac{1}{n}\sum_{i=1}^{n} f(x_i,y_j)\log_2 \frac{f(x_i,y_j)}{f(x_i)f(y_j)} \tag{11}$$

where f denotes the estimated density based on a sample of n observations of (x, y). The Parzen window approach is a simple KDE in which the estimator for f is given by

$$\hat{f}(x) = \frac{1}{n}\sum_{i=1}^{n} K_h(x-x_i) \tag{12}$$

where $\hat{f}(x)$ denotes the estimate of the pdf at x, $x_i\{i = 1,\dots,n\}$ denotes the samples observations of X, and K_h is the kernel function where h denotes the kernel bandwidth (or, smoothing parameter). A common choice for K_h is the Gaussian kernel

$$K_h = \frac{1}{(\sqrt{2\pi}h)^d\sqrt{|\Sigma|}}\exp\left(\frac{-\|x-x_i\|}{2h^2}\right) \tag{13}$$

where d denotes the number of dimensions of X, $\Sigma=\{\sigma_{ij}\}$ is the sample covariance matrix and $\|x-x_i\|$ is the Mahalanobis distance metric given by

$$\|x-x_i\| = (x-x_i)^T\Sigma^{-1}(x-x_i). \tag{14}$$

The kernel expression in (12) is used with (13) and (14) to produce the kernel estimator as defined below

$$\hat{f}(x) = \frac{1}{n(\sqrt{2\pi}h)^d\sqrt{|\Sigma|}}\sum_{i=1}^{n}\exp\left(\frac{-\|x-x_i\|}{2h^2}\right) \tag{15}$$

The performance of the kernel estimator, in terms of accuracy, is dependent more on the choice of bandwidth as opposed to choice of kernel itself [26][50]. The optimal choice of bandwidth depends on the distribution of the data samples. In [25][26][51] the Gaussian reference bandwidth, h_G, for MI estimation is adopted as an efficient choice. The Gaussian reference bandwidth is determined using the following rule proposed by Silvermann [52]

$$h_G = \left(\frac{1}{d+2}\right)^{1/(d+4)} \sigma n^{-1/d+4} \tag{16}$$

where σ is the standard deviation of the data samples. The MI calculation can be easily extended to the multivariate case where the response/output variable Y is dependent on multiple input variables. For example, two input variables X and Z. Given X the uncertainty is reduced by a certain amount and the partial mutual information is defined as the further reduction in the uncertainty surrounding Y that is gained by the additional mutual observation of Z. Partial MI (PMI) is analogous to the partial correlation coefficient, $R`_{ZY.X}$, which quantifies the linear dependence of Y on variable Z that is not accounted for by the input variable X. This is normally calculated by filtering Y and Z via regression on X to obtain some residuals, u and v, respectively [26]. Pearson's correlation can be used to estimate X. PMI can be applied in a similar way to estimate the arbitrary dependence between variables. Using the KDE approach an estimator for the regression of Y on X is written as

$$\hat{m}_Y(x) = E[y|X=x] = \frac{1}{n}\frac{\sum_{i=1}^{n} y_i K_h(x-x_i)}{\sum_{i=1}^{n} K_h(x-x_i)} \tag{17}$$

where $\hat{m}_Y(x)$ is the regression estimator; n is the number of observed values (y_i; x_i); K_h is given as in (12) and $E[y|X=x]$ denotes the conditional expectation of y given an observed x. An estimator $\hat{m}_Z(x)$ can be similarly constructed, and the residuals u and v estimated using the expressions

$$y = Y - \hat{m}_Y(X) \tag{18}$$

and

$$u = Z - \hat{m}_Z(X) \tag{19}$$

Using these residuals the PMI can then be calculated as

$$I'_{ZY.X} = I(v;u) \tag{20}$$

where the subscript notation $I'_{ZY.X}$ or $I(Z;Y|X)$ can be used. PMI allows for the evaluation of variables taking into account any information already provided by a given variable X.

Given a candidate set C, and output variable, Y, the PMI based input variable selection (PMIS) algorithm proceeds at each iteration by finding the candidate, C_s, that maximises the PMI with respect to the output variable, conditional on the inputs that have been previously selected. The statistical significance of the PMI estimated for C_s can be assessed based on the confidence bounds drawn from the distribution generated by a bootstrap loop. If the input is significant, C_{s_s} is added to S and the selection continues; otherwise there are no more significant candidates remaining and the algorithm is terminated [25][26].

The size of the bootstrap, B, is important in the implementation of PMIS since it can influence both the accuracy and overall computational efficiency of the algorithm. May et al [26] discuss the implications for selecting the bootstrap size in terms of the accuracy – computational efficiency trade-off and present a new approach which does not rely on a bootstrap or direct comparison with the critical value of MI (as is necessary with some of the other approaches compared, such as the tabulated critical values approach [26]). In [26] the use of the Hampel test criterion is suggested as a termination criterion.

3.5 Hampel Test Criterion

Outlier detection methods are robust statistical methods for determining whether a given value, x, is significantly different from another within a set of values X. In the case of PMIS, having identified the most relevant candidate, the outlier, it is necessary to determine whether this candidate is statistically significantly greater than the others and to keep this candidate if it is. The Z-test is a commonly adopted approach for outlier detection where the deviation of a single observation is compared with the sample mean of all observations. Based on the 3σ rule for Gaussian distributions, outliers lie greater than three standard deviations from the population mean and therefore an observed value with a Z-score greater than 3 is generally considered to be an outlier. The Z-test can be particularly sensitive when a population contains multiple outliers. One very distant outlier could disrupt the distribution of the population (mean and variance) resulting in other outliers not being identified i.e., hiding and masking outliers. The sensitivity of outlier detection methods to masking is determined based on the proportion of outliers that must be present to significantly alter the data distribution, referred to as the breakdown point, which is $1/n$ for the Z-test since only one sufficiently large outlier will cause the test to breakdown [26].

Since the candidate set of variables in the PMIS method is likely to contain more than one relevant variable (analogous to outliers in the aforementioned outlier test) a modified Z-score is necessary to improve the robustness of the test. The Hampel distance test proposed in [53] and compared in [26] is based on the population median. Because the Hampel distance test breakdown point is 2/n it is considered to be one of the most robust outlier tests when the data contains multiple outliers. To calculate the Hampel distance, the absolute deviation from the median for all candidates is calculated as follows

$$d_j = \left| I_{C_j Y \cdot S} - I_{C_j Y \cdot S}^{(50)} \right| \tag{21}$$

where d_j denotes the absolute deviation and $I_{C_j Y \cdot S}^{(50)}$ is the medium PMI for candidate set C. Taking $d_j^{(50)}$ as the median absolute deviation (MAD), the Hampel distance (modified Z-score) for candidate C_j is

$$Z_j = \frac{d_j}{1.4826 d_j^{(50)}}. \tag{22}$$

The factor of 1.4826 scales the distance such that the rule Z>3 can be applied, as in the case of the conventional Z-test [26]. The value Z_s is determined for candidate C_s and if $Z_s > 3$, the candidate is selected and added to S; otherwise the forward selection algorithm is terminated as described in the following subsection.

PMIS algorithm using Hampel distance criterion

1: Let $S \rightarrow \phi$ (Initialisation)
2:   While $S \neq \phi$ (Forward Selection)
3:   Construct kernel regression estimator $\hat{m}_Y(S)$
4:   Calculate residual output $u = Y - \hat{m}_Y(S)$
5:   For each $C_j \in C$
6:       Construct kernel regression estimator $\hat{m}_{C_j}(S)$
7:       Calculate residual output $u = C_j - \hat{m}_{C_j}(S)$
8:       Estimate $I(v;u)$
9:   Find candidate C_s (and v_s) that maximises $I(v;u)$
10: Estimate Z_s for C_s
11: If $Z_s > 3$ (Selection/Termination)
12:      Move C_s to S
13: Else
14:      Break
15: Return

Using PMIS the optimal selection of time delayed EEG signal samples which minimise the uncertainty about a future output (prediction) can be estimated.

3.6 Using PMIS to Optimize the NTSPP Framework

For every channel, in the case of the BCI presented in sections 2 and 3.1, there is assumed to be an optimal selection of input variables (EEG time series embedding), that will enable accurate prediction for the channel and accurate specialisation of a neural network for that channel. The optimal selection however is likely to differ depending on the class of data (motor imagery) being assessed. So, for a 3 channel system with 2 classes of data there is assumed to be at least 6 optimal embedding configurations, one for each channel per class. When applying a BCI that involves time embedding the EEG for prediction, as is the case for the NTSPP framework, the optimal embedding for both classes cannot be applied simultaneously in the online BCI as the class is unknown a priori and a decision has to be made given data from 3 channels. It is therefore necessary to decide which embedding should be applied, not necessarily to maximise the prediction accuracy for both classes, but to maximise the specialisation for the networks in such a way that the difference between signals predicted for both classes is maximal i.e., separability is maximised. In the case where 3 channels and 2 classes are available,

Table 2 Different combinations of embedding dimension for three channels. For any configuration each can use the embedding parameters that are optimal for either class '1' or '2' but not both.

Configuration	C3	C4	C_z
A	1	1	1
B	1	1	2
C	1	2	1
D	1	2	2
E	2	1	1
F	2	1	2
G	2	2	1
H	2	2	2

$C \in \{1,2\}$, there are 2^3 possible configurations for deciding which of the channel's assumed optimal embedding configuration should be used as shown in Table 2.

From Table 2, if configuration A is selected then the embedding values selected for class 1 on all channels would be used whereas if configuration D is selected the embedding parameters chosen for class 1 would be used for channel C3 and those chosen for class 2 would be used for channels C4 and Cz. As this chapter presents the first assessment of this approach, a heuristic based approach was adopted to determine the best configuration. The following section describes the complete BCI setup and parameters optimization procedure.

3.7 Parameters Optimization and BCI Setup

In motor imagery BCIs, the parameter search space and the available data can be extensive, particularly when there are multiple stages of signal processing, therefore a phased approach to parameter selection is conducted. In the proposed BCI setup it is necessary to find of the optimal combination of lagged input variables (embedding parameters) to train the predictor networks. An inner-outer cross-validation (CV) is performed, where all other BCI parameters are optimized for each of the embedding configurations including the optimal subject-specific frequency bands (shown in Fig. 4). In the outer fold, NTSPP is trained on up to 10 trials randomly selected from each class (2 seconds of event related data from each trial resulting in 2500 samples for each channel/class) using standard time series embedding parameters: embedding ($\Delta=6$) and time lag ($\tau=1$). The trained networks then predict all the data from the training folds to produce a surrogate set of trials containing only EEG predictions. No parameter tuning is necessary at this stage as the SOFNN adapts autonomously to the signals [23]. The 4 training folds from the outer splits are then split into 5 folds on which an inner 5-fold cross validation is performed. Firstly, the time point of maximum separability is found for the inner data and, (if necessary, channel selection can be performed), both using the R^2 correlation analysis with a standard 8-26 Hz band [40]. Using the information regarding the optimal time point, a 2 second window of data around the time point of maximum separability is taken from 10 randomly chosen trials from each class and PMIS is applied to find the optimal embedding for each channel and each class. Using one of the combinations shown in table 1 the NTSPP framework is retrained with these embedding parameters, again using 10 randomly chosen trials from the outer training folds but in this case using the 2 second window of data around the time point of maximum separability (using the most separable data segments increases network specialization). A new surrogate dataset for the outer training fold data is obtained and the 4 training folds from the outer splits are then split into 5 folds on which another inner 5-fold cross validation is performed. Firstly, the time point of maximum separability is found for the inner data using the R^2 correlation analysis with a standard 8-26 Hz band [40]. Using the best time point and best channels from the correlation analysis, a PSO based search is conducted to identify the optimal frequency bands where CSP, feature extraction and classification is performed to determine classification accuracy levels on each of the folds for each of the bands selected by PSO and tested using 4 CSP surrogate channels [40]. After each frequency band is tested on the test fold, PSO swarm particles communicate the accuracy levels to one another and the algorithm converges, identifying the optimal band for that test fold much quicker than searching the complete space of all the possible bands (cf. Fig. 4(a) for a graphical representation of a PSO search). After optimal bands for each of the inner folds have been identified the finally selected band is the average classification accuracy (CA) of the 5 bands weighted by the CA of the test fold as illustrated in Fig. 4(b). NTSPP-SF-CSP is then applied on the outer fold training set, where a feature set is extracted and an LDA classifier is trained at every time point across the trials and tested for that point on the outer test folds. The average across the five-folds is used to identify the optimal number of CSPs (between 1-3 from each side of W) and the final time point of maximum separation for the corresponding combination of PMIS selected lagged input variables.

There are eight combinations of selected lag variable combinations as shown in Table 2 therefore the above process is conducted for each of the combinations. At the end of this process the embedding configuration which provides the best mean accuracy is known however it is then necessary to select which channel-specific embedding works best i.e., from the outer cross validation PMIS is applied for each of the 5 training folds and each time the exact embedding parameters may differ for each channel. To obtain the best setup for cross session tests (to ensure generalization to the unseen testing data) the complete system is retrained a further 5 times on all the training data using the chosen parameters for each fold and tested on the training data 5 times. In this case the lag combination for each of the 5 folds is tested with other BCI parameters selected in the cross-validation. The embedding parameters configuration and parameter combination which provides the highest mean accuracy across the 5 tests on the complete training data set is used for cross session tests. In the case where two tests produce the same results on one of the training data tests the parameter setup that achieves the average best accuracy across 8 lag configurations for a particular training test, corresponding to the best average lag configuration across the 5 training tests, is used to determine the best setup. The system is finally tested on the unseen test/evaluation data (as given for the BCI competition outlined in section 2). All parameter optimization is expedited using the Matlab® Parallel Processing Toolbox and a high performance computing (HPC) cluster with 384 cores. Subject analysis and 5-fold cross validations were run in parallel along with parallelization of parts of the kernel regression estimation and PMI calculation for selecting input variables and multiple cross validation tests.

4 Results

The objective of this research was to improve the NTSPP framework, which is a predictive framework involving training a mixture of experts on EEG data produced for two classes of motor imagery recorded from three channels. The hypothesis is that specializing the networks for a particular motor imagery (class) leads to improvement in the separability of the predicted output i.e., when the mixture of networks produce predictions for an unknown class of motor imagery, networks trained on that particular class of motor imagery should predict the data sufficiently accurately and differently compared to the other networks which are trained on the other class of motor imagery. Maximizing the difference in the prediction for each class of motor imagery ultimately should lead to better classification accuracy (BCI performance) when features are extracted from the predicted signals and classified i.e., when all other components are merged with the predictive framework. The hypothesis is based on the observation that dynamics of the EEG differs across channels and between motor imageries. The 2nd hypothesis is thus that each channel will have different and optimal embedding parameters which will optimize prediction performance and enable network specialization. Therefore, selecting these optimal embedding parameters will result in improved NTSPP performance and thus improved BCI performance. For example, if PMIS selected $x(t-1)$, $x(t-3)$, and $x(t-5)$ as the best predictors for channel C3 for class 1 and $x(t-1)$, $x(t-2)$, $x(t-3)$, and $x(t-10)$ for the same channel but for class 2, training the networks for class 1 and class 2 on only one of these embedding combinations for this channel, for example, the embedding parameters for class 1, then the class 2 network would not be able to specialize/train on the same channel using class 2 data as accurately, as the optimal embedding parameters for that channel are not been utilised i.e., for each channel only of the two networks trained on the channel is specialized to predict the channel whilst the other is not.

To test both hypotheses the overall BCI performance and the selected embedding parameters for each channel and class that produce the optimal BCI performance are assessed. BCI performance with NTSPP and PMIS is compared with the case were no NTSPP is performed (only CSP and SF) and where NTSPP is performed with a standard embedding/time lag setup (NTSPP6).

Table 3 The optimal time series embedding parameters for each channel and for each class for all subjects. The optimal configuration for NTSPP is shown in column 2 (corresponding to table 2) and in bold in wide columns 3 and 4.

S	NTSPP Conf.	Class 1 (Left Hand Motor Imagery)			Class 2 (Right Hand Motor Imagery)		
		C3	C4	Cz	C3	C4	Cz
1	C 1,2,1	**x(t-1),x(t-10)**	x(t-1), x(t-9)	**x(t-1), x(t-3)**	x(t-1), x(t-5)	**x(t-1)**	x(t-1),x(t-10)
2	D 1,2,2	**x(t-1), x(t-2)**	x(t-1),x(t-2)	x(t-1),x(t-2) x(t-5),x(t-6)	x(t-1)	**x(t-1),x(t-2)**	**x(t-1)**
3	A 1,1,1	**x(t-1),x(t-2), x(t-4),x(t-5), x(t-6),x(t-7), x(t-10)**	**x(t-1),x(t-2), x(t-3),x(t-4), x(t-5),x(t-6)**	**x(t-1),x(t-2), x(t-3),x(t-5)**	x(t-1),x(t-2), x(t-3),x(t-4), x(t-5),x(t-6)	x(t-1),x(t-2), x(t-3),x(t-4), x(t-5),x(t-6)	x(t-1),x(t-2), x(t-3),x(t-10)
4	B 1,1,2	**x(t-1),x(t-10)**	x(t-1)	x(t-1),x(t-10)	x(t-1), x(t-7), x(t-8)	**x(t-1)**	**x(t-1),x(t-10)**
5	B 1,1,2	**x(t-1)**	x(t-1),x(t-2), x(t-3),x(t-4)	x(t-1), x(t-2)	x(t-1),x(t-2), x(t-3),x(t-4), x(t-5),x(t-7)	x(t-1), x(t-2)	**x(t-1),x(t-2), x(t-7),x(t-8)**
6	E 2,1,1	x(t-1),x(t-2), x(t-6),x(t-8), x(t-9)	**x(t-1), x(t-2), x(t-3),x(t-4), x(t-5)**	**x(t-1),x(t-2), x(t-3)**	**x(t-1), x(t-2), x(t-3), x(t-5)**	x(t-1), x(t-2), x(t-3), x(t-4), x(t-5),x(t-10)	x(t-1),x(t-2), x(t-3),x(t-7), x(t-8)
7	A 1,1,1	**x(t-1)**	**x(t-1)**	**x(t-5),x(t-6)**	x(t-1)	x(t-1)	x(t-6)
8	B 1,1,2	**x(t-1)**	**x(t-1), x(t-8)**	x(t-1)	x(t-1)	x(t-1),x(t-10)	**x(t-10)**
9	A 1,1,1	**x(t-1)**	**x(t-1)**	**x(t-1)**	x(t-1)	x(t-1)	x(t-2), x(t-3)
10	C 1,2,1	**x(t-1)**	x(t-1), x(t-2)	**x(t-1), x(t-2), x(t-3)**	x(t-1)	**x(t-1)**	x(t-1),x(t-2), x(t-3)
11	D 1,2,2	**x(t-1),x(t-2), x(t-3),x(t-4), x(t-5),x(t-6), x(t-8),x(t-9)**	x(t-1)	x(t-1),x(t-2), x(t-3),x(t-4), x(t-5),x(t-6)	x(t-1),x(t-2)	**x(t-1),x(t-2), x(t-5)**	**x(t-1),x(t-2), x(t-7),x(t-10)**
12	C 1,2,1	**x(t-1)**	x(t-1)	**x(t-1)**	x(t-1),x(t-10)	**x(t-1),x(t-10)**	x(t-1),x(t-8), x(t-9),x(t-10)
13	H 2,2,2	x(t-1)	x(t-1),x(t-8)	x(t-1),x(t-2), x(t-8)	**x(t-1),x(t-7), x(t-8)**	**x(t-1)**	**x(t-1),x(t-7), x(t-8)**
14*	G 2,2,1	x(t-1),x(t-2), x(t-9)	x(t-1),x(t-2)	**x(t-1),x(t-7), x(t-8),x(t-10)**	**x(t-1),x(t-2), x(t-3),x(t-4), x(t-5)**	**x(t-1),x(t-2), x(t-3),x(t-4), x(t-5)**	x(t-1),x(t-4)
15	A 1,1,1	**x(t-1),x(t-2)**	**x(t-1),x(t-2)**	**x(t-1),x(t-2)**	x(t-1)	x(t-1)	x(t-1),x(t-2),
16	H 2,2,2	x(t-1)	x(t-1), x(t-2), x(t-4),x(t-10)	x(t-1), x(t-2), x(t-3),x(t-10)	**x(t-1)**	**x(t-1), x(t-2), x(t-10)**	**x(t-1)**
17	H 2,2,2	x(t-1)	x(t-1)	x(t-1)	**x(t-1)**	**x(t-1)**	x(t-1),x(t-10)
18	G 2,2,1	x(t-9)	x(t-1)	**x(t-1)**	**x(t-1), x(t-9)**	**x(t-1)**	x(t-1)

4.1 PMIS Selected Embedding Parameters

Table 3 shows the PMIS selected embedding/lag parameters for each of the 3 channels for each class. The best configuration in terms of which embedding/lag parameters were used for each channel are shown in bold (as outlined only one embedding/lag setup can be used for each channel – the best training accuracy during system optimization as outlined above is used to determine which setup is used, either class 1 or class 2). As can be seen, across all subjects there are different embedding/lag parameters chosen using PMIS. Within subjects there are different embedding/lag parameters for each channel and for each class. Using the selection method outlined above, the best configuration of embedding/lag parameters differ significantly across subjects. Colum 2 shows the lag configurations corresponding to Table 2. For subjects 3, 7, 15, 16 and 17 the best configuration is based on the embedding/lag combinations for one class across all signals. The optimal setup could be derived based on one class only for a number of other subjects, as the embedding/lag combination is identical for some channels regardless of class e.g., for subject channel Cz parameters could have been used for both classes and likewise for subject 18 (subject 4 could have used all class 1 parameters and subject 18 could have used all class 2 parameters). This selection process could have chosen either in these cases. For the remaining subjects however a mixture of parameters have been selected between the two classes, emphasizing the need to assess all embedding parameters combination on a channel-, class- and subject-specific basis. The following subsection outlines how these subjects performed in terms of overall classification accuracy using these embedding configurations in the NTSPP framework.

4.2 BCI Performance

The cross validation (CV) performances for all subjects are presented in Figure 6(a) whilst the cross session (x-Session) single-trial performances are presented in Figure 6(b). The average performances across subjects are presented in Figure 6(c). In the majority of cases NTSPP-PMIS provides the best cross-validation performance (within session) for all subjects (Fig. 6(a)). In some cases NTSPP6 outperforms NTSPP-PMIS. Subject 10 is the only case where there is a significant drop in CV performance given by NTSPP-PMIS compared to No-NTSPP. There is a slight drop in CV performance for subjects 17 and 18 but overall the CV results indicate a slight improvement in the average within-session performance for NTSPP6 and a greater improvement for NTSPP-PMIS across all subjects compared to No-NTSPP. The average across subjects is shown in Fig. 6(c). The averages are compared across all subjects as well as across the two competition groupings as both datasets have different attributes which influence performance[2]. In terms of the CV average there are slight improvements given by NTSPP-PMIS which are statistically significant ($p<0.05$) as shown in Table 4 where the results of two statistical tests are presented. The parametric statistical test repeated measures ANOVA (which is akin to a t-test (related) for two groups) [54] and the Wilcoxon signed rank test [55], a non-parametric statistical test, are used for clarity (the results indicate from both tests are similar and correlated). The improvement given by NTSPP6 over No-NTSPP is not significant in the CV test but is more significant (not statistically) in the cross-session tests on all subjects. The cross session performance difference between No-NTSPP and NTSPP-PMIS is statistically significant however NTSPP–PMIS is not shown to be statistically better than NTSPP6 in the cross session tests. The performance of the BEST NTSPP-PMIS cross session results are presented here to show what is theoretically possible with

---

[2] For dataset 2B (subjects 1-9), the data used for training are from a 3rd feedback session after 2 sessions without feedback and the testing data is from two further feedback sessions whereas the dataset for 2A (subjects 10-18) are trained and tested on 2 sessions with no feedback and within these sessions subjects performed another 2 motor imageries (4 class data acquisition; there is also a significant difference in the number or trials performed for both groups (cf. section 2 for further details).

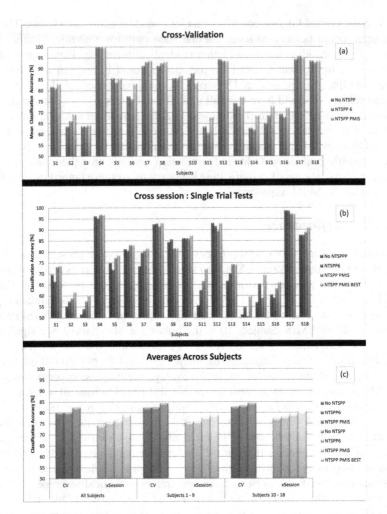

Fig. 5 BCI performance results for all subjects and approaches: (a) average cross-validation classification accuracy; (b) cross-session (x-Session) classification accuracy (c) mean accuracies across all subjects and two groups of subjects (groups based on datasets 2A and 2B). Results are presented for 3 methods: No NTSPP, NTSPP6 (standard embedding dimension 6 and time lag 1) and NTSPP with embedding selected using PMIS and configured according to Table 3. In figures (a) and (b) the absolute best performing NTSPP PMIS embedding setup is shown for information only (this best setup is the accuracy that could have been obtained if the absolute best embedding setup was determined from the training data i.e., in some cases the setup chosen did not provide the best generalization performance on the test data.

the proposed NTSPP-PMIS approach if the parameters can be selected appropriately from the training data (i.e., these results were generated by applying all possible NTSPP-PMIS configurations across the sessions and viewing the best results). The results show that the possible best performances are statistically better than those produced using the proposed heuristic configuration optimization method for NTSPP-PMIS. The significance of the results is discussed in the following section.

Table 4 Results of statistical tests comparing the average performance across subjects (All, S1-S9 and S10-S19) for each of the methods/approaches (p-values), where the average performances are shown in Figure 6(c). $p<0.05$ indicates a statistical difference in the performance produced by the methods compared. Results of two statistical tests (parametric and non-parametric) are shown for comparison and verification.

Methods	CV	Cross Session		
Repeated Measures ANOVA	**All**	**All**	**S1-9**	**S10-18**
No NTSPP vs NTSPP6	0.8714	0.1033	0.6149	0.1094
No NTSPP vs NTSPP-PMIS	**0.0046**	**0.0264**	0.0566	0.1953
NNTSPP6 vs NTSPP-PMIS	**0.0093**	0.2823	0.1271	1
NTSPP PMIS vs BEST	-	**0.0052**	**0.0177**	**0.0151**
Wilcoxon Signed Rank Test				
No NTSPP vs NTSPP6	0.7925	0.1640	0.7969	0.1250
No NTSPP vs NTSPP-PMIS	**0.0019**	**0.0331**	0.0742	0.2656
NTSPP6 vs NTSPP-PMIS	**0.0083**	0.2366	0.1289	0.9688
NTSPP PMIS vs BEST	-	**0.0002**	**0.0313**	**0.0156**

5 Discussion

This study presents for the first time the use of partial mutual information input variable selection (PMIS) for selecting channel-, class- and subject specific embedding parameters from EEG time-series. The results presented in section 4.1 show that, depending on the particular brain state (class), the channel-specific embedding varies and is subject-specific. Past studies have investigated the optimal subject-specific embedding parameters for BCI [15][21] but focused on using one set for all channels with the same embedding parameters being selected for both classes i.e., the embedding was optimized based on overall classification performance without first selecting embedding parameters for particular channels. The results presented here show the variability in the brain and intra- and inter-subject differences in EEG dynamics. It is therefore recommended to optimize the channel specific embedding parameters when attempting to make predictions about a future brain state, be it one step or multiple steps ahead.

Although, in this study, the aim is not to exploit the use of advanced prediction of future brain states to reduce system latency (cf. next section for a discussion on how this may be potentially beneficial in BCI), the NTSPP approach is based on EEG time-series prediction and the results clearly demonstrate that there are improvements given by the NTSPP framework when channel- and class-specific embedding configurations are deployed for each subject. The results show that within session cross-validation differences between NTSPP-PMIS and NTSPP6 or No NTSPP are statistically significant and that the cross session performance difference between NTSPP-PMIS and No-NTSPP is statistically significant. In previous work it was shown that in some cases, depending on the number of channels and classes being investigated, NTSPP6 provides significant improvement over No NTSPP whereas in this work the cross session differences are not shown to be statistically different, although an improvement is observable. The approach presented in [22] involved only testing four frequency bands between 8-24Hz for the EEG spectral filter whereas in this work the frequency bands are subject-specifically tuned in the setup with a fine resolution using PSO and then applied along with NTSPP6[3] i.e., the results presented here suggest that NTSPP6 is less effective when the subject-specific bands are tuned. However, as shown here, when we deploy PMIS embedding selection there are improvements even with optimized frequency bands in the spectral filter. With the proposed parameter configuration using PMIS and a heuristic as well as computationally intelligent search methods (PSO and SOFNN) for other parameter combinations/settings, NTSPP-PMIS can generalize reasonably well across sessions. This therefore is a positive indication that the use of NTSPP can indeed improve performance of the BCI. The BEST results (as outlined these are identified after viewing the testing performance across all embedding configurations) show there can be even greater gains provided by the NTSPP-PMIS framework if the parameter optimization approach is further improved to ensure better generalization. The following subsection outlines why the approach used is suboptimal and other limitations of this study.

5.1 Limitations

In terms of PMIS and NTSPP, only 10 trials randomly selected from the available trials from each class are used to, firstly, identify the optimal embedding for each channel and then to train the SOFNNs in the NTSPP framework. Using more trials in the PMIS setup, only using trials which are highly separable i.e., omitting trials which are less separable, may improve the specificity and accuracy of the PMIS algorithm. Likewise, for NTSPP and the SOFNN training, using more trials and only those that are most separable, along with PMIS selected embedding as described may enhance the specialization of the networks leading to increased difference in the prediction for both classes and enhanced separability. The SOFNN is deployed in this framework using standard hyper parameters, identified based on a study of a small number of subjects [37]. It is highly probable that fine tuning the SOFNN parameters to suit the channel-specific embedding will also lead to greater specialization. In addition, the data segments within the event-related portion of trials on which PMIS and the SOFNN are deployed could be fine-tuned and assessed more closely using smaller or larger segments around the time point of maximum separation in trials (in this work a 2s window around the max separation point was used). Using more trials from which to select data may also improve specialization. In this study only 2 seconds of the data was used from 10 randomly selected trials resulting in 2500 samples for PMI selection and SOFNN

---

[3] Even though the same subjects are analyzed the data splits in [22] are also different (based on the feedback and non-feedback sessions) and therefore results presented here are not directly comparable to those presented in [22].

training. This is a low number of trials relative to the amount of data available however results in a significant number of samples on which to train multiple networks, multiple times in a cross validation. Simply selecting data more instinctively could have a major impact on PMIS and the SOFNNs in the NTSPP framework without any additional parameter optimization.

Improvements to other elements of the BCI are possible and ongoing however it is desirable to keep parameter tuning in the BCI setup to a minimum or indeed in any application therefore there is always the aim to ensure the system can be setup quickly using an auto-calibrating approach, hence the reason for using computational intelligence based approaches such as the SOFNN which can adapt and tune its weights and structure automatically during the learning process, and the use of PSO to select optimal frequency bands quickly and efficiently. CSP is not only used here to improve separability but to help identify redundancy in the signals. The use of linear classifiers for easy training and adaptation is necessary but classifier performance can be improved to account for inter-session variability and sensorimotor learning as the subject endeavors to improve BCI performance (research is ongoing in this area [56][57]). Future work will involve investigating the parameters that are providing the best cross-session performances and developing an optimized and efficient framework where optimal performance and cross session generalization is guaranteed. For example, subject 14 in this work was poorly performing regardless of the method deployed but NTSPP-PMIS failed completely (~50% classification accuracy) given the parameters selected on the training data whereas the BEST performance shows that NTSPP-PMIS parameters could have been much better for this subject had the training data being more carefully used to setup the system. These issues are currently being investigated along with other potential benefits of the NTSPP framework as outlined in the following subsection.

6 Conclusions and Future Work

This chapter has shown for the first time that partial mutual information input variable selection (PMIS) can be used to select embedding parameters for EEG time series prediction and by selecting channel-, class- and subject-specific embedding parameters predictive performance and over all classification of EEG data can be improved for a two class EEG-based BCI using the NTSPP framework. The PMIS approach can be improved by using more data and further assessment of the criteria for considering whether a particular embedded sample of the time series provides information about the predicted input. This may be improved using the bootstrapping or Akaike Information Criterion as compared by May et al [26] however the approach used here, involving the Hampel distance criterion, is efficient. By exploring better parameters for the PMIS approach, the NTSPP setup and the complete BCI it is expected that the BCI presented here can be improved significantly. This work provides evidence of this potential. The PMIS approach will also aid in the investigation of other BCI configurations involving the NTSPP framework, for example, multiclass systems and multiple channel EEG montages. Previous work has already shown that the performance gains provided by the

NTSPP framework are greater when multiple channels and multiple classes are used [22][55]. Channel- and class-specific embedding is likely to further increase that improvement. NTSPP has also been shown to have the capacity to reduce the latency involved in motor imagery BCIs involving continuous classification; producing higher signal separability faster (i.e., earlier in the trial) by predicting the EEG times series multiple steps ahead [59]. This has the potential to reduce the time required (latency) for a subject to exceed a threshold with the continuous classifier output, as the NTSPP predicts multiple steps ahead in time characteristics in the data which are more separable. Features can then be extracted from the predicted separable segments of the data before that separability actually is produced by the sensorimotor activity. A preliminary study of this is presented in [59]. Again, that preliminary study used standard embedding parameters. For multiple-step-ahead prediction the prediction error increases as the prediction horizon increases and therefore PMIS embedding parameter selection will be even more pertinent and can be exploited in such a multi-step-ahead NTSPP framework. Further work will be carried out to verify if combining CSP and SF with the multiple-step-ahead prediction NTSPP framework and PMIS has potential for improved accuracy and information transfer rate in BCI. It may also be possible that PMIS selected EEG embedding parameters can be used as class predictors i.e., the optimal selected embedding parameters can be selected on a trial-by-trial basis using PMIS and used as signal features. The investigation would involve determining if such features provided sufficient inter class variability and intra class correlation to enable reliable discrimination of brain states.

In summary, this work shows how a range of traditional signal processing tools can be combined with multiple computational intelligence based approaches to develop a more autonomous parameter optimization setup and ultimately a more accurate BCI. Finally, the novel developments in signal processing and embedding selection using PMIS will be integrated with our real-time BCI, when sufficiently validated, for application in assistive technologies and entertainment for the physically impaired [4][5][6][8] and rehabilitation [7].

References

[1] Wolpaw, J.R., Birbaumer, N., McFarland, D.J., Pfurtscheller, G., Vaughan, T.M.: Brain-computer interfaces for communication and control. J. Clinical Neurophysiology 113, 767–791 (2002)

[2] Kubler, A., Kotchoubey, B., Kaiser, J., Wolpaw, J.R., Birbaumer, N.: Brain-Computer communication: unlocking the locked-in. Psychological Bulletin 127(3), 358–375 (2001)

[3] Pfurtscheller, G., Guger, C., Muller, G., Krausz, G., Neuper, C.: Brain oscillations control hand orthosis in a tetraplegic. Neurosci. Lett. 292, 211–214 (2000)

[4] Coyle, D., Satti, A., Stow, J., McCreadie, K., Carroll, A., McElligott, J.: Operating a Brain Computer Interface: Able Bodied vs. Physically Impaired Performance. In: Proc. of the Recent Advances in Assistive Technology & Engineering Conference (2011)

[5] Stow, J., Coyle, D., Carroll, A., Satti, A., McElligott, J.: Achievable Brain Computer Communication through Short Intensive Motor Imagery Training despite Long Term Spinal Cord Injury. In: Proc. of the Annual IICN Registrar's Prize in Neuroscience (2011)

[6] Coyle, D., Carroll, A., Stow, J., McCann, A., Ally, A., McElligott, J.: Enabling Control in the Minimally Conscious State in a Single Session with a Three Channel BCI. In: Proc. of the 1st International DECODER Workshop (2012)

[7] Prasad, G., Herman, P., Coyle, D., McDonough, S., Crosbie, J.: Applying a brain-computer interface to support motor imagery practice in people with stroke for upper limb recovery: a feasibility study. J. Neuroeng. Rehab. 7(60), 1–17 (2011)

[8] Coyle, D., Garcia, J., Satti, A., McGinnity, T.M.: EEG-based Continuous Control of a Game using a 3 Channel Motor Imagery BCI. In: IEEE Symposium Series on Computa-tional Intelligence, pp. 88–93 (2011)

[9] Enzinger, C., Ropele, S., Fazekas, F., Loitfelder, M., Gorani, F., Seifert, T., Reiter, G., Neuper, C., Pfurtscheller, G., Muller-Putz, G.: Brain motor system function in a patient with complete spinal cord injury following extensive brain–computer interface training. Exp. Brain Res. 190, 215–223 (2008)

[10] Chatrian, G.E., Peterson, M., Lazarte, J.A.: The blocking of the rolandic wicket rhythm and some central changes related to movement. Electroencephalogr. Clin. Neurophysiol. 11, 497–510 (1959)

[11] Pfurtscheller, G., Neuper, C., Flotzinger, D., Pregenzer, M.: EEG-based discrimina-tion between imagination of right and left hand movement. Electroencephalography and Clinical Neurophysiology 113(6), 642–651 (1997)

[12] Felzer, T., Freisleben, B.: Analyzing EEG signals using the probability estimated guarded neural classifier. IEEE Trans. on Neural Sys. and Rehab. Eng. 11(2), 361–371 (2003)

[13] Anderson, C., Sijercic, Z.: Classification of EEG signals from four subjects during five mental tasks. In: Proc of the Conference on Eng. Applications in Neural Net-works (EANN 1996), pp. 407–414 (1996)

[14] Muller, K.-R., Anderson, C.W., Birch, G.E.: Linear and nonlinear methods for brain-computer interfaces. IEEE Trans. on Neural Systems and Rehab. Eng. 11(2), 165–169 (2003)

[15] Schlogl, A., Flotzinger, D., Pfurtscheller, G.: Adaptive autoregressive modelling used for single-trial EEG classification. Biomedizinische Technik, Band 42, 162–167 (1997)

[16] Forney, E., Anderson, C.W.: Classification of EEG during Imagined Mental Tasks by Forecasting with Elman Recurrent Neural Networks. In: Proceedings of the Interna-tional Joint Conference on Neural Networks, pp. 2749–2755 (2011)

[17] Pfurtscheller, G., Neuper, C., Schlogl, A., Lugger, K.: Separability of EEG signals recorded during right and left motor imagery using adaptive autoregressive parame-ters. IEEE Transactions on Rehabilitation Engineering 6(3), 316–324 (1998)

[18] Schloegl, A.: The electroencephalogram and the adaptive autoregressive model: theory and applications. Shaker Verlag, Aachen (2000)

[19] Kohlmorgen, J., Müller, K.-R., Rittweger, J., Pawelzik, K.: Identification of non-stationary dynamics in physiological recordings. Biological Cybernetics 83(1), 73–84 (2000)

[20] Haselsteiner, E., Pfurtscheller, G.: Using Time-Dependent NNs for EEG classifica-tion. IEEE Trans. on Rehab. Eng. 8(4), 457–462 (2000)

[21] Coyle, D., Prasad, G., McGinnity, T.M.: A time-series prediction approach for feature extraction in a brain-computer interface. IEEE Transactions on Neural Systems and Rehabilitation Engineering 13(4), 461–467 (2005)

[22] Coyle, D.: Neural network based auto association and time-series prediction for bio-signal processing in brain-computer interfaces. IEEE Computational Intelligence Magazine 4(4), 47–59 (2009)

[23] Coyle, D., Prasad, G., McGinnity, T.M.: Faster self-organizing fuzzy neural network training and a hyperparameter analysis for a brain-computer interface. IEEE Transactions on Systems, Man and Cybernetics (Part B) 39(6), 1458–1471 (2009)

[24] Coyle, D., Prasad, G., McGinnity, T.M.: Improving the separability of multiple feature types for a brain-computer interface by neural time-series prediction preprocessing. Biomedical Signal Processing and Control, 196–204 (2010)

[25] Sharma, A.: Seasonal to inter annual rainfall probabilistic forecasts for improved water supply management: part 1 – a strategy for system predictor identification. Journal of Hydrology 239, 232–239 (2000)

[26] May, R.J., Maier, H.R., Dandy, G.C., Gayani Fernando, T.M.K.: Non-linear variable selection for artificial neural networks using partial mutual information. Environmental Modelling and Software 23, 1312–1326 (2008)

[27] Blankertz, et al.: BCI Competition III and IV (2005), http://www.bbci.de/competition/

[28] Blankertz, et al.: The BCI competition. III: Validating alternative approaches to actual BCI problems. IEEE Trans. Neural. Syst. Rehabil. Eng. 14, 153–159 (2006)

[29] Schlogl, A., Lee, F., Birschof, H., Pfurtscheller, G.: Characterization of four-class motor imagery EEG data for the BCI-competition 2005. J. of Neural Engineering 2, L.14–L.22 (2005)

[30] Schlogl, et al.: BCI-Competition IV (Dataset 2A and 2B) (2008), http://www.bbci.de/competition/iv/desc_2b.pdf, http://www.bbci.de/competition/iv/desc_2a.pdf

[31] Leeb, R., Lee, F., Keinrath, C., Scherer, R., Bischof, H., Pfurtscheller, G.: Brain-computer communication: motivation, aim, and impact of exploring a virtual apartment. IEEE Transactions on Neural Systems and Rehabilitation Engineering 15, 473–482 (2007)

[32] Schlogl, A., Keinrath, C., Zimmermann, D., Scherer, R., Leeb, R., Pfurtscheller, G.: A fully automated correction method for EOG artifacts in EEG recordings. Clin. Neuro-Phys. 118(1), 98–104 (2007)

[33] Hornik, K., Stinchcombe, M., White, H.: Multilayer feedforward networks are universal approximators. Neural Networks 2, 359–366 (1989)

[34] Jang, J.S.R.: Neuro-Fuzzy & Soft Computing. Prentice-Hall (1997)

[35] Leng, G.: Algorithmic Developments for Self-Organising Fuzzy Neural Networks. PhD Dissertation, University of Ulster (2003)

[36] Prasad, G., McGinnity, T.M., Leng, G., Coyle, D.: On-line identification of self-organizing fuzzy neural networks for modelling time-varying complex systems. In: Plamen, et al. (eds.) Evolving Intelligent Systems, pp. 302–324. John Wiley, NY (2010)

[37] Coyle, D., Prasad, G., McGinnity, T.M.: Faster Self-organising Fuzzy Neural Network Training and Improved Autonomy with Time-Delayed Synapses for Locally Recurrent Learning. In: Temel (ed.) System and Circuit Design for Biologically-Inspired Learning, pp. 156–183. IGI-Global (2010)

[38] Ramouser, H., Muller-Gerking, J., Pfurtscheller, G.: Optimal spatial filtering of single trial EEG during imagined hand movement. IEEE Trans. on Rehab. Eng. 8(4), 441–446 (2000)

[39] Blankertz, B., Tomioka, R., Lemm, S., Kawanabe, M., Müller, K.R.: Optimizing spatial filters for robust EEG Analysis. IEEE Signal Processing Magazine, 41–56 (2008)

[40] Satti, A., Coyle, D., Prasad, G.: Spatio-spectral & temporal parameter searching using class correlation analysis and particle swarm optimization for a brain computer interface. In: Proc. of the 2009 IEEE Systems, Man and Cybernetics Conference, pp. 1731–1735 (2009)

[41] Kennedy, J., Eberhart, R.: Particle swarm optimization. In: Proceedings IEEE International Conference on Neural Networks, vol. 1, pp. 1942–1948 (1995)

[42] Herman, P., Prasad, G., McGinnity, T.M., Coyle, D.: Comparative analysis of spectral approaches to feature extraction for EEG-based motor imagery classification. IEEE Transactions on Neural Systems and Rehabilitation Engineering 16(4), 317–326 (2008)

[43] Coyle, D., Prasad, G., McGinnity, T.M.: A time-frequency approach to feature extraction for a brain-computer interface with a comparative analysis of performance measures. EURASIP JASP, Trends in Brain-Computer Interfaces (special issue) 19, 3141–3151 (2005)

[44] Coyle, D., Prasad, G., McGinnity, T.M., Herman, P.: Estimating the predictability of EEG recorded over the motor cortex using information theoretic functionals. In: Proceedings of the 2nd International Brain-Computer Interface Workshop and Training Course, Biomedizinische Technik, pp. 43–44 (2004)

[45] Fraser, A.M.: Information and Entropy in Strange Attractors. IEEE Trans. on Info. Theory. 35(2), 245–262 (1989)

[46] Palus, M., Pecen, L., Pivka, D.: Estimating predictability: The redundancy and surrogate data method. Neural Network World 5(4), 537–552 (1995)

[47] Palus, M.: Testing for nonlinearity using redundancies: Quantitative and qualitative aspects. Physica D, 186–205 (1995)

[48] Williams, G.P.: Chaos Theory Tamed. Taylor and Francis, London (1997)

[49] Shannon, C.E., Weaver, W.: The mathematical theory of communication. University of Illinois Press (1963)

[50] Scott, D.W.: Multivariate Density Estimation: Theory, Practice and Visualisation. John Wiley and Sons, New York (1992)

[51] Chow, T.W.S., Huang, D.: Estimating optimal feature subsets using efficient estimation of high-dimensional mutual information. IEEE Transactions on Neural Networks 16(1), 213–224 (2005)

[52] Silverman, B.W.: Density Estimation for Statistics and Data Analysis. Chapman and Hall, London (1986)

[53] Davies, L., Gather, U.: The identification of multiple outliers. Journal of the American Statistical Association 88(423), 782–792 (1993)

[54] Zar, J.H.: Biostatistical Analysis, 4th edn., pp. 255–259. Upper Saddle River, New-Jersey (1999)

[55] Greene, J., D'Oliveira, M.: Learning to use statistical tests in psychology. Open University Press (1982)

[56] Satti, A., Guan, C., Coyle, D., Prasad, G.: A covariate shift minimisation method to alleviate non-stationarity effects for an adaptive brain-computer interface. In: 20th International Conference Pattern Recognition, August 23-26, pp. 105–108 (2010)

[57] Krusienski, D.J., Grosse-Wentrup, M., Galan, F., Coyle, D., Miller, K.J., Forney, E., Anderson, C.W.: Critical Issues in Brain Computer Interface Research. Journal of Neural Engineering 8, 025002 (8pp) (2011)

[58] Coyle, D., McGinnity, T.M., Prasad, G.: A multi-class brain-computer interface with SOFNN-based prediction preprocessing. In: IEEE World Congress on Computational Intelligence, pp. 3695–3702 (2008)

[59] Coyle, D., Prasad, G., McGinnity, T.M.: Improving information transfer rates of a brain-computer interface by self-organising fuzzy neural network-based multi-step-ahead time-series prediction. In: Proceedings of the 3rd IEEE Systems, Man and Cybernetics (UK&RI Chapter) Conference, pp. 230–235 (2004)

Chapter 13
How to Describe and Propagate Uncertainty When Processing Time Series: Metrological and Computational Challenges, with Potential Applications to Environmental Studies

Christian Servin, Martine Ceberio, Aline Jaimes,
Craig Tweedie, and Vladik Kreinovich

Abstract. Time series comes from measurements, and often, measurement inaccuracy needs to be taken into account, especially in such volatile application areas as meteorology and economics. Traditionally, when we deal with an individual measurement or with a sample of measurement results, we subdivide a measurement error into random and systematic components: systematic error does not change from measurement to measurement while random errors corresponding to different measurements are independent. In time series, when we measure the same quantity at different times, we can also have correlation between measurement errors corresponding to nearby moments of time. To capture this correlation, environmental science researchers proposed to consider the third type of measurement errors: periodic. This extended classification of measurement error may seem ad hoc at first glance, but it leads to a good description of the actual errors. In this paper, we provide a theoretical explanation for this semi-empirical classification, and we show how to efficiently propagate all types of uncertainty via computations.

1 Formulation of the Problem

In many applications areas – e.g., in meteorology, in financial analysis – the value of the important variable (temperature, stock price, etc.) changes with time. In order to adequately predict the corresponding value, we need to analyze the observed time series and to make a prediction based on this analysis; see, e.g., [3, 20].

All the values that form the time series come from measurements or from expert estimates. Neither measurements nor expert estimates are 100% accurate, especially in such volatile application areas as meteorology and economics. Thus, the

Christian Servin · Martine Ceberio · Aline Jaimes · Craig Tweedie · Vladik Kreinovich
Cyber-ShARE Center,
University of Texas at El Paso,
El Paso, TX 79968, USA
e-mail: vladik@utep.edu

W. Pedrycz & S.-M. Chen (Eds.): Time Series Analysis, Model. & Applications, ISRL 47, pp. 279–299.
DOI: 10.1007/978-3-642-33439-9_13   © Springer-Verlag Berlin Heidelberg 2013

actual values of the corresponding variables are, in general, slightly different from the observed values x_t. These measurement uncertainties affects the result of data processing.

For example, in meteorological and environmental applications, we measure, at different locations, temperature, humidity, wind speed and direction, flows of carbon dioxide and water between the soil and atmosphere, intensity of the sunlight, reflectivity of the plants, plant surface, etc. Based on these *local* measurement results, we estimate the *regional* characteristics such as the carbon fluxes describing the region as a whole – and then use these estimates for predictions. These predictions range from short-term meteorological predictions of weather to short-term environmental predictions of the distribution and survival of different ecosystems and species to long-term predictions of climate change; see, e.g., [1, 12]. Many of these quantities are difficult to measure accurately: for example, the random effects of turbulence and the resulting rapidly changing wind speeds and directions strongly affect our ability to accurately measure carbon dioxide and water flows; see, e.g., [18]. The resulting measurement inaccuracy is one of the main reasons why it is difficult to forecast meteorological, ecological, and climatological phenomena.

It is therefore desirable to describe how the corresponding measurement uncertainty affects the result of data processing. In this paper, we analyze this problem, describe the related challenges, and show how these challenges can be overcome.

2   Traditional Approach to Measurement Errors

When we are interested in the value x of some quantity that we can measure directly, we apply an appropriate measuring instrument and get the measurement result $\widetilde{x}$. In the ideal world, the measurement result $\widetilde{x}$ is exactly equal to the desired value x. In practice, however, there is noise, there are imperfection, there are other factors which influence the measurement result. As a consequence, the measurement result $\widetilde{x}$ is, in general, different from the actual (unknown) value x of the quantity of interest, and the *measurement error* $\Delta x \overset{\text{def}}{=} \widetilde{x} - x$ is different from 0.

Because of this, if we repeatedly measure the same quantity by the same measuring instrument, we get, in general, slightly different results. Some of these results are more frequent, some less frequent. For each interval of possible values, we can find the frequency with which the measurement result gets into this interval; at first, some of these frequencies change a lot with each new measurement, but eventually, once we have a large number of measurements, these frequencies stabilize – and become *probabilities* of different values of $\widetilde{x}$ and, correspondingly, probabilities of different values of measurement error Δx. In other words, the measurement error becomes a *random variable*.

Usually, it is assumed that random variables corresponding to different measurement errors are statistically independent from each other. In statistics, independence of two events A and B means that the probability of A does not depend on B, i.e., that the conditional probability $P(A|B)$ of A under condition B is equal to the unconditional probability $P(A)$ of the event A.

The probability $P(A)$ of the event A can be estimated as the ratio $\dfrac{N(A)}{N}$ of the number of cases $N(A)$ when the event A occurred to the total number N of observed cases. Similarly, the probability $P(B)$ of the event B can be estimated as the ratio $\dfrac{N(B)}{N}$ of the number of cases $N(B)$ when the event A occurred to the total number N of observed cases, and the probability $P(A\,\&\,B)$ of both events A and B can be estimated as the ratio $\dfrac{N(A\,\&\,B)}{N}$ of the number of cases $N(A\,\&\,B)$ when both events A and B occurred to the total number N of observed cases. In contrast, to estimate the conditional probability of A given B, we must only take into account cases when B was observed. As a result, we get an estimate $P(A\,|\,B) \approx \dfrac{N(A\,\&\,B)}{N(B)}$. Since $P(A\,\&\,B) \approx \dfrac{N(A\,\&\,B)}{N}$ and $P(B) \approx \dfrac{N(B)}{N}$, we conclude that $N(A\,\&\,B) \approx P(A\,\&\,B) \cdot N$ and $N(B) \approx P(B) \cdot N$ and therefore, $P(A\,|\,B) \approx \dfrac{P(A\,\&\,B) \cdot N}{P(B) \cdot N} = \dfrac{P(A\,\&\,B)}{P(B)}$, so $P(A\,|\,B) \approx \dfrac{P(A\,\&\,B)}{P(B)}$. The larger the sample, the more accurate are these estimates, so in the limit when N tends to infinity, we get the equality $P(A\,|\,B) = \dfrac{P(A\,\&\,B)}{P(B)}$, i.e., equivalently, $P(A\,\&\,B) = P(A\,|\,B) \cdot P(B)$. For independent events, $P(A\,|\,B) = P(A)$ and thus, $P(A\,\&\,B) = P(A) \cdot P(B)$.

So, under the independence assumption, if we have two different series of measurements, resulting in measurement errors Δx and Δy, then the probability $P(\Delta x \in [\underline{x}, \overline{x}] \,\&\, \Delta y \in [\underline{y}, \overline{y}])$ that Δx is in an interval $[\underline{x}, \overline{x}]$ *and* Δy is in an interval $[\underline{y}, \overline{y}]$ is equal to the product of the two probabilities:

- the probability $P(\Delta x \in [\underline{x}, \overline{x}])$ that Δx is in the interval $[\underline{x}, \overline{x}]$, and
- the probability $P(\Delta y \in [\underline{y}, \overline{y}])$ that Δy is in the interval $[\underline{y}, \overline{y}]$:

$$P(\Delta x \in [\underline{x}, \overline{x}] \,\&\, \Delta y \in [\underline{y}, \overline{y}]) = P(\Delta x \in [\underline{x}, \overline{x}]) \cdot P(\Delta y \in [\underline{y}, \overline{y}]).$$

Usually in metrology, the measurement error is divided into two components (see, e.g., [16]):

- the *systematic* error component, which is defined as the expected value (mean) $E(\Delta x)$ of the measurement errors, and
- the *random* error component which is defined as the difference $\Delta x - E(\Delta x)$ between the measurement error Δx and its systematic component $E(\Delta x)$.

Systematic error component is usually described by the upper bound Δ_s on its absolute value: $|E(\Delta x)| \leq \Delta_s$, while the random error is usually described by its mean square value

$$\sigma = \sqrt{E\left[(\Delta x - E(\Delta x))^2\right]}.$$

In statistical terms, $\sigma = \sqrt{V}$ is the *standard deviation* of the random variable Δx, i.e., the square root of the *variance* $V = E\left[(\Delta x - E(\Delta x))^2\right]$.

The practical meaning of these components – and the practical difference between them – can be described if, in order to improve measurement accuracy, we repeatedly measure the same quantity several times. Once we have several results $\widetilde{x}^{(1)}, \ldots, \widetilde{x}^{(M)}$ of measuring the same (unknown) quantity x, we can then take the arithmetic average

$$\widetilde{x} = \frac{\widetilde{x}^{(1)} + \ldots + \widetilde{x}^{(M)}}{M}$$

as the new estimate.

One can easily see that the measurement error $\Delta x = \widetilde{x} - x$ corresponding to this new estimate is equal to the average of the measurement errors $\Delta x^{(k)} = \widetilde{x}^{(k)} - x$ corresponding to individual measurements:

$$\Delta x = \frac{\Delta x^{(1)} + \ldots + \Delta x^{(M)}}{M}.$$

What are the systematic and random error components of this estimate? Let us start with the systematic error component, i.e., in mathematical terms, with the mean. It is known that the mean of the sum is equal to the sum of the means, and that when we divide a random variable by a constant, its mean is divided by the same constant. All M measurements are performed by the same measuring instrument with the same systematic error $E\left(\Delta x^{(1)}\right) = \ldots = E\left(\Delta x^{(M)}\right)$. Thus, for the sum $\Delta x^{(1)} + \ldots + \Delta x^{(M)}$, the mean is equal to

$$E\left(\Delta x^{(1)} + \ldots + \Delta x^{(M)}\right) = E\left(\Delta x^{(1)}\right) + \ldots + E\left(\Delta x^{(M)}\right) = M \cdot E\left(\Delta x^{(k)}\right).$$

Therefore, the mean of the ratio Δx (which is obtained by dividing the above sum by M) is M times smaller than the mean of the sum, i.e., equal to $E(\Delta x) = E\left(\Delta x^{(k)}\right)$. In other words, the systematic error component does not decrease if we simply repeat the measurements.

In contrast, the random component decreases, or, to be precise, its standard deviation decreases. Indeed, for independent random variables, the variance of the sum is equal to the sum of the variances, and when we divide a random variable by a constant, the variance is divided by the square of this constant. The variance $V = \sigma^2$ of each random error component is equal to $V^{(1)} = \ldots = V^{(M)}$; thus, the variance of the sum $\Delta x^{(1)} + \ldots + \Delta x^{(M)}$ is equal to the sum of these variances, i.e., to

$$V\left[\Delta x^{(1)} + \ldots + \Delta x^{(M)}\right] = V^{(1)} + \ldots + V^{(M)} = M \cdot \left(\sigma^{(k)}\right)^2.$$

Therefore, the variance of the ratio Δx (which is obtained by dividing the above sum by M) is M^2 times smaller than the variance of the sum, i.e., equal to $\dfrac{\left(\sigma^{(k)}\right)^2}{M}$. So, the standard deviation σ (which is the square root of this variance) is equal to $\dfrac{\sigma^{(k)}}{\sqrt{M}}$. In other words, the more times we repeat the measurement, the smaller the resulting random error.

So, when we repeat the same measurement several times, the random error disappears, and the only remaining error component is the systematic error.

3 The Traditional Metrological Approach Does Not Work Well for Time Series

In the traditional approach, we represent the measurement error as the sum of two components:

- a *systematic* component which is *the same* for all measurements, and
- a *random* component which is *independent* for different measurements.

When we process time series, this decomposition is insufficient: e.g., usually, there are strong correlations between measurement errors corresponding to consequent measurements.

To achieve a better representation of measurement errors, researchers in environmental science have proposed a semi-empirical idea of introducing the third component of measurement error: the *seasonal* (*periodic*) component; see, e.g., [14].

For example, a seasonal error component can represent errors that only happen in spring (this is where the name of this error component comes from), or errors that only happen at night, etc.

From the purely mathematical viewpoint, we can have periodic error components corresponding to all possible frequencies. However, from the physical viewpoint, it makes sense to concentrate on the components with physically meaningful frequencies – and with frequencies which are multiples of these frequencies, e.g., double or triple the daily or yearly frequencies.

For example, in environmental observations, it makes sense to concentrate on daily and yearly periodic errors. If we are interested in the effect of human activity, then we need to add weekly errors – since human activity periodically changes from weekdays to weekends.

The idea of using three components of measurement error works extremely well – which leads to two related challenges:

- A *metrological* challenge: how can we explain this success? What is the foundation of this idea?
- A *computational* challenge: how can we efficiently describe this new error component and how can we efficiently propagate it through computations?

In this paper, we address both challenges:

- we provide a theoretical justification for the semi-heuristic idea of the third error component, and
- we show a natural way for efficiently describing this error component, and show how to efficiently propagate different error components through computations.

4 First Result: A Theoretical Explanation of the Three-Component Model of Measurement Error

Our objective is to analyze measurement errors $\Delta x(t)$ corresponding to time series. Namely, we want to represent a generic measurement error as a linear combination of several error components.

This division into components can be described on different levels of granularity. Let us consider the level where the granules are the smallest, i.e., where each granule corresponds to a finite-dimensional linear space, i.e., to the linear space whose elements can be determined by finitely many parameters.

Each component of the measurement error is thus described by a finite-dimensional linear space L, i.e., by the set of all the functions of the type $x(t) = c_1 \cdot x_1(t) + \ldots + c_n \cdot x_n(t)$, where $x_1(t), \ldots, x_n(t)$ are given functions, and $c_1, \ldots, c_n$ are arbitrary constants.

In most applications, observed signals continuously (and even smoothly) depend on time, so we will assume that all the functions $x_i(t)$ are smooth (differentiable).

Also, usually, there is an upper bound on the measurement error, so we will assume that each of the the functions $x_i(t)$ are bounded by a constant.

Finally, for a long series of observations, we can choose a starting point arbitrarily. If instead of the original starting point, we take a starting point which is t_0 seconds earlier, then each moment of time which was originally described as moment t is not described as moment $t + t_0$. Then, for describing measurement errors, instead of the original function $x(t)$, we have a new function $x_{t_0}(t)$ for which $x_{t_0}(t + t_0) = x(t + t_0)$. It is reasonable to require that the linear space that describes a component of the measurement error does not not change simply because we changed the starting point. Thus, we arrive at the following definitions.

Definition 1. We say that a function $x(t)$ of one variable is *bounded* if there exists a constant M for which $|x(t)| \leq M$ for all t.

Definition 2. We say that a class F of functions of one variable is *shift-invariant* if for every function $x(t) \in F$ and for every real number t_0, the function $x(t + t_0)$ also belongs to the class F.

Definition 3. By an *error component* we mean a shift-invariant finite-dimensional linear space of functions

$$L = \{c_1 \cdot x_1(t) + \ldots + c_n \cdot x_n(t)\},$$

where $x_1(t), \ldots, x_n(t)$ are given bounded smooth functions and c_i are arbitrary numbers.

Theorem 1. *Every error component is a linear combination of the functions*

$$x(t) = \sin(\omega \cdot t) \text{ and } x(t) = \cos(\omega \cdot t).$$

Proof.

1°. Let us first use the assumption that the linear space L is shift-invariant.

For every i from 1 to n, the corresponding function $x_i(t)$ belongs to the linear space L. Since the error component is shift-invariant, we can conclude that for every real number t_0, the function $x_i(t + t_0)$ also belongs to the same linear space. Thus, for every i from 1 to n and for every t_0, there exist values $c_1, \ldots, c_n$ (possibly depending on i and on t_0) for which

$$x_i(t + t_0) = c_{i1}(t_0) \cdot x_1(t) + \ldots + c_{in}(t_0) \cdot x_n(t). \tag{1}$$

2°. We know that the functions $x_1(t), \ldots, x_n(t)$ are smooth. Let us use the equation (1) to prove that the functions $c_{ij}(t_0)$ are also smooth (differentiable).

Indeed, if we substitute n different values $t_1, \ldots, t_n$ into the equation (1), then we get a system of n linear equations with n unknowns to determine n values $c_{i1}(t_0)$, $\ldots, c_{in}(t_0)$:

$$x_i(t_1 + t_0) = c_{i1}(t_0) \cdot x_1(t_1) + \ldots + c_{in}(t_0) \cdot x_n(t_1);$$

$$\ldots$$

$$x_i(t_n + t_0) = c_{i1}(t_0) \cdot x_1(t_n) + \ldots + c_{in}(t_0) \cdot x_n(t_n).$$

The solution of a system of linear equations – as determined by the Cramer's rule – is a smooth function of all the coefficients and right-hand sides. Since all the right-hand sides $x_i(t_j + t_0)$ are smooth functions of t_0 and since all the coefficients $x_i(t_j)$ are constants (and thus, are also smooth), we conclude that each dependence $c_{ij}(t_0)$ is indeed smooth.

3°. Now that we know that all the functions $x_i(t)$ and $c_{ij}(t_0)$ are differentiable, we can differentiate both sides of the equation (1) with respect to t_0 and then take $t_0 = 0$. As a result, we get the following systems of n differential equations with n unknown functions $x_1(t), \ldots, x_n(t)$:

$$\dot{x}_i(t) = c_{i1} \cdot x_1(t) + \ldots + c_{in} \cdot x_n(t),$$

where $\dot{x}_i(t)$ denotes derivative over time, and c_{ij} denoted the value of the corresponding derivative $\dot{c}_{ij}$ when $t_0 = 0$.

4°. We have shown that the functions $x_1(t), \ldots, x_n(t)$ satisfy a system of linear differential equations with constant coefficients.

It is known that a general solution of such system of equations is a linear combination of functions of the type $t^k \cdot \exp(\lambda \cdot t)$, where k is a natural number (non-negative integer), and λ is a complex number. Specifically, λ is an eigenvalue of the matrix c_{ij}, and the value $k > 0$ appears when we have a degenerate eigenvalue, i.e., an eigenvalue for which there are several linearly independent eigenvectors.

5°. Every complex number λ has the form $a + i \cdot \omega$, where a is its real part and ω is its imaginary part. So:

$$\exp(\lambda \cdot t) = \exp(a \cdot t) \cdot \cos(\omega \cdot t) + i \cdot \exp(a \cdot t) \cdot \sin(\omega \cdot t).$$

Thus, every function $x_i(t)$ can be represented as a linear combination of expressions of the types $t^k \cdot \exp(a \cdot t) \cdot \cos(\omega \cdot t)$ and $t^k \cdot \exp(a \cdot t) \cdot \sin(\omega \cdot t)$.

6°. Now, we can use the requirement that the functions $x_i(t)$ are bounded.

6.1°. Because of this requirement, we cannot have $a \neq 0$:

- for $a > 0$, the function is unbounded for $t \to +\infty$, while
- for $a < 0$, the function is unbounded for $t \to -\infty$.

So, we must have $a = 0$.

6.2°. Similarly, if $k > 0$, the corresponding function is unbounded. Thus, we must have $k = 0$.

7°. Thus, every function $x_i(t)$ is a linear combination of the trigonometric functions $x(t) = \sin(\omega \cdot t)$ and $x(t) = \cos(\omega \cdot t)$.
    The theorem is proven.

What are the practical conclusions of this result? We have concluded that the measurement error $\Delta x(t)$ can be described as a linear combination of sines and cosines corresponding to different frequencies ω.

In practice, depending on the relation between the frequency ω and the frequency f with which we perform measurements, we can distinguish between small, medium, and large frequencies:

- frequencies ω for which $\omega \ll f$ are *small*;
- frequencies ω for which $\omega \gg f$ are *large*, and
- all other frequencies ω are medium.

Let us consider these three types of frequencies one by one.

When the frequency ω is low, the corresponding values $\cos(\omega \cdot t)$ and $\sin(\omega \cdot t)$ practically do not change with time: the change period is much larger than the usual observation period.

Thus, we can identify low-frequency components with *systematic* error component – the error component that practically does not change with time.

When the frequency ω is high, $\omega \gg f$, the phases of the values $\cos(\omega \cdot t_i)$ and $\cos(\omega \cdot t_{i+1})$ (or, alternatively,
$\sin(\omega \cdot t_i)$ and $\sin(\omega \cdot t_{i+1})$) corresponding to the two sequential measurements t_i and t_{i+1} differ so much that for all practical purposes, the resulting values of cosine or sine functions are independent.

Thus, high-frequency components can be identified with *random* error component – the error component for which measurement errors corresponding to different measurements are independent.

In contrast to the cases of low and high frequencies, where the periodicity of the corresponding cosine and sine functions is difficult to observe, components $\cos(\omega \cdot t)$ and $\sin(\omega \cdot t)$ corresponding to medium frequencies ω are observably periodic.

It is therefore reasonable to identify medium-frequency error components with *seasonal* (*periodic*) components of the measurement error.

This conclusion explains why, in addition to the original physically meaningful frequencies, it is also reasonable to consider their multiples:

- We know that the corresponding error component is a periodic function of time, with the physically meaningful period T_0.
- It is known that every periodic function can be explained into Fourier series, i.e., represented as a linear combination of sines and cosines with frequencies ω which are multiples of the basic frequency $\omega_0 = \dfrac{2\pi}{T_0}$ corresponding to the period T_0.

Thus, we have indeed provided a justification to the semi-empirical three-component model of measurement error.

5 Periodic Error Component: Technical Details

In the above section, we explained that the periodic error component is as fundamental as the more traditional systematic and random error components. It is therefore necessary to extend the usual analysis of error components and their propagation to this new type of measurement errors.

For systematic and random error components we know:

- how to describe reasonable bounds on this error component, and
- how to estimate this error component when we calibrate the measuring instrument.

Specifically, the random error component is characterized by its standard deviation σ, while a systematic error component s is characterized by the upper bound Δ: $|s| \leq \Delta$.

The standard deviation σ of the measuring instrument can be estimated if we repeatedly measure the same quantity x by this instrument. Then, the desired standard deviation can be estimated as the sample standard deviation of the corresponding measurement results $\widetilde{x}^{(1)}, \ldots, \widetilde{x}^{(M)}$:

$$\sigma \approx \sqrt{\frac{1}{M} \cdot \sum_{k=1}^{M} \left(\widetilde{x}^{(k)} - E \right)^2},$$

where $E = \dfrac{1}{M} \cdot \sum_{k=1}^{M} \widetilde{x}^{(k)}$.

To estimate the systematic error component, it is not enough to have the given measuring instrument, we also need to *calibrate* the measuring instrument, i.e., to measure the same quantity x with an additional much more accurate ("standard") measuring instrument – whose measurement result $\widetilde{x}_s$ is assumed to be very close to the actual value x of the measured quantity. Here, $E \approx E(\widetilde{x})$ and $\widetilde{x}_s \approx x$, so the difference $E - x_s$ is approximately equal to $E(\widetilde{x}) - x = E(\widetilde{x} - x) = E(\Delta x)$. Thus, this difference $E - \tilde{x}_s$ can be used as a good approximation to the systematic error component.

Since we want to also take into account the periodic error component, it is desirable to provide answers to the above two questions for the periodic error component as well.

How can we describe reasonable bounds for each part of the periodic error component? For each frequency ω, the corresponding linear combination

$$a_c \cdot \cos(\omega \cdot t) + a_s \cdot \sin(\omega \cdot t)$$

can be equivalently represented as $A \cdot \cos(\omega \cdot t + \varphi)$. This is the form that we will use for describing the periodic error component.

Similarly to the systematic error, for the amplitude A, we will assume that we know the upper bound P: $|A| \leq P$.

For phase φ, it is natural to impose a requirement that the probability distribution of phase be invariant with respect to shift $t \to t + t_0$. When time is thus shifted, the phase is also shifted by $\varphi_0 = \omega \cdot t_0$. Thus, the requirement leads to the conclusion that the probability distribution for the phase be shift-invariant, i.e., that the corresponding probability density function $\rho(\varphi)$ is shift-invariant $\rho(\varphi) = \rho(\varphi + \varphi_0)$ for every possible shift φ_0. This means that this probability density function must be constant, i.e., that the phase φ is uniformly distributed on the interval $[0, 2\pi]$.

How can we estimate the periodic error component when calibrating a measuring instrument? When we compare the results of measuring the time series by our measuring instrument and by a standard measuring instrument, we get a sequence of differences $\tilde{x}(t) - \tilde{x}_s(t)$ that approximates the actual measurement errors $\Delta x(t)$.

Periodic error components are sinusoidal components corresponding to several frequencies. In data processing, there is a known procedure for representing each sequence as a linear combination of sinusoids of different frequency – Fourier transform. To find the periodic components, it is therefore reasonable to perform a Fourier Transform; the amplitudes of the Fourier transform corresponding to physically meaningful frequencies (and their multiples) ω will then serve as estimates for the amplitude of the corresponding periodic measurement error component.

Computing Fourier transform is fast: there is a known Fact Fourier Transform (FFT) algorithm for this computation; see, e.g., [2].

In this process, there is a still a computational challenge. Indeed, while the standard measuring instrument is reasonably accurate and its measurement results $\tilde{x}_s(t)$ provide a good approximation to the actual values $x(t)$, these results are still somewhat different from the actual values $x(t)$. Hence, the observed differences $\tilde{x}(t) - \tilde{x}_s(t)$ are only approximately equal to the measurement errors $\Delta x(t) = \tilde{x}(t) - x(t)$. When we apply FFT in a straightforward way, this approximation error sometimes leads to drastic over-estimation of the results; see, e.g., [4, 13]. Because of this fact, many researchers replaced FFT by much slower – but more accurate – error estimation algorithms.

In our paper [13], we showed how we can modify the FFT techniques so that we get (almost) exact error bounds while being (almost) as fast as the original FFT. So, to estimate the periodic error component, we need to use thus modified FFT algorithm.

6 Because of Our Justification, the Three-Component Model of Approximation Error Can Also Be Applied to Expert Estimates

In many practical situations, the measurement results are not sufficient to make reasonable conclusions. We need to supplement measurement results with the knowledge of experts. The use of expert knowledge in processing data is one of the important aspects of *computational intelligence*.

For example, when a medical doctor makes a diagnosis and/or prescribes medicine, he or she is usually not following an algorithm that inputs the patients stats and outputs the name of the disease and the dosage of the corresponding medicine. If medicine was that straightforward, there would have been no need for skilled medical doctors. A good doctor also uses his/her experience, his/her intuition. Similarly, in environmental research, we *measure* temperature, humidity, etc. However, to make meaningful conclusions, it is necessary to supplement these measurement results with *expert estimates* of, e.g., amount of leaves on the bushes ("low", "medium", "high"), state of the leaves – and many other characteristics which are difficult to measure but which can be easily estimated by an expert.

We have mentioned that in data processing, it is important to take into account the uncertainty of measurement results. Expert estimates are usually even much less accurate than measurement results. So, it is even more important to take into account the uncertainty of expert estimates.

The main idea behind most methods for dealing with uncertainty of expert estimates is to treat an expert as a measuring instrument and use the corresponding metrological techniques.

One of the main techniques for describing expert uncertainty is *fuzzy techniques*; see, e.g., [9, 15]. While these techniques are not exactly probabilistic, many fuzzy techniques are similar to the probabilistic ones.

For example, one of the most widely used methods of determining the (fuzzy) degree of belief $\mu_P(x)$ that a certain value x satisfies the property P (e.g., that a certain temperature is low) is to poll several experts and take, as $\mu_P(x)$, the proportion of those who thing that x satisfies this property.

Good news is that in our analysis of the error components, we never used the fact that this error comes from measurements. We can therefore apply the exact same analysis to the approximation error of the expert estimates.

Thus, while our main current emphasis is on measurement results and measurement uncertainty, it is desirable to apply the same three-component decomposition to inaccuracies of expert estimates as well.

7 How to Propagate Uncertainty in the Three-Component Model

In the previous sections, we analyzed how to *describe* the uncertainty related to measurements and/or expert estimates. Some quantities can be indeed directly measured or estimates. However, there are many quantities of interest which cannot be directly measured or estimated.

An example of such a quantity is a carbon flux that describes the exchange of carbon between the soil and the atmosphere; see, e.g., [12]. It is difficult to measure this flux directly. Instead, we measure the humidity, wind and concentration of different gases at different height of a special meteorological tower, and then use the results of these measurements to process the data.

In general, for many quantities y, it is not easy (or even impossible) to measure them directly. Instead, we measure related quantities $x_1, \ldots, x_n$, and use the known relation $y = f(x_1, \ldots, x_n)$ between x_i and y to estimate the desired quantity y.

Since measurements come with uncertainty, the resulting estimate is, in general, somewhat different from the actual value of the desired quantity – even when the relation $y = f(x_1, \ldots, x_n)$ is known exactly. It is therefore desirable to *propagate* this uncertainty, i.e., to find out how accurate is the estimate based on (approximate) measurement results.

In practical applications, many inputs to the data processing algorithm come from the same sensor at different moments of time. In other words, as inputs, we have the results $\tilde{x}_i(t_{ij})$ of measuring the values $x_i(t_{ij})$ by the i-th sensor at the j-th moment of time $t_{ij} = t_0 + j \cdot \Delta t_i$, where t_0 is the starting moment of all the measurements, and Δt_i is the time interval between the two consecutive measurements performed by the i-th sensor.

The desired quantity y depends on all these values:

$$y = f(x_1(t_{11}), x_1(t_{12}), \ldots, x_2(t_{21}), x_2(t_{22}), \ldots, x_n(t_{n1}), x_n(t_{n2}), \ldots).$$

Instead of the actual values $x_i(t_{ij})$, we only know the measurement results $\tilde{x}_i(t_{ij})$, results which differ from the actual values by the corresponding measurement errors $\Delta x_i(t_{ij})$:

$$\tilde{x}_i(t_{ij}) = x_i(t_{ij}) + \Delta x_i(t_{ij}).$$

After applying the data processing algorithm f to the measurement results $\tilde{x}_i(t_{ij})$, we get the estimate $\tilde{y}$ for the desired quantity y:

$$\tilde{y} = f(\tilde{x}_1(t_{11}), \tilde{x}_1(t_{12}), \ldots, \tilde{x}_n(t_{n1}), \tilde{x}_n(t_{n2}), \ldots).$$

We are interested in estimating the difference

$$\Delta y = \tilde{y} - y = f(\tilde{x}_1(t_{11}), \tilde{x}_1(t_{12}), \ldots, \tilde{x}_n(t_{n1}), \tilde{x}_n(t_{n2}), \ldots) -$$

$$f(x_1(t_{11}), x_1(t_{12}), \ldots, x_n(t_{n1}), x_n(t_{n2}), \ldots).$$

We know that the actual (unknown) value $x_i(t_{ij})$ of each measured quantity is equal to

$$x_i(t_{ij}) = \widetilde{x}_i(t_{ij}) - \Delta x_i(t_{ij}).$$

Thus, the desired difference has the form

$$\Delta y = f(\widetilde{x}_1(t_{11}), \ldots, \widetilde{x}_n(t_{n1}), \widetilde{x}_n(t_{n2}), \ldots) -$$

$$f(\widetilde{x}_1(t_{11}) - \Delta x_1(t_{11}), \ldots, \widetilde{x}_n(t_{n1}) - \Delta x_n(t_{n1}), \widetilde{x}_n(t_{n2}) - \Delta x_n(t_{n2}), \ldots).$$

Our objective is to estimate this difference based on the known information about the measurement errors $\Delta x_i(t_{ij})$.

Measurement errors are usually relatively small, so terms quadratic and of higher order in terms of $\Delta x_i(t_{ij})$ can be safely ignored.

For example, if the measurement error is 10%, its square is 1% which is much much smaller than 10%. If we measure with a higher accuracy, e.g., of 1%, then the square of this value is 0.01% which is even mich more smaller than the error itself.

Thus, we can *linearize* the above formula, i.e., expand the dependence of Δy on $\Delta x_i(t_{ij})$ in Taylor series and keep only linear terms in this expansion. As a result, we arrive at the following formula:

$$\Delta y = \sum_i \sum_j C_{ij} \cdot \Delta x_i(t_{ij}),$$

where C_{ij} denotes the corresponding partial derivative $\dfrac{\partial y}{\partial x_i(t_{ij})}$.

As a result of this linearization, we can consider all three components separately. Indeed, we know that each measurement errors $\Delta x_i(t_{ij})$ consists of three components: systematic component s_i, random component r_{ij}, and periodic component(s) $A_{\ell i} \cdot \cos(\omega_\ell \cdot t_{ij} + \varphi_{\ell i})$ corresponding to different physically meaningful frequencies (and their multiples) ω_ℓ:

$$\Delta x_i(t_{ij}) = s_i + r_{ij} + \sum_\ell A_{\ell i} \cdot \cos(\omega_\ell \cdot t_{ij} + \varphi_{\ell i}).$$

The dependence of Δy on the measurement errors $\Delta x_i(t_{ij})$ is linear. Thus, we can represent Δy as the sum of different components coming from, correspondingly, systematic, random, and periodic errors:

$$\Delta y = \Delta y_s + \Delta y_r + \sum_\ell \Delta y_{p\ell},$$

where

$$\Delta y_s = \sum_i \sum_j C_{ij} \cdot s_i;$$

$$\Delta y_r = \sum_i \sum_j C_{ij} \cdot r_{ij};$$

$$\Delta y_{p\ell} = \sum_i \sum_j C_{ij} \cdot A_{\ell i} \cdot \cos(\omega_\ell \cdot t_{ij} + \varphi_{\ell i}).$$

So, it is indeed sufficient to estimate the effect of all three types of measurement error components separately.

In these estimations, we will make a natural assumption: that measurement errors corresponding to different time series are independent. Indeed, as we have mentioned earlier,

- while measurement errors corresponding to measurement by the same sensor at consecutive moments of time are correlated,
- measurement errors corresponding to different sensors usually come from different factors and are, therefore, largely independent.

Because of this assumption, we arrive at the following algorithms for estimating different components of Δy.

Propagating random component is the traditional part of error propagation. A natural way to describe the resulting error Δy_r is to use simulations (i.e., a so-called Monte-Carlo approach).

By definition of the random error component, the values r_{ij} and r_{ik} corresponding to measurements by the same i-th sensor at different moments of time t_{ij} and $t_{ij'}$ are independent. We are also assuming that the values r_{ij} and $r_{i'j'}$ corresponding to different sensors are independent. Thus, all the values r_{ij} corresponding to different pairs (i, j) are independent.

There are many such values, since each sensor performs the measurements with a high frequency – e.g., one reading every second or every minute. The value Δy_r is thus a linear combination of a large number of independent random variables. It is known that under reasonable conditions, the probability distribution of such a combination tends to normal; this is what is known as the Central Limit Theorem – one of the main reasons why normal distributions are ubiquitous in nature; see, e.g., [19].

A normal distribution is uniquely determined by its mean and standard deviation. We know that each measurement error r_{ij} has mean 0 and a known standard deviation σ_i corresponding to measurements of the i-th sensor. The mean of the linear combination is equal to the linear combination of means. Thus, the mean of Δy_r is 0. The standard deviation can be obtained if we repeatedly simulate random errors and take a standard deviation of the corresponding empirical values $\Delta y_r^{(1)}$, $\Delta y_r^{(2)}$, ... Thus, we arrive at the following algorithm.

Propagating random component: algorithm. The random component Δy_r is normally distributed with zero mean. Its standard deviation can be obtained as follows:

- First, we apply the algorithm f to the measurement results $\tilde{x}_i(t_{ij})$ and get the estimate $\tilde{y}$.
- Then, for $k = 1, \ldots, N$, we do the following:

  - simulate the random errors $r_{ij}^{(k)}$ as independent random variables (e.g., Gaussian) with 0 mean and standard deviation σ_i;

- form simulated values $x_i^{(k)}(t_{ij}) = \widetilde{x}_i(t_{ij}) - r_{ij}^{(k)}$;
- substitute the simulated values $x_i^{(k)}(t_{ij})$ into the data processing algorithm f and get the result $y^{(k)}$.

- Finally, we estimate the standard deviation σ of the random component Δy_r as

$$\sigma = \sqrt{\frac{1}{N} \cdot \sum_{k=1}^{N} \left(y^{(k)} - \widehat{y}\right)^2}.$$

Mathematical comment. The proof that this algorithm produces a correct result easily follows from the fact that for simulated values, the difference $y^{(k)} - \widehat{y}$ has the form $\sum_i \sum_j C_{ij} \cdot r_{ij}^{(k)}$ and thus, has the exact same distribution as $\Delta y_r = \sum_i \sum_j C_{ij} \cdot \Delta x_i(t_{ij})$; see, e.g., [10].

Metrological comment. In some practical situations, instead of the standard deviations $\sigma_i = \sqrt{E[(\Delta x)^2]}$ that describe the *absolute* accuracy, practitioners often describe *relative* accuracy δ_i such as 5% or 10%. In this case, the standard deviation σ_i can be obtained as $\sigma_i = \delta_i \cdot m_i$, i.e., by multiplying the given value δ_i and the mean square value of the signal

$$m_i = \sqrt{\frac{1}{T_i} \cdot \sum_j (\widetilde{x}_i(t_{ij}))^2},$$

where T_i is the total number of measurements performed by the i-th sensor.

Let us now consider the problem of propagating systematic component. By definition, the systematic component Δy_s of the resulting error Δy is equal to $\Delta y_s = \sum_i \sum_j C_{ij} \cdot s_i$. If we combine terms corresponding to different j, we conclude that $\Delta y_s = \sum_i K_i \cdot s_i$, where $K_i \overset{\text{def}}{=} \sum_j C_{ij}$.

The values K_i can be explicitly described. Namely, one can easily see that if for some small value $\delta > 0$, for this sensor i, we take $\Delta x_i(t_{ij}) = \delta$ for all j, and for all other sensors i', we take $\Delta x_{i'}(t_{i'j}) = 0$, then the resulting increase in y will be exactly equal to $\delta \cdot K_i$.

Once we have determined the coefficients K_i, we need to find out the smallest and the largest possible value of the sum $\Delta y_s = \sum_i K_i \cdot s_i$. Each parameter s_i can take any value between $-\Delta_{si}$ and Δ_{si}, and these parameters are independent. Thus, the sum is the largest when each term $K_i \cdot s_i$ is the largest.

Each term is a linear function of s_i. A linear function is increasing or decreasing depending on whether the coefficient K_i is positive or negative.

- When $K_i \geq 0$, the linear function $K_i \cdot s_i$ is increasing and thus, its largest possible value is attained when s_i attains its largest possible value Δ_{si}. Thus, this largest possible value is equal to $K_i \cdot \Delta_{si}$.

- When $K_i \leq 0$, the linear function $K_i \cdot s_i$ is decreasing and thus, its largest possible value is attained when s_i attains its smallest possible value $-\Delta_{si}$. Thus, this largest possible value is equal to $-K_i \cdot \Delta_{si}$.

In both cases, the largest possible value is equal to $|K_i| \cdot \Delta_{si}$ and thus, the largest possible value Δ_s of the sum Δy_s is equal to $\Delta_s \overset{\text{def}}{=} \sum_i |K_i| \cdot \Delta_{si}$. Similarly, one can prove that the smallest possible value of Δy_s is equal to $-\Delta_s$.

Thus, we arrive at the following algorithm for computing the upper bound Δ_s on the systematic component Δy_s.

Propagating systematic component: algorithm. The largest possible value Δ_s of the systematic component Δy_s can be obtained as follows:

- First, we apply the algorithm f to the measurement results $\widetilde{x}_i(t_{ij})$ and get the estimate $\widetilde{y}$.
- Then, we select a small value δ and for each sensor i, we do the following:

  - for this sensor i, we take $x_i^{(i)}(t_{ij}) = \widetilde{x}_i(t_{ij}) + \delta$ for all moments j;
  - for all other sensors $i' \neq i$, we take $x_{i'}^{(i)}(t_{i'j}) = \widetilde{x}_i(t_{i'j})$;
  - substitute the resulting values $x_{i'}^{(i)}(t_{i'j})$ into the data processing algorithm f and get the result $y^{(i)}$.

- Finally, we estimate the desired bound Δ_s on the systematic component Δy_s as

$$\Delta_s = \sum_i \left| \frac{y^{(i)} - \widetilde{y}}{\delta} \right| \cdot \Delta_{si}.$$

Metrological comment. In some practical situations, instead of the *absolute* bound Δ_{si} on the systematic error of the i-th sensor, practitioners often describe *relative* accuracy δ_i such as 5% or 10%. In this case, a reasonable way to describe the absolute bound is to determine it as $\Delta_{si} = \delta_i \cdot a_i$, i.e., by multiplying the given value δ_i and the mean absolute value of the signal

$$a_i = \frac{1}{T_i} \cdot \sum_j |\widetilde{x}_i(t_{ij})|.$$

Numerical example. Let us consider a simple case when we are estimating the difference between the average temperatures at two nearby locations. For example, we may be estimating the effect of a tree canopy on soil temperature, by comparing the temperature at a forest location with the temperature at a nearby clearance location. Alternatively, we can be estimating the effect of elevation of the temperature by comparing the temperatures at different elevations. In this case, we use the same frequency $\Delta t_1 = \Delta t_2$ for both sensors, so $t_{1j} = t_{2j} = t_j$. The difference in average temperatures is defined as

$$y = f(x_1(t_0), x_2(t_0), x_1(t_1), \ldots, x_2(t_1), \ldots, x_1(t_n), x_2(t_n)) =$$

$$\frac{x_1(t_0) + \ldots + x_1(t_n)}{n+1} - \frac{x_2(t_0) + \ldots + x_2(t_n)}{n+1}.$$

Let us assume that the know upper bound on the systematic error of the first sensor is $\Delta_{s1} = 0.1$, and the upper bound on the systematic error of the second sensor is $\Delta_{s2} = 0.2$. We perform measurements at three moments of time $t = 0, 1, 2$. During these three moments of time, the first sensor measured temperatures $\tilde{x}_1(t_0) = 20.0$, $\tilde{x}_1(t_1) = 21.9$, and $\tilde{x}_1(t_2) = 18.7$, and the second second measured temperatures $\tilde{x}_2(t_0) = 22.4$, $\tilde{x}_2(t_1) = 23.5$, and $\tilde{x}_2(t_2) = 21.0$. In this case, the estimate $\tilde{y}$ for the desired difference y between average temperatures is equal to

$$\tilde{y} = \frac{20.0 + 21.9 + 18.7}{3} - \frac{22.4 + 23.5 + 21.0}{3} = 20.2 - 22.3 = -2.1.$$

According to our algorithm, we first select a small value δ, e.g., $\delta = 0.1$.

Then, we modify the results of the first sensor while keeping the results of the second sensor unchanged. As a result, we get $x_1^{(1)}(t_0) = \tilde{x}_1(t_0) + \delta = 20.0 + 0.1 = 20.1$, and similarly $x_1^{(1)}(t_1) = 22.0$ and $x_1^{(1)}(t_2) = 18.8$; we also get $x_2^{(1)}(t_0) = \tilde{x}_2(t_0) = 22.4$, and similarly $x_2^{(1)}(t_1) = 23.5$ and $x_2^{(1)}(t_2) = 21.0$. For thus modified values, we get

$$y^{(1)} = \frac{x_1^{(1)}(t_0) + x_1^{(1)}(t_1) + x_1^{(1)}(t_2)}{3} - \frac{x_2^{(1)}(t_0) + x_2^{(1)}(t_1) + x_2^{(1)}(t_2)}{3} =$$

$$\frac{20.1 + 22.0 + 18.8}{3} - \frac{22.3 + 23.5 + 21.0}{3} = 20.3 - 22.3 = -2.0.$$

Similarly, we modify the results of the second sensor while keeping the results of the first sensor unchanged. As a result, we get $x_1^{(2)}(t_0) = \tilde{x}_1(t_0) = 20.0$, and similarly $x_1^{(2)}(t_1) = 21.9$ and $x_1^{(2)}(t_2) = 18.7$; we also get $x_2^{(2)}(t_0) = \tilde{x}_2(t_0) + \delta = 22.4 + 0.1 = 22.5$, and similarly $x_2^{(2)}(t_1) = 23.6$ and $x_2^{(2)}(t_2) = 21.1$. For thus modified values, we get

$$y^{(2)} = \frac{x_1^{(2)}(t_0) + x_1^{(2)}(t_1) + x_1^{(2)}(t_2)}{3} - \frac{x_2^{(2)}(t_0) + x_2^{(2)}(t_1) + x_2^{(2)}(t_2)}{3} =$$

$$\frac{20.0 + 21.9 + 18.7}{3} - \frac{22.4 + 23.6 + 21.1}{3} = 20.2 - 22.4 = -2.2.$$

Finally, we estimate the desired bound Δ_s on the systematic component $\Delta_s y$ as

$$\Delta_s = \frac{|y^{(1)} - \tilde{y}|}{\delta} \cdot \Delta_{s1} + \frac{|y^{(2)} - \tilde{y}|}{\delta} \cdot \Delta_{s2} =$$

$$\frac{|(-2.0) - (-2.1)|}{0.1} \cdot 0.1 + \frac{|(-2.2) - (-2.1)|}{0.1} \cdot 0.3 = 1 \cdot 0.1 + 1 \cdot 0.3 = 0.4.$$

Finally, let us consider the problem of propagating the periodic components. By definition, the periodic-induced component $\Delta y_{p\ell}$ of the resulting error Δy is equal to

$$\Delta y_{p\ell} = \sum_i \sum_j C_{ij} \cdot A_{\ell i} \cdot \cos(\omega_\ell \cdot t_{ij} + \varphi_{\ell i}),$$

i.e., to

$$\Delta y_{p\ell} = \sum_i \sum_j C_{ij} \cdot A_{\ell i} \cdot (\cos(\omega_\ell \cdot t_{ij}) \cdot \cos(\varphi_{\ell i}) - \sin(\omega_\ell \cdot t_{ij}) \cdot \sin(\varphi_{\ell i})).$$

By combining the terms corresponding to different j, we conclude that

$$\Delta y_{p\ell} = \sum_i A_{\ell i} \cdot K_{ci} \cdot \cos(\varphi_{\ell i}) + \sum_i A_{\ell i} \cdot K_{si} \cdot \sin(\varphi_{\ell i}),$$

where $K_{ci} \overset{\text{def}}{=} \sum_j C_{ij} \cdot \cos(\omega_\ell \cdot t_{ij})$ and $K_{si} \overset{\text{def}}{=} \sum_j C_{ij} \cdot \sin(\omega_\ell \cdot t_{ij})$.

The values K_{ci} and K_{si} can be explicitly described. Namely:

- One can easily see that if for some small value $\delta > 0$, for this sensor i, we take $\Delta x_i(t_{ij}) = \delta \cdot \cos(\omega_\ell \cdot t_{ij})$ for all j, and for all other sensors i', we take $\Delta x_{i'}(t_{i'j}) = 0$, then the resulting increase in y will be exactly equal to $\delta \cdot K_{ci}$.
- Similarly, if for this sensor i, we take $\Delta x_i(t_{ij}) = \delta \cdot \sin(\omega_\ell \cdot t_{ij})$ for all j, and for all other sensors i', we take $\Delta x_{i'}(t_{i'j}) = 0$, then the resulting increase in y will be exactly equal to $\delta \cdot K_{si}$.

Once we have determined the coefficients K_{ci} and K_{si}, we need to describe the probability distribution of the sum $\Delta y_{p\ell} = \sum_i A_{\ell i} \cdot K_{ci} \cdot \cos(\varphi_{\ell i}) + \sum_i A_{\ell i} \cdot K_{si} \cdot \sin(\varphi_{\ell i})$. We assumed that all φ_i are independent (and uniformly distributed). Thus, for the case of multiple sensors, we can apply the Central Limit Theorem and conclude that the distribution of the sum $\Delta y_{p\ell}$ is close to normal.

In general, normal distribution is uniquely determined by its first two moments: mean and variance (or, equivalently, standard deviation). The mean of each sine and cosine term is 0, so the mean of the sum $\Delta y_{p\ell}$ is zero as well. Since the terms corresponding to different sensors are independent, the variance of the sum is equal to the sum of the variances of individual terms. For each i, the mean of the square

$$(A_{\ell i} \cdot K_{ci} \cdot \cos(\varphi_{\ell i}) + A_{\ell i} \cdot K_{si} \cdot \sin(\varphi_{\ell i}))^2 =$$

$$A_{\ell i}^2 \cdot (K_{ci}^2 \cdot \cos^2(\varphi_{\ell i}) + K_{si}^2 \cdot \sin(\varphi_{\ell i}) + 2 \cdot K_{ci} \cdot K_{si} \cdot \cos(\varphi_{\ell i}) \cdot \sin(\varphi_{\ell i}))$$

is equal to $\dfrac{1}{2} \cdot A_{\ell i}^2 \cdot (K_{ci}^2 + K_{si}^2)$. Thus, the variance of the sum is equal to

$$\frac{1}{2} \cdot \sum_i A_{\ell i}^2 \cdot (K_{ci}^2 + K_{si}^2).$$

Each amplitude $A_{\ell i}$ can take any value from 0 to the known bound $P_{\ell i}$. The above expression monotonically increases with $A_{\ell i}$, and thus, it attains its largest value when $A_{\ell i}$ takes the largest value $P_{\ell i}$. Thus, the largest possible value of the variance is equal to $\dfrac{1}{2} \cdot \sum_i P_{\ell i}^2 \cdot (K_{ci}^2 + K_{si}^2)$.

Thus, we arrive at the following algorithm for computing the upper bound $\sigma_{p\ell}$ of the standard deviation of the periodic-induced component $\Delta y_{p\ell}$ on the approximation error Δy.

Propagating periodic-induced component: algorithm. The upper bound $\sigma_{p\ell}$ on the standard deviation of the periodic-induced component $\Delta y_{p\ell}$ can be obtained as follows:

- First, we apply the algorithm f to the measurement results $\widetilde{x}_i(t_{ij})$ and get the estimate $\widetilde{y}$.
- Then, we select a small value δ and for each sensor i, we do the following:

  - for this sensor i, we take $x_i^{(ci)}(t_{ij}) = \widetilde{x}_i(t_{ij}) + \delta \cdot \cos(\omega_\ell \cdot t_{ij})$ for all moments j;
  - for all other sensors $i' \neq i$, we take $x_{i'}^{(ci)}(t_{i'j}) = \widetilde{x}_i(t_{i'j})$;
  - substitute the resulting values $x_{i'}^{(ci)}(t_{i'j})$ into the data processing algorithm f and get the result $y^{(ci)}$;
  - then, for this sensor i, we take $x_i^{(si)}(t_{ij}) = \widetilde{x}_i(t_{ij}) + \delta \cdot \sin(\omega_\ell \cdot t_{ij})$ for all moments j;
  - for all other sensors $i' \neq i$, we take $x_{i'}^{(si)}(t_{i'j}) = \widetilde{x}_i(t_{i'j})$;
  - substitute the resulting values $x_{i'}^{(si)}(t_{i'j})$ into the data processing algorithm f and get the result $y^{(si)}$.

- Finally, we estimate the desired bound $\sigma_{p\ell}$ as

$$\sigma_{p\ell} = \sqrt{\frac{1}{2} \cdot \sum_i P_{\ell i}^2 \cdot \left(\left(\frac{y^{(ci)} - \widetilde{y}}{\delta} \right)^2 + \left(\frac{y^{(si)} - \widetilde{y}}{\delta} \right)^2 \right)}.$$

Metrological comment. In some practical situations, instead of the *absolute* bound $P_{\ell i}$ on the amplitude of the corresponding periodic error components, practitioners often describe *relative* accuracy $\delta_{\ell i}$ such as 5% or 10%. In this case, a reasonable way to describe the absolute bound is to determine it as $\sigma_i = \delta_i \cdot m_i$, i.e., by multiplying the given value δ_i and the mean square value of the signal

$$m_i = \sqrt{\frac{1}{T_i} \cdot \sum_j (\widetilde{x}_i(t_{ij}))^2}.$$

Example. To test our algorithm, we have applied it to compute the corresponding error component in the problem of estimating carbon and water fluxes described in the paper [14], where such the notion of a periodic error component was first introduced. Our numerical results are comparable with the conclusions of that paper. In the future, we plan to apply all the above algorithms to the results obtained by the sensors on the Jornada Experimental Range Eddy covariance tower and on the nearby robotic tram, and by the affiliated stationary sensors [5, 6, 7, 8, 11, 17].

8 Conclusion

In many application areas, it is necessary to process time series. In this processing, it is necessary to take into account uncertainty with which we know the corresponding values. Traditionally, measurement uncertainty has been classified into systematic and random components. However, for time series, this classification is often not sufficient, especially in the analysis of seasonal meteorological and environmental time series. To describe real-life measurement uncertainty more accurately, researchers have come up with a semi-empirical idea of introducing a new type of measurement uncertainty – that corresponds to periodic errors. In this paper, we provide a mathematical justification for this new error component, and describe efficient algorithms for propagating the resulting three-component uncertainty.

Acknowledgements. This work was supported in part by the National Science Foundation grants HRD-0734825 (Cyber-ShARE Center of Excellence) and DUE-0926721, and by Grant 1 T36 GM078000-01 from the National Institutes of Health.

The authors are greatly thankful to the anonymous referees for valuable suggestions.

References

1. Aubinet, M., Vesala, T., Papale, D. (eds.): Eddy Covariance – A Practical Guide to Measurement and Data Analysis. Springer, Hiedelberg (2012)
2. Cormen, T.H., Leiserson, C.E., Rivest, R.L., Stein, C.: Introduction to Algorithms. MIT Press, Cambridge (2009)
3. Cryer, J.D., Chan, K.-S.: Time Series Analysis. Springer, New York (2010)
4. Garloff, J.: Zur intervallmässigen Durchführung der schnellen Fourier-Transformation. ZAMM 60, T291–T292 (1980)
5. Herrera, J.: A robotic tram system used for understanding the controls of Carbon, water, of energy land-atmosphere exchange at Jornada Experimental Range. Abstracts of the 18th Symposium of the Jornada Basin Long Term Ecological Research Program, Las Cruces, New Mexico, July 15 (2010)
6. Jaimes, A.: Net ecosystem exchanges of Carbon, water and energy in creosote vegetation cover in Jornada Experimental Range. Abstracts of the 18th Symposium of the Jornada Basin Long Term Ecological Research Program, Las Cruces, New Mexico, July 15 (2010)
7. Jaimes, A., Tweedie, C.E., Peters, D.C., Herrera, J., Cody, R.: GIS-tool to optimize site selection for establishing an eddy covariance and robotic tram system at the Jornada Experimental Range. Abstracts of the 18th Symposium of the Jornada Basin Long Term Ecological Research Program, Las Cruces, New Mexico, July 15 (2010)
8. Jaimes, A., Tweedie, C.E., Peters, D.C., Ramirez, G., Brady, J., Gamon, J., Herrera, J., Gonzalez, L.: A new site for measuring multi-scale land-atmosphere Carbon, water and energy exchange at the Jornada Experimental Range. Abstracts of the 18th Symposium of the Jornada Basin Long Term Ecological Research Program, Las Cruces, New Mexico, July 15 (2010)
9. Klir, G.J., Yuan, B.: Fuzzy Sets and Fuzzy Logic. Prentice Hall, Upper Saddle River (1995)

10. Kreinovich, V.: Interval computations and interval-related statistical techniques: tools for estimating uncertainty of the results of data processing and indirect measurements. In: Pavese, F., Forbes, A.B. (eds.) Data Modeling for Metrology and Testing in Measurement Science, pp. 117–145. Birkhauser-Springer, Boston (2009)

11. Laney, C., Cody, R., Gallegos, I., Gamon, J., Gandara, A., Gates, A., Gonzalez, L., Herrera, J., Jaimes, A., Kassan, A., Kreinovich, V., Nebesky, O., Pinheiro da Silva, P., Ramirez, G., Salayandia, L., Tweedie, C.: A cyberinfrastructure for integrating data from an eddy covariance tower, robotic tram system for measuring hyperspectral reflectance, and a network of phenostations and phenocams at a Chihuahuan Desert research site. Abstracts of the FLUXNET and Remote Sensing Open Workshop: Towards Upscaling Flux Information from Towers to the Globe, Berkeley, California, June 7-9, p. 48 (2011)

12. Lee, X., Massman, W., Law, B.: Handbook of Micrometeorology – A Guide for Surface Flux Measurements. Springer, Heidelberg (2011)

13. Liu, G., Kreinovich, V.: Fast convolution and fast Fourier transform under interval and fuzzy uncertainty. Journal of Computer and System Sciences 76(1), 63–76 (2010)

14. Moncrieff, J.B., Malhi, Y., Leuning, R.: The propagation of errors in long-term measurements of land-atmospheric fluxes of carbon and water. Global Change Biology 2, 231–240 (1996)

15. Nguyen, H.T., Walker, E.A.: First Course In Fuzzy Logic. CRC Press, Boca Raton (2006)

16. Rabinovich, S.: Measurement Errors and Uncertainties: Theory and Practice (2005)

17. Ramirez, G.: Quality data in light sensor network in Jornada Experimental Range. Abstracts of the 18th Symposium of the Jornada Basin Long Term Ecological Research Program, Las Cruces, New Mexico, July 15 (2010)

18. Richardson, A.D., et al.: Uncertainty quanitication. In: Aubinet, M., Vesala, T., Papale, D. (eds.) Eddy Covariance – A Practical Guide to Measurement and Data Analysis, pp. 173–209. Springer, Heidelberg (2012)

19. Sheskin, D.J.: Handbook of Parametric and Nonparametric Statistical Procedures. Chapman and Hall/CRC Press, Boca Raton (2011)

20. Shumway, R.H., Stoffer, D.S.: Time Series Analysis and Its Applications. Springer, New York (2010)

Chapter 14
Building a Rough Sets-Based Prediction Model of Tick-Wise Stock Price Fluctuations

Yoshiyuki Matsumoto and Junzo Watada

Abstract. Rough sets enable us to mine knowledge in the form of IF-THEN decision rules from a data repository, a database, a web base, and others. Decision rules are used to reason, estimate, evaluate, and forecast. The objective of this paper is to build the rough sets-based model for analysis of time series data with tick-wise price fluctuations where knowledge granules are mined from the data set of tick-wise price fluctuations. We show how a method based on rough sets helps acquire the knowledge from time-series data. The method enables us to obtain IF-THEN type rules for forecasting stock prices.

1 Introduction

Changes in economic time series data influence corporate profits. Therefore, various prediction analysis methods have been proposed. In methods for analyzing stock prices and currency exchange rates, analyses of the graphical movement of price time series data and fundamental analyses based on the corporate performance and economical environment are widely employed. Matsumoto and Watada applied a chaos-based method to analyze and estimate time series data [11]. The objective of the paper is to mine knowledge in the form of rules from economic time series data based on rough sets theory and to apply the rules to forecast problems.

In rough sets theory [13][14] the concept of discernibility plays a pivotal role in analyzing rough sets. When we deal with objects using finite values of finite attributes, different objects result in their indiscernibility because of similar patterns of the attribute values.

Yoshiyuki Matsumoto
Faculty of Economics, Shimonoseki City University,
2-1-1, Daigaku-cho, Shimonoseki, Yamaguchi 751-8510, Japan
e-mail: matsumoto@shimonoseki-cu.ac.jp

Junzo Watada
Graduate School of Information, Production and Systems, Waseda University,
2-7 Hibikino, Wakamatsu-ku, Kitakyushu, Fukuoka 808-0135, Japan
e-mail: junzow@osb.att.ne.jp

W. Pedrycz & S.-M. Chen (Eds.): Time Series Analysis, Model. & Applications, ISRL 47, pp. 301–329.
DOI: 10.1007/978-3-642-33439-9_14   © Springer-Verlag Berlin Heidelberg 2013

This theory enables us to mine knowledge granules. We can apply the decision rule to reason, estimate, evaluate, and forecast unknown objects. In this paper, the rough sets model is used to analyze time series data of tick-wise price fluctuations where the knowledge granules are mined from the data set of tick-wise price fluctuations.

For example, when we describe a tomato and an apple using the two attributes of color and shape, both have the same values of being red and round. In this case, the tomato and the apple are indiscernible from one another. The rough sets analysis provides a way to mine a minimal number of rules that can discern objects based on discernible relations.

We explain a process of reduction to obtain a minimal number of attributes for discerning among all given objects and the decision rules to discern among all classes when including all objects. Then, employing rough sets theory, we propose a method of analysis and forecasting of economic time series data.

The chapter has the following structure. First, the basic concepts are given in Section 2. The prediction problem of up and down movements of stock price is explained in Section 3. In Section 4, the changes in terms are discussed from the perspective of knowledge acquisition. In Section 5, we build rough sets model based on intraday trading. Finally, we give concluding remarks in Section 6.

2 Rough Set Theory

A rough set is especially useful for domains where the data collected are imprecise and/or incomplete about the domain objects. It provides a powerful tool for a data analysis and data mining of imprecise and ambiguous data. A reduction is the minimal set of attributes that preserves the indispensability relation, that is, the classification power of the original dataset [19]. Rough set theory has many advantages, such as providing efficient algorithms for finding hidden patterns in data, finding minimal sets of data (data reduction), evaluating the significance of data, and generating the minimal sets of decision rules from data. It is easy to understand and to offer a straightforward interpretation of the results [3]. These advantages can simplify analyses, which is why many applications use a rough set approach as their research method. The rough set theory is of fundamental importance in artificial intelligence and cognitive science, especially in the areas of machine learning, knowledge acquisition, decision analysis, knowledge discovery from databases, expert systems, decision support systems, inductive reasoning, and pattern recognition [2][1][9].

Rough set theory has been applied to the management of many issues, including expert systems, empirical study of materials data [18], machine diagnosis [4], travel demand analysis [7], web screen design [6], IRIS data classification [8], business failure prediction, solving linear programs, data mining [21] and α-RST [16]. Another paper discusses the preference-order of the attribute criteria needed to extend the original rough set theory, such as sorting, choice and ranking problems [17], the insurance market [15], and unifying rough set theory with fuzzy theory [5]. Rough set theory provides a simple way to analyze data and reduct information.

2.1 Information Systems

Generally, an information system denoted *IS* is defined as $IS = (U,A)$, where universe U consists of finite objects and is named a universe and A is a finite set of n attributes $\{a_1, a_2, \cdots, a_n\}$. Each attribute $i\{1,2,\cdots,n\}$ belongs to set A, that is, $a \in A$. An object $\omega(\omega \in U)$ has a value $f_a(\omega)$ for each attribute, which is defined as $f_a : U \to V_a$. f_a means that object ω in the U has a value $f_a(\Omega) \subset\in V_a$ for attribute $a \in A$, where V_a is a set of values of attribute $a \in A$. It is called a domain of attribute a.

2.2 Core and Reduct of Attributes

Core and reduct attribute sets, $COR(B)$ and $RED(P)$, respectively, are two fundamental concepts of a rough set. The reduct set is a minimal subset of attributes that can realize the same object classification as the full set of attributes. The core set is common to all reducts [5]. The reduct set of attributes can remove the superfluous and redundant attributes and provide the decision maker simple and clear information. There may be more than one reduct set of attributes. If the set of attributes is dependent, we are interested in identifying all of the possible minimal subsets of attributes that have the same number of elementary sets (called the reducts) [5]. The reduct set of attributes does not affect the decision-making process, and the core attributes are the most important attributes in decision making. If a set of attributes is indispensable, it is called the core. [5]

$$RED(P) \subseteq A, COR(B) = \cap RED(P). \tag{1}$$

2.3 Decision Rules

Decision rules can also be regarded as a set of decision (classification) rules of the form $a_k to d_j$, where a_k means that the attribute a_k has value 1, d_j denotes the decision attributes and the symbol *'to'* denotes propositional implication. In the decision rule $\theta \to \psi$, formulae θ and ψ are called the condition (premise) and decision (conclusion), respectively [10]. For the decision rules we can minimize the set of attributes, reduce the superfluous attributes and classify elements into different groups. In this way we can have many decision rules, where each rule shows meaningful attributes. The stronger rule will encompass more objects and the strength of each decision rule indicates the appropriateness of the rules.

IT-THEN type Decision rules enable us to understand the latent structure decision. This is the most different from many machine learning mechanism. Even we can apply these rules in human decision making after the knowledge acquisition was succeeded.

In this chapter, the decision rules obtained from tick-wise stock price fluctuation data will be applied to forecast the future movement of stock price.

3   Forecast of Up and Down Movements of TOPIX

3.1   A Rough Set Approach to Analyzing Time-Series Data

In this section, a rough set analysis is used to analyze time series data; the focal time
series data and the changes of related data are also analyzed due to the influence on
the focal data. Generally speaking, data treated in a rough set analysis are categorical. In this paper, the change of the value is calculated from its previous single
period value and from two categories: plus and minus are defined by increases and
decreases, respectively. Such categorical data are analyzed by a rough set method.
For instance, when the information of the three past periods is analyzed, let us select
the upward or downward movements from the first to third periods for a conditional
attribute and the present change for a decision attribute. That is, the present change
is decided using the increasing and decreasing movement in the three past periods
as shown in Table 1.

When employing other time series data that may influence the decision attribute,
such time series data are additionally taken as a conditional attribute, and the present
movement is decided depending on these attributes, as shown in Table 2.

Table 1 Only one attribute time-series data

No	Conditional attribute			Decision attribute
	1 period previous	2 period previous	3 period previous	present period
1	+	-	-	-
2	+	-	+	+
3	+	-	-	-
4	+	+	-	+
5	-	-	+	-

Table 2 Including related data

No.	Conditional attribute						Decision attribute
	Target data			Related data			
	1	2	3	1	2	3	
	pp	pp	pp	pp	pp	pp	Present Period
1	+	—	—	—	—	—	—
2	+	+	—	+	+	+	+
3	+	—	—	+	—	—	—
4	+	—	+	—	—	+	+
5	—	+	—	—	+	+	—

PP: previous period

3.2 Analysis of TOPIX

The method described above is employed to analyze the TOPIX time series data. The Dollar-Yen exchange rates, NY Dow-Jones Industrial Average of 30 stocks (DJIA) and NASDAQ Index are used as a related time series rules data. Let us forecast the changes of TOPIX by means of the knowledge acquisition from these changes. The data employed are monthly values from 1995 to 2003. From the first half of the data, 50 samples are used for knowledge acquisition, and from the latter half of the data, 50 samples are used for verifying the model. The increasing and decreasing movements from 6-month periods (half a year) are employed in the knowledge acquisition. That is, these changes of increasing and decreasing movements from the first to sixth periods in the past are taken to obtain the respective conditional attributes. However, the change of the present period is taken to obtain the decision attributes. An analysis was performed for four combinations of the above-mentioned data: 1) TOPIX, 2) TOPIX and Dol-lar-Yen exchange rates, 3) TOPIX and NY Dow-Jones Industrial Average, and 4) TOPIX and NASDAQ index. For the first case, the changes of TOPIX are calculated from the first period to the sixth period, and for the other cases, the changes are calculated, and the conditional attributes are found for the other data and TOPIX.

3.3 Results

The minimal decision rules acquired by means of a rough set analysis are shown in Tables 3 to 10, where $x(\ )$ denotes a change of TOPIX, $y(\ )$ denotes related data such as the Dollar-Yen exchange rate, NY Dow-Jones Industrial Average, and NASDAQ Index. The number in the parenthesis denotes the number of changes in a period.

Table 3 shows that the decision attributes have a $+$ movement after applying the rules obtained from only the TOPIX time series data. Rules with a greater value of C.I. are more reliable. Table 4 shows that decision attributes have a $-$ movement after implementing the rules obtained from only the time series data of TOPIX. Table 5 shows that the decision attributes have a $+$ movement after applying the rules obtained from the time series data of both TOPIX and the Dollar-Yen exchange rate. In this case, the number of obtained rules is large. Therefore, Table 3 illustrates only the top 30 C.I. values. Table 6 shows the top 30 C.I. values from the decision attributes that have a $-$ movement after applying the rules obtained from only time series data of both TOPIX and the Dollar-Yen exchange rates. Table 7 and Table 8 show the rules obtained from the TOPIX and NY Dow-Jones Industrial Average, and Table 9 and Table 10 show the rules obtained from the TOPIX and NASDAQ indices.

Table 11 illustrates the forecasted results based on these rules. Using the three top rules for the C.I. value, the 50 values from the latter half of the data set are forecasted. With respect to the C.I. values of the obtained rules, the rule obtained using the related data is better than the one obtained from only using TOPIX. This result shows that the related data could acquire better rules that cover a wider range.

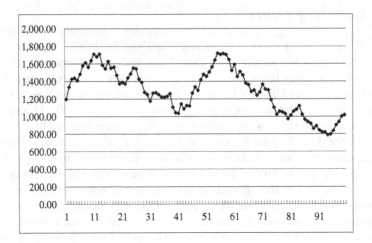

Fig. 1 TOPIX

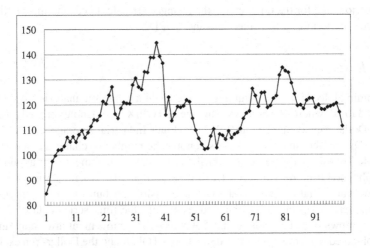

Fig. 2 Dollar-Yen exchange rates

The best C.I. value was observed when using the rule of (−) movement based on TOPIX and the Dollar-Yen Exchange Rate, which covers 40% of the entire range.

With respect the forecasted results using the rules obtained, it is better to use the related data than to use only the TOPIX times series data. Considering the result of all increasing and decreasing movements, the NY Dow-Jones Industrial Average shows the best result in forecasting among all of the combinations. With respect to a single direction, the NASDAQ index in the increasing direction has the most precise forecasting among all of the combinations in both directions, but its decreasing direction produced the worst result.

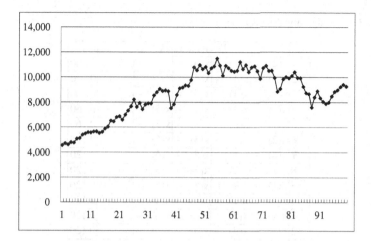

Fig. 3 NY Dow-Johns Industrial Average

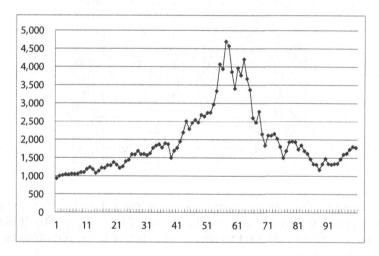

Fig. 4 NASDAQ index

Table 12 shows the results obtained by forecasting using the decision rules acquired from the decision rule analysis. It was frequently observed that the forecasting precision was worse than the result using the 3 rules of the highest C.I. values. Because decision rules with a low C.I. value are employed in forecasting, the forecasting precision should be worse. Nevertheless, the number of objects that fit the obtained decision rules is larger in the case of the decision rule analysis. That is, even though the forecasting precision is worse, the number of forecastable objects increases.

Table 3 Case using only TOPIX (rise)

No.	Decision rules										C.I.
1	x(2)	-	x(3)	-	x(6)	-					0.200
2	x(1)	+	x(4)	+	x(6)	-					0.160
3	x(1)	-	x(4)	-	x(5)	+	x(6)	-			0.160
4	x(1)	-	x(3)	+	x(4)	-	x(5)	+			0.160
5	x(1)	-	x(2)	+	x(4)	-	x(5)	+			0.160
6	x(2)	+	x(3)	-	x(4)	-	x(5)	+			0.120
7	x(3)	-	x(4)	-	x(5)	+	x(6)	-			0.120
8	x(1)	+	x(2)	-	x(3)	+	x(5)	+			0.120
9	x(1)	+	x(2)	-	x(5)	+	x(6)	-			0.120
10	x(1)	-	x(2)	-	x(3)	+	x(4)	-			0.120
11	x(1)	-	x(2)	-	x(4)	-	x(6)	-			0.120
12	x(2)	-	x(3)	+	x(4)	-	x(5)	+			0.120
13	x(1)	-	x(3)	-	x(4)	-	x(6)	-			0.120
14	x(1)	-	x(3)	+	x(4)	-	x(6)	+			0.080
15	x(1)	-	x(2)	+	x(3)	+	x(5)	+			0.080
16	x(1)	-	x(3)	+	x(5)	+	x(6)	+			0.080
17	x(1)	+	x(3)	+	x(4)	+	x(5)	+			0.080
18	x(1)	+	x(2)	-	x(4)	+	x(5)	+			0.080
19	x(2)	-	x(4)	-	x(5)	+	x(6)	-			0.080
20	x(2)	+	x(3)	+	x(4)	+	x(6)	-			0.040
21	x(2)	-	x(3)	+	x(5)	+	x(6)	+			0.040
22	x(1)	-	x(2)	+	x(5)	+	x(6)	+			0.040
23	x(1)	+	x(3)	-	x(5)	+	x(6)	-			0.040
24	x(1)	-	x(2)	-	x(3)	+	x(6)	+			0.040
25	x(1)	+	x(2)	-	x(3)	-	x(4)	-	x(5)	-	0.040

In the case in which we use three rules with higher C.I. values, there are one-third less objects that fit the rules than the number of total objects (50). However, this number is 80% of the total number of objects that fit the rules obtained by decision rule analysis.

3.4 Remarks

In this section, we proposed a method based on a rough set to analyze time series data, TOPIX time series data and forecasted future changes. We employed data related to TOPIX, the Dollar-Yen Exchange Rate, Dow-Jones Industrial Average of 30 stocks and NASDAQ index. For these data, the decision rules are acquired in terms of a rough set theory analysis. By employing rules with higher C.I. values, the related data could obtain better results than TOPIX without any related data.

Table 4 Case using only TOPIX (down)

No.	Decision rules											C.I.
1	x(1)	+	x(3)	-	x(5)	-	x(6)	+				0.160
2	x(1)	-	x(2)	-	x(3)	+	x(4)	+				0.120
3	x(1)	-	x(2)	-	x(3)	-	x(6)	+				0.120
4	x(2)	-	x(3)	-	x(4)	+	x(6)	+				0.120
5	x(1)	+	x(2)	+	x(4)	-	x(5)	-				0.120
6	x(2)	+	x(3)	+	x(4)	-	x(5)	-				0.120
7	x(1)	+	x(2)	+	x(4)	-	x(6)	+				0.120
8	x(1)	+	x(2)	+	x(5)	-	x(6)	+				0.120
9	x(2)	-	x(3)	-	x(5)	-	x(6)	+				0.120
10	x(3)	-	x(4)	+	x(5)	-	x(6)	+				0.120
11	x(2)	+	x(4)	-	x(5)	-	x(6)	-				0.120
12	x(1)	+	x(2)	+	x(3)	+						0.120
13	x(1)	-	x(3)	-	x(5)	+	x(6)	+				0.080
14	x(1)	-	x(2)	-	x(4)	+	x(5)	-				0.080
15	x(2)	-	x(3)	+	x(4)	+	x(5)	-	x(6)	-	0.080
16	x(1)	-	x(3)	-	x(4)	+	x(6)	+				0.040
17	x(2)	+	x(4)	+	x(5)	+	x(6)	-				0.040
18	x(1)	-	x(4)	+	x(5)	+	x(6)	+				0.040
19	x(1)	-	x(2)	-	x(4)	+	x(6)	+				0.040
20	x(2)	-	x(4)	+	x(5)	+	x(6)	+				0.040
21	x(1)	-	x(2)	-	x(5)	-	x(6)	+				0.040
22	x(1)	+	x(3)	+	x(5)	+	x(6)	+				0.040
23	x(1)	-	x(3)	+	x(4)	+	x(5)	+				0.040
24	x(1)	-	x(2)	+	x(4)	+	x(5)	+				0.040
25	x(1)	+	x(3)	+	x(5)	-	x(6)	-				0.040
26	x(1)	-	x(2)	-	x(3)	-	x(4)	-	x(5)	+	0.040

The combination of TOPIX with the Dow-Jones Industrial Average resulted in forecasting with the highest precision. Additionally, we forecasted the present values using rules obtained by decision rule analysis. Even if the forecasting precision was worse than in the case of using three rules with the highest C.I. values, the number of objects that fit the rules is greater than that in the case of using C.I. values. Therefore, it is effective when we forecast data that do not fit rules with high C.I. values. In other words, the forecasting can be mutually compensated if we forecast time series data by using rules with higher C.I. values when the objects fit such rules and by using rules obtained by decision rule analysis otherwise. In this application, we employed two categories of increasing and decreasing movements of time series data. If we utilize additional categories, it may be possible to obtain additional knowledge. Obtaining decision rules that cover entire states should also be analyzed.

Table 5 Case of TOPIX & Dollar-Yen (rise)

No.	Decision rules							C.I.
1	x(3)	-	x(6)	-	y(6)	+		0.280
2	x(1)	+	x(6)	-	y(2)	+		0.240
3	x(6)	-	y(1)	+	y(2)	+	y(4) +	0.240
4	x(5)	+	x(6)	-	y(1)	+	y(3) +	0.240
5	x(5)	+	x(6)	-	y(1)	+	y(4) +	0.240
6	x(4)	-	x(6)	-	y(1)	+	y(2) +	0.240
7	x(6)	-	y(1)	+	y(3)	+	y(6) +	0.240
8	x(1)	-	y(1)	+	y(3)	+	y(6) +	0.240
9	x(1)	-	x(4)	-	x(5)	+	y(1) +	0.240
10	x(4)	-	y(2)	+	y(4)	-		0.240
11	x(3)	-	x(4)	-	y(2)	+		0.240
12	y(2)	+	y(5)	-	y(6)	-		0.200
13	x(2)	-	x(3)	-	x(6)	-		0.200
14	x(6)	-	y(1)	+	y(2)	+	y(6) -	0.200
15	x(1)	-	x(4)	-	y(1)	+	y(2) +	0.200
16	x(2)	-	x(4)	-	y(2)	+	y(3) -	0.200
17	x(4)	-	y(1)	+	y(2)	+	y(3) -	0.200
18	x(2)	-	x(4)	-	x(6)	-	y(2) +	0.200
19	x(5)	-	x(6)	-	y(1)	+	y(2) +	0.200
20	x(3)	-	x(5)	-	x(6)	-	y(2) +	0.200
21	x(3)	-	x(6)	-	y(1)	+	y(2) +	0.200
22	x(6)	-	y(1)	+	y(2)	+	y(3) +	0.200
23	x(3)	-	x(6)	-	y(1)	+	y(5) -	0.200
24	x(3)	-	x(5)	+	y(1)	+	y(3) +	0.200
25	x(4)	-	x(5)	+	x(6)	-	y(1) +	0.200
26	x(4)	-	x(5)	+	x(6)	-	y(2) +	0.200
27	x(2)	+	x(6)	-	y(1)	+	y(2) +	0.200
28	x(3)	+	x(4)	+	y(3)	+	y(6) +	0.200
29	x(4)	+	x(6)	-	y(3)	+	y(6) +	0.200
30	x(6)	-	y(1)	+	y(4)	+	y(6) +	0.200

4 Regression-Based Knowledge Acquisition and Difference among Terms

4.1 Analysis by Means of Regression Line

In general, rough sets analysis deals with categorical data. Therefore, the objective of this section is to obtain a regression line for the time series data to forecast and to employ the increasing and decreasing trends of the regression line using the condition attributes. For example, when we analyze the past six fiscal terms, we obtain the

Table 6 Case of TOPIX & Dollar-Yen (down)

No.	Decision rules								C.I.
1	x(5)	-	y(2)	-					0.400
2	y(2)	-	y(6)	-					0.400
3	x(3)	-	y(5)	+	y(6)	-			0.240
4	x(3)	-	x(6)	+	y(2)	-			0.240
5	y(1)	+	y(2)	-	y(3)	-			0.200
6	x(4)	+	y(3)	-	y(5)	+			0.200
7	x(4)	+	y(5)	+	y(6)	-			0.200
8	x(1)	+	x(4)	-	y(2)	-			0.200
9	x(2)	+	x(5)	-	x(6)	+	y(1)	+	0.200
10	x(3)	-	x(5)	-	x(6)	+	y(1)	+	0.200
11	x(1)	-	x(5)	-	y(5)	+	y(6)	-	0.200
12	x(2)	+	x(5)	-	y(5)	+	y(6)	-	0.200
13	x(4)	-	y(2)	-	y(3)	-			0.160
14	x(6)	+	y(2)	-	y(3)	-			0.160
15	x(4)	+	x(6)	+	y(3)	-			0.160
16	x(1)	+	x(3)	+	y(5)	-			0.160
17	x(3)	+	x(4)	+	y(3)	-			0.160
18	x(1)	-	y(3)	+	y(6)	-			0.160
19	x(3)	-	y(2)	-	y(5)	+			0.160
20	x(1)	+	x(2)	+	y(2)	-			0.160
21	y(1)	-	y(2)	-	y(3)	+			0.160
22	x(1)	+	y(2)	-	y(5)	-			0.160
23	y(2)	-	y(3)	-	y(5)	+			0.160
24	x(1)	+	x(3)	-	y(2)	-			0.160
25	x(5)	-	y(3)	+	y(5)	+	y(6)	-	0.160
26	x(3)	-	x(6)	+	y(3)	-	y(5)	+	0.160
27	x(3)	-	x(6)	+	y(1)	+	y(3)	-	0.160
28	x(4)	+	y(2)	+	y(3)	-	y(6)	+	0.160
29	x(3)	-	x(4)	-	x(6)	+	y(1)	+	0.160
30	x(1)	+	y(4)	+	y(5)	-	y(6)	+	0.160

trends of the regression line for all six terms, the former three terms and the latter three terms. The trend a is determined as follows:

$$a = \frac{\sum\limits_{i=1}^{n}(X_i - \overline{X})(Y_i - \overline{Y})}{\sum\limits_{i=1}^{n}(x_i - \overline{X})^2} \qquad (2)$$

The obtained trend is employed as a condition attribute for the rough sets analysis. That is, with respect to each of the total, former and latter parts, we forecasted whether each part has an increasing or a decreasing trend, depending on such data.

Table 7 Case of TOPIX & DJIA (rise)

No.	Decision rules					
1	x(6) -	y(2) +	y(4) +	y(6) +		0.280
2	x(6) -	y(4) +	y(5) +	y(6) +		0.280
3	x(4) -	x(5) +	y(3) +	y(4) +		0.240
4	x(2) -	x(3) +	y(2) -	y(3) +		0.240
5	x(1) -	x(4) -	x(5) +	y(3) +		0.240
6	x(2) -	x(3) -	x(6) -			0.200
7	x(1) +	x(2) -	y(2) -			0.200
8	x(3) +	x(5) +	y(2) -			0.200
9	x(4) -	x(5) +	y(2) -			0.200
10	x(3) -	x(4) -	x(6) -	y(2) +		0.200
11	x(2) -	x(3) +	x(5) +	y(1) +		0.200
12	x(2) -	x(5) +	x(6) -	y(1) +		0.200
13	x(3) +	x(5) +	y(1) +	y(4) +		0.200
14	x(3) -	x(6) -	y(1) +	y(2) +		0.200
15	x(4) -	x(5) +	y(1) +	y(4) +		0.200
16	x(2) -	x(4) -	y(2) -	y(4) +		0.200
17	x(2) -	x(6) -	y(3) +	y(4) +	y(6) +	0.200
18	x(1) -	x(4) -	x(6) -	y(2) +	y(6) +	0.200
19	x(1) +	x(4) +	x(6) -			0.160
20	x(2) -	x(4) -	x(6) -	y(6) +		0.160
21	x(1) +	x(2) -	x(4) +	y(3) +		0.160
22	x(1) +	x(4) +	y(1) +	y(3) +		0.160
23	x(1) +	x(3) -	x(6) -	y(2) +		0.160
24	x(1) +	x(3) -	x(6) -	y(5) +		0.160
25	x(3) -	x(6) -	y(1) +	y(5) +		0.160
26	x(3) -	x(4) -	x(6) -	y(5) +		0.160
27	x(5) +	x(6) -	y(4) +	y(6) +		0.160
28	x(6) -	y(1) -	y(4) +	y(6) +		0.160
29	x(4) -	x(6) -	y(1) -	y(2) +		0.160
30	x(4) -	y(1) -	y(2) +	y(3) +		0.160

In the case shown in Fig. 5, the condition attribute shows that the total trend indicates +, the former part has a plus trend and the latter part indicates a minus trend. Such values are evaluated with respect to each of the samples; we then mine the knowledge using rough sets analysis. Table 14 illustrates this process. According to this process, we analyzed the trend of each of the past data points and forecast the increasing and decreasing movements of the present term depending on the data.

4.2 Hybrid Method Rough Sets Analysis with Regression

In this research, we employ linear regression to analyze time series data and to obtain rules included in the trends of time series data through rough set analysis. The data analyzed here are time series data sets of the stock price index (TOPIX) of

Table 8 Case of TOPIX & DJIA (down)

No.	Decision rules								C.I.
1	x(6)	+	y(2)	+	y(5)	+			0.360
2	x(5)	-	x(6)	+	y(5)	+			0.320
3	x(3)	-	x(6)	+	y(5)	+	y(6)	+	0.320
4	x(4)	+	x(6)	+	y(5)	+			0.240
5	x(2)	-	x(6)	+	y(2)	+			0.240
6	x(6)	+	y(3)	-	y(5)	+			0.240
7	x(1)	+	x(5)	-	x(6)	+	y(2)	+	0.240
8	x(1)	+	x(6)	+	y(5)	+	y(6)	+	0.240
9	x(5)	-	x(6)	+	y(3)	-			0.200
10	x(4)	-	x(6)	+	y(2)	+	y(6)	+	0.200
11	x(3)	-	x(5)	-	x(6)	+	y(6)	+	0.200
12	x(3)	-	x(6)	+	y(3)	+	y(6)	+	0.200
13	x(1)	+	x(2)	+	x(6)	+	y(5)	+	0.200
14	x(2)	+	x(4)	-	x(5)	-	y(5)	+	0.200
15	x(1)	+	x(4)	-	y(1)	+	y(2)	+	0.200
16	x(2)	+	x(4)	-	x(5)	-	y(6)	+	0.200
17	x(2)	-	x(6)	+	y(5)	+	y(6)	+	0.200
18	x(3)	-	x(4)	-	x(6)	+	y(6)	+	0.200
19	x(2)	-	x(3)	-	x(6)	+	y(6)	+	0.200
20	x(2)	-	x(6)	+	y(3)	-			0.160
21	x(1)	-	x(3)	-	x(6)	+	y(6)	+	0.160
22	x(2)	+	x(3)	-	x(6)	+	y(5)	+	0.160
23	x(1)	+	x(2)	+	x(4)	-	y(1)	+	0.160
24	x(1)	+	x(2)	+	x(5)	-	y(6)	+	0.160
25	x(1)	+	x(2)	+	y(1)	+	y(3)	+	0.160
26	x(1)	+	x(3)	+	x(4)	-	y(2)	+	0.160
27	x(1)	+	x(4)	-	x(6)	+	y(2)	+	0.160
28	x(1)	+	x(6)	+	y(2)	+	y(3)	+	0.160
29	x(1)	+	x(5)	-	y(3)	-	y(6)	+	0.160
30	x(1)	+	y(3)	-	y(5)	+	y(6)	+	0.160

the Tokyo stock exchange. The objective of the paper is to mine the knowledge from obtained trends included in the past data and to apply the knowledge to forecast the present trend. All data employed are divided into three six-year terms (one from 1987 to 1992, a second from 1993 to 1998, and a third from 1999 to 2004). Let us scrutinize the difference of the obtained rules of knowledge among these terms. The data employed in mining are the trend data before one year (twelve months). The data from the past twelve months are analyzed with respect to seven terms, including the whole term, the former and latter six-month terms and four three-month terms. Then, increasing and decreasing trends are analyzed, and these trends are taken as condition attributes, and the rules of knowledge are obtained through a rough set analysis. In this case, the present increasing and decreasing trends are taken as a decision attribute. In this paper, each of the three kinds of terms is employed in data mining, and the obtained rules of knowledge are compared.

Table 9 Case of TOPIX & NASDAQ (rise)

No.	Decision rules					C.I.
1	x(6) -	y(1) +	y(5) +	y(6) +		0.240
2	x(6) -	y(2) +	y(5) +	y(6) +		0.240
3	x(5) +	x(6) -	y(1) +	y(5) +		0.240
4	x(1) +	x(2) -	y(1) +	y(4) +	y(6) +	0.240
5	x(2) -	x(3) -	x(6) -			0.200
6	x(3) -	x(6) -	y(4) -			0.200
7	x(1) -	x(5) +	y(4) -			0.200
8	x(2) -	x(6) -	y(5) +	y(6) +		0.200
9	x(2) -	x(3) +	x(4) -	y(6) +		0.200
10	x(4) +	x(6) -	y(1) +	y(5) +		0.200
11	x(3) -	x(6) -	y(1) +	y(5) +		0.200
12	x(3) -	x(6) -	y(2) +	y(3) +		0.200
13	x(4) -	x(5) +	y(1) +	y(4) +		0.200
14	x(1) +	x(5) +	y(4) +	y(5) +		0.200
15	x(4) -	x(5) +	x(6) -	y(5) +		0.200
16	x(6) -	y(1) -	y(2) +	y(6) +		0.200
17	x(2) -	x(3) +	x(5) +	y(1) +		0.200
18	x(1) +	x(2) -	y(2) +	y(6) +		0.200
19	x(2) -	x(5) +	x(6) +	y(1) +		0.200
20	x(1) +	y(2) -	y(4) +	y(6) +		0.200
21	x(5) +	x(6) -	y(4) -	y(5) +		0.200
22	x(1) -	x(4) -	x(5) +	y(3) +		0.200
23	x(2) -	x(4) -	y(1) +	y(4) +	y(6) +	0.200
24	x(2) -	x(3) +	y(1) +	y(5) +	y(6) +	0.200
25	x(1) +	x(4) +	x(6) -			0.160
26	x(2) -	x(6) -	y(4) -			0.160
27	x(3) -	y(4) -	y(6) -			0.160
28	x(3) -	y(3) +	y(6) -			0.160
29	x(3) -	y(1) +	y(6) -			0.160
30	x(2) -	x(4) -	x(6) -	y(6) +		0.160

4.3　Rules Obtained in Each Term

Let us show the rules obtained in each of three terms in Tables 15 to 20, where we show the cases of increasing and decreasing trends are different for the present term, where + and − denote increasing and decreasing situations, respectively. As the number of rules obtained is large, as shown in Table 21, we showed 10 rules from the top according to the covering index value (C.I.), which shows the rate of the number of covering objects over all of the objects, where the covering objects denotes the same value for the decision attribute. The covering objects indicate that the rule can be applicable.

Table 15 and Table 16 illustrate the rules obtained from 1987-1992. The bubble economy and the crash of the bubble economy are included in the term. Table 15

Table 10 Case of TOPIX & NASDAQ (down)

No.	\multicolumn{10}{Decision rules}										C.I.
1	x(5)	-	x(6)	+	y(5)	+					0.280
2	x(3)	-	x(5)	-	x(6)	+	y(6)	+			0.240
3	x(1)	+	x(5)	-	x(6)	+	y(2)	+			0.240
4	x(1)	+	y(2)	+	y(5)	-					0.200
5	x(2)	-	x(6)	+	y(2)	+					0.200
6	x(4)	-	x(6)	+	y(2)	+	y(6)	+			0.200
7	x(6)	+	y(2)	+	y(3)	+	y(6)	+			0.200
8	x(1)	+	x(6)	+	y(1)	+	y(2)	+			0.200
9	x(2)	+	x(4)	-	x(5)	-	y(5)	+			0.200
10	x(2)	+	x(4)	-	x(5)	-	y(6)	+			0.200
11	x(3)	-	x(6)	+	y(1)	+	y(2)	+	y(6)	+	0.200
12	x(1)	+	x(5)	-	y(2)	+	y(3)	-	y(6)	+	0.200
13	x(1)	+	x(3)	-	x(5)	-	y(3)	-	y(6)	+	0.200
14	x(6)	+	y(4)	-							0.160
15	x(1)	+	y(4)	-	y(6)	+					0.160
16	x(1)	+	x(3)	-	y(5)	-					0.160
17	x(3)	+	x(5)	-	y(6)	-					0.160
18	x(6)	-	y(1)	+	y(5)	-					0.160
19	x(2)	+	x(6)	+	y(3)	+	y(6)	+			0.160
20	x(1)	+	x(4)	-	x(5)	-	y(2)	+			0.160
21	x(1)	+	x(2)	+	x(5)	-	y(4)	+			0.160
22	x(1)	+	x(2)	+	x(5)	-	y(6)	+			0.160
23	x(1)	+	x(2)	+	x(6)	+	y(1)	+			0.160
24	x(1)	+	x(6)	+	y(2)	+	y(3)	+			0.160
25	x(1)	-	x(3)	-	x(6)	+	y(5)	+			0.160
26	x(1)	-	x(3)	-	x(6)	+	y(6)	+			0.160
27	x(2)	+	x(4)	-	x(6)	+	y(6)	+			0.160
28	x(3)	-	x(4)	+	x(6)	+	y(1)	+			0.160
29	x(1)	+	x(3)	-	x(5)	-	x(6)	+			0.160
30	x(3)	-	x(5)	-	x(6)	+	y(3)	-			0.160

Table 11 Forecasted Results

Using only TOPIX (rise)	47.4%
Using only TOPIX (down)	58.3%
Using TOPIX & Dollar-yen (rise)	51.6%
Using TOPIX & Dollar-yen (down)	56.0%
Using TOPIX & DJIA (rise)	80.0%
Using TOPIX & DJIA (down)	53.8%
Using TOPIX & NASDAQ (rise)	91.7%
Using TOPIX & NASDAQ (down)	33.3%

illustrates that the present term shows an increase when the entire past year shows an increasing TOPIX trend. However, each quarter showed a decreasing trend. When the entire term shows an increasing trend, although each quarter partly shows a

Table 12 Forecasted Results(Decision Rule Analysis method)

Using only TOPIX (rise)	45.5%
Using only TOPIX (down)	43.3%
Using TOPIX & Dollar-yen (rise)	52.6%
Using TOPIX & Dollar-yen (down)	50.0%
Using TOPIX & DJIA (rise)	70.0%
Using TOPIX & DJIA (down)	39.0%
Using TOPIX & NASDAQ (rise)	65.6%
Using TOPIX & NASDAQ (down)	41.0%

Table 13 Conform Rate

Using only Top 3 rules	32.3%
Using Decision Rule Analysis method	77.3%

Table 14 Trends of data

No.	Condition Attribute (Trend)			Decision Attribute
	Total Term	Former Term	Latter Term	Present Term
1	+	+	—	—
2	+	—	+	+
3	+	—	—	—
4	+	—	—	+
5	—	—	+	—

decreasing one, in such cases, it is highly possible that the present term increases the TOPIX. Table 16 shows that the fourth quarter has a more pronounced decreasing trend, while the second and third quarters have more increasing trends. Therefore, the fourth quarter changes from an increasing trend into a decreasing trend, and there is a continuous decreasing trend starting from the fourth quarter.

Table 17 and Table 18 illustrate the rules obtained in the term from 1993 to 1998. During this term, after the bubble economy burst, that is, after the asset-inflated economy collapsed, the stock prices depressed. The rules obtained from this term show a higher C.I. than those from the other terms and a larger volume. It can be emphasized that the rules obtained in this term show more reliability than those from other terms. Table 18 depicts a general decreasing trend (over the past one year), although no other evidence is apparent. There are few rules with the same condition attributes. Therefore, the rules obtained show a large variety.

Table 19 and Table 20 illustrate the rules obtained in the term from 1999 to 2004. The term includes the IT bubble economy. Table 19 indicates increasing trends in general (in the past one year), which also increase in the latter part and in the third quarter, which means that, when an increasing trend is observed during this half-year, the situation will continue to the present term. Table 20 illustrates that both the former and latter halves of the year are increasing, and the fourth quarter is shown

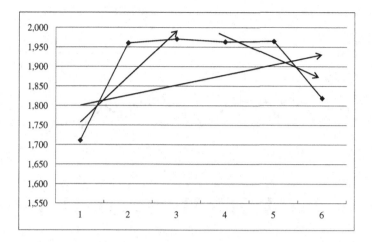

Fig. 5 Trends by Linear Regression

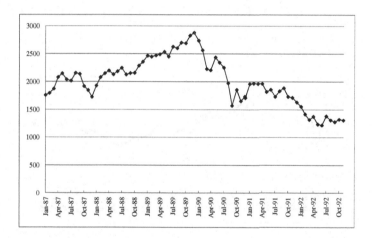

Fig. 6 TOPIX monthly data from 1987 to 1992

as decreasing. Even if the increasing situation appeared in the past year, in the case in which the portion immediately prior is shown to be decreasing, there is a high possibility that the present term continues to decrease.

4.4 Comparison of Each Term

Tables 22 to 27 illustrate the comparison of the obtained rules using a rough sets analysis. The rules of the decision attribute obtained in each of three different types of terms forecast the increasing and decreasing movements in the present term.

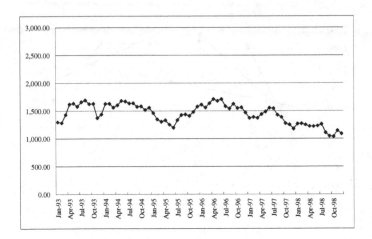

Fig. 7 TOPIX monthly data from 1993 to 1998

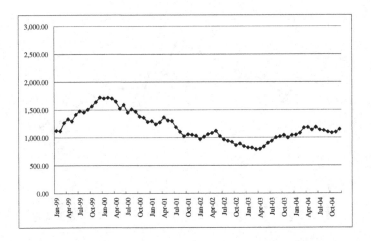

Fig. 8 TOPIX monthly data from 1999 to 2004

Carefully, we verified whether the terms have the same rules as those for different terms. That is, we checked whether the rules obtained in a term can be applicable to another terms. "Increasing/decreasing" denotes that the decision attribute illustrates the increasing and decreasing nature of the present term, respectively. The decision attribute shows a similar movement in the compared term, that is, increasing and increasing or decreasing and decreasing. However, the opposite situation means the decision attribute shows the other movement to be in the opposite direction to the compared term, that is, increasing and decreasing or decreasing and increasing. The number of rules illustrates the ratio of the corresponding rules over all the rules. "C.I." indicates the total C.I. of the corresponding rules.

Table 15 Rules (increasing state) in 1987-1992

No.	Whole	Former	Latter	First Quarter	Second Quarter	Third Quarter	Fourth Quarter	C.I.
1		-	+	+		+		0.111
2	+				-		-	0.083
3		+		-		-		0.083
4	+	-			-			0.083
5	+	-		+				0.083
6			+		-		-	0.083
7	+				-	-		0.083
8	+	-	+				-	0.083
9		+			-	-	+	0.083
10	+		+	-			-	0.083

Table 16 Rules (decreasing state) in 1987-1992

No.	Whole	Former	Latter	First Quarter	Second Quarter	Third Quarter	Fourth Quarter	C.I.
1				+	+	+	-	0.111
2			-		+	+	-	0.111
3	+			-	+		+	0.083
4				-	+	+	+	0.083
5		-		-	+	+		0.083
6	-	+				-	-	0.083
7		+			-	-	-	0.083
8		-		-		+		0.056
9			+	+	+		-	0.056
10		+	+	+			-	0.056

Table 22 and Table 23 compare the obtained rules in the term from 1987 to 1992. These tables explain that the knowledge obtained in the term from 1987 to 1992 cannot be applicable to the term from 1993 to 1998. Although the C.I. of the rules with the same decision attribute is approximately 10%, the C.I. of the rules with the opposite direction movement is approximately 20% to 30%. In the term from 1987 to 1992, the rules of increasing in the present term will be changed to the rules of decreasing trends. The term from 1987 to 1992 includes the bubble economy and the collapse of the bubble economy. It is clear that knowledge obtained from this term will not be applicable to the term from 1993 to 1998. Compared with the term from 1999 to 2004, the C.I. of the rules of the decision attribute with the same decreasing trend is greater than those with a different trend. However, with respect to the rules of increasing trends, the C.I. of the rules with the opposite trends is greater than those with the same trend, demonstrating that the obtained knowledge is not applicable. However, it should be noted that the knowledge in this term is

Table 17 Rules (increasing state) in 1993-1998

No.	Whole	Former	Latter	First Quarter	Second Quarter	Third Quarter	Fourth Quarter	C.I.
1			-	+	+	-		0.171
2	+				-	+	+	0.114
3		+		+	-	+	-	0.114
4		+	+	+			-	0.086
5	+			+	+		-	0.086
6	+			-	-	+		0.086
7				+	+	-	-	0.086
8	+				+	-	-	0.086
9	+		-			+	-	0.086
10	+		-		+	-		0.086

Table 18 Rules (decreasing state) in 1993-1998

No.	Whole	Former	Latter	First Quarter	Second Quarter	Third Quarter	Fourth Quarter	C.I.
1				+	-		+	0.189
2				+		+	+	0.189
3		-	+	+				0.162
4		-		+		+		0.135
5	-	+	-	-				0.135
6	-		+	+			+	0.108
7		+	-	-		-		0.108
8	-				+	-	-	0.108
9			-		+	-	-	0.108
10	-	+	-			-	-	0.108

better than that obtained from the 1993 to 1998 term because the 1999 to 2004 term includes the IT bubble economy, and therefore, the knowledge is coincident with the term of the bubble economy.

Table 24 and Table 25 show the comparison of the rules obtained from the 1993 to 1998 term. The comparison with the term from 1987 to 1992 shows that the increase resulting from the same movement is much greater than that from the opposite movement. The obtained knowledge in the term shows workability. However, the decrease of the rules of the opposite movement shows more incident, that is, the knowledge obtained is not workable. As the term from 1987 to 1992 includes a great portion of the bubble economy, it is easy to identify increasing movement but difficult to forecast decreasing movement. The comparison of the term from 1999 to 2004 shows that the increasing movement has a larger C.I. for opposite movement, but the decreasing movement has a larger C.I. for the same movement. Because the

Table 19 Rules (increasing state) in 1999-2004

No.	Whole	Former	Latter	First Quarter	Second Quarter	Third Quarter	Fourth Quarter	C.I.
1				-	+	+		0.167
2	+		+	-	+			0.167
3	+			-	+		+	0.139
4	+	-			+			0.083
5		-			+	+		0.083
6		+	+	-	+			0.083
7	-					+	+	0.056
8	+				-		-	0.056
9	+	+		+	-			0.056
10			+	+	-	+		0.056

Table 20 Rules (decreasing state) in 1999-2004

No.	Whole	Former	Latter	First Quarter	Second Quarter	Third Quarter	Fourth Quarter	C.I.
1		+			+	+	-	0.139
2				+	+	+	-	0.139
3	-	+			-			0.111
4	-		+				-	0.083
5	-	+		+				0.083
6			+	+	+		-	0.083
7	+	+			+		-	0.083
8	+			-	-			0.056
9	+				-		+	0.056
10		+				-	-	0.056

Table 21 Number of rules in each term

Term	Up/Down	No. of Rules
1987-1992	Up	46
	Down	26
1993-1998	Up	52
	Down	35
1999-2004	Up	33
	Down	37

term from 1999 to 2004 includes the period after the bubble economy burst, it is easier to forecast decreasing movements.

Table 26 and Table 27 illustrate the comparison of the obtained rules in the term from 1999 to 2004. The comparison with the term from 1987 to 1992 illustrates that

Table 22 Comparison of the term from 1987 to 1992 (increasing)

Term	Increasing/decreasing	Number of rules	C.I.
1993-1998	The same increasing	6.5%	7.89%
	Contrary decreasing	26.1%	27.63%
1999-2004	The same increasing	23.9%	28.95%
	Contrary decreasing	30.4%	27.63%

Table 23 Comparison of the term from 1987 to 1992 (decreasing)

Term	Increasing/decreasing	Number of rules	C.I.
1993-1998	The same decreasing	15.4%	8.70%
	Contrary increasing	23.1%	21.74%
1999-2004	The same decreasing	38.5%	41.30%
	Contrary increasing	23.1%	28.26%

Table 24 Comparison of the term from 1993 to 1998 (increasing)

Term	Increasing/decreasing	Number of rules	C.I.
1987-1992	The same increasing	38.5%	32.69%
	Contrary decreasing	11.5%	10.58%
1999-2004	The same increasing	28.8%	24.04%
	Contrary decreasing	38.5%	34.62%

Table 25 Comparison with the term from 1993 to 1998 (decreasing)

Term	Increasing/decreasing	Number of rules	C.I.
1987-1992	The same decreasing	22.9%	10.47%
	Contrary increasing	40.0%	29.07%
1999-2004	The same decreasing	42.9%	27.91%
	Contrary increasing	8.6%	6.98%

Table 26 Comparison with the term from 1999 to 2004 (increasing)

Term	Increasing/decreasing	Number of rules	C.I.
1987-1992	The same increasing	30.3%	19.30%
	Contrary decreasing	3.0%	8.77%
1993-1998	The same increasing	6.1%	3.51%
	Contrary decreasing	12.1%	7.02%

knowledge of increasing movement is more workable, but knowledge of decreasing movement shows an opposite movement and is unhelpful. It is the same reason as for the term from 1993 to 1998. We can understand the increasing movement in

Table 27 Comparison with the term from 1999 to 2004 (decreasing)

Term	Increasing/decreasing	Number of rules	C.I.
1987-1992	The same decreasing	13.5%	16.90%
	Contrary increasing	35.1%	25.35
1993-1998	The same decreasing	24.3%	19.72%
	Contrary increasing	10.8%	8.45%

the bubble economy is easy to forecast, whereas the decreasing movement is not. The comparison with the term from 1993 to 1998 does not show that the increasing movement has a higher C.I., but the decreasing movement indicates an appropriate C.I. to some extent.

4.5　Remarks

In this paper, we analyzed economical time series data using a rough sets model, and TOPIX time series data were employed to forecast future trends. To obtain forecasting knowledge, we divided all of the years into three terms: 1987 to 1992, 1993 to 1998 and 1999 to 2004. Then, we acquired the rules of knowledge for each term and compared them among the different terms. The result of each term is different depending on the term. One reason is because the terms included notable economical states, such as a bubble economy, economy recession and IT bubble economy, which influenced stock prices. Nevertheless, it shows the difficulty of mining unified knowledge to forecast stock price movements. However, even if the knowledge is obtained from a different term, it showed that such knowledge can still be employed in forecasting. It should be emphasized from this fact that we can separate the knowledge into permanent or general knowledge, which can be applicable to any term, and the specific or limited knowledge, which can be employed to only a special term.

5　The Rough Sets Model Based on Intraday Trading Data

5.1　Intraday Trading Data

The intraday trading data are the record of all of the trading transactions that occur in the market for all of the stocks and the yen and dollar currency trading in the exchange market. The intraday trading data of stocks include the stock code, traded market, date, traded time, and traded amount in units of minutes and can be employed to various stock analyses. 28 exemplifies the intraday trading data of stocks, which included the trading change records of Fuji Heavy Industry Stock Prices on June 2, 2008. It shows the records containing all of the traded transactions in units of minutes in the market The stock price does not change in each transaction. The five transactions occurred between 9:10 and 9:11, but the stock price was fixed at 495

Table 28 Intratrading data

Date	Stock Number	Time	Price	Amount
20080602	72700	909	497	1000
20080602	72700	909	496	3000
20080602	72700	910	495	11000
20080602	72700	910	495	7000
20080602	72700	911	495	5000
20080602	72700	911	495	4000
20080602	72700	911	495	1000

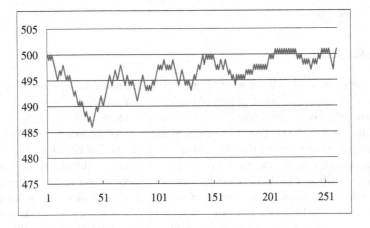

Fig. 9 Intraday trading without duplications

JPY. It seems very reasonable to consider these transactions as one trade. Therefore, in this paper, we considered same continuous price transactions as a single transaction. Fig. 9 illustrates the data after removing duplicated transactions with the same price.

5.2 Intraday Trading Analysis by the Rough Sets Model

The objective of this section is to mine the knowledge from intraday trading transaction data of stocks using a rough sets analysis. The employed data are intraday trading data without any duplicated trading but with the same continuous price shown in Fig. 9. The conditional attributes express how many times intraday trading occurred prior to the focal stock timing. Based on the frequency of changes, we mined the trading knowledge. For example, when we identified an increase 6 times and a decrease 2 times out of the last 8 trading transactions, the change frequency was denoted as +4. When we identified an increase 3 times and a decrease 5 times out of the last 8 trading transactions, the change frequency was denoted as -2. The forecasted value (decision attribute) estimates how many times the future price will increase or decrease. Similar to the past change, we intend to evaluate how many times the price will increase or decrease.

Table 29 and Table 30 show the data employed in the analysis. For example, let us consider the past change frequency of a stock shown in Table 29; the present is denoted No. 7 with a price of 494 JPY, and previously, No. 6 had a price of 495 JPY. The present price decreased from the 1 time previous price. Then, the conditional attribute shows that the "1 time previous price" is "-1." With respect to No. 5 with a price of 496 JPY, as the prices at No. 6 and at No. 7 decreased continuously, the conditional attribute "2 times previous price" is "-2." In the same way, we calculated how many times the change occurred for "4 times previous change," "8 times previous change" and "16 times previous change." In terms of these conditional attributes, we acquired knowledge on future intratrading changes. We take this value for a decision attribute and mine the forecasting knowledge. The objective of this

Table 29 The past change frequency

No.	Price	Previous times				
		1	2	4	8	16
1	496	-1	-2	0	0	-4
2	495	-1	-2	-2	0	-4
3	496	1	0	-2	0	-4
4	495	-1	0	-2	-2	-4
5	496	1	0	0	0	-4
6	495	-1	0	0	-2	-4
7	494	-1	-2	-2	-4	-4
8	493	-1	-2	-2	-4	-4
9	492	-1	-2	-4	-4	-4
10	493	1	0	-2	-2	-4

Table 30 Future expected change frequency

No	Price	Times in future		
		1 time	2 times	4 times
1	496	-1	0	0
2	495	1	0	0
3	496	-1	0	-2
4	495	1	0	-2
5	496	-1	-2	-4
6	495	-1	-2	-2
7	494	-1	-2	-2
8	493	-1	0	-2
9	492	1	0	-2
10	493	-1	-2	-2

paper is to retrieve the rules on future change frequency based on past experiences. The future expected change frequency can be evaluated in the same way. For example, let the present state be at No. 5. The prices at No. 5 and at No. 6 are 496 JPY and 495 JPY, respectively. From No. 5, the 1 time future change frequency is "-1." In the same way, as the prices from No.7 to No. 9 decreased, the 2 times future intraday trading change frequency is "-2" and the 4 times future change is "-4." Let us use this value as a decision attribute and mine the knowledge on forecasting. The objective of this paper is to retrieve the knowledge about future change frequency based on past experiences.

5.3   Knowledge Acquisition by Minimal Decision Rule

We employed intraday trading data of the stock price of Fuji Heavy Industry Corporation for one month in June 2008. We used the past change frequency of the stock price as the conditional attributes and the future change frequency of the stock price as a decision attribute. Then, we applied the rough sets analysis to obtain the minimal decision rules. In this research, we acquired rules of the stock prices increasing 4 times continuously or decreasing. Table 31 to Table 34 show some of the obtained rules. Table 31 and Table 32 show the rules obtained for "+4" in the 4 times future change, which indicated that the stock prices increased 4 times continuously. Table 31 shows a necessity rule, and the cases that satisfied the rule in which the 4 times future change continuously occurred. Table 32 shows a possibility rule, and the cases that satisfied the rule in which the 4 times future change can continuously occur. Table 32 shows only the top seven rules with the largest covering rates.

Table 31 indicates that in June 2008, almost all of the changes were decreasing. The rules in No. 1 show that the price decreased in 10 of 16 total cases. The rules in No. 2 illustrate that 8 cases decreased. We can observe that the nearest values did not change and that after the stock price continuously decreased, it continuously increased. In Table 32, we find that the decreasing trend of a stock price is major in the past. In the No. 3 and No. 4 cases, there are increasing trends. Even if we have an increasing change in the past, it is possible that a stock price will continuously increase. In the same way, Table 33 shows the necessity rules of 4 times continuously decreasing, and Table 34 illustrates the possibility rules of 4 times continuously decreasing. Table 34 indicates the top 5 rules with the largest covering rates. In Table 33, the increasing trends prevail. The latest trading transactions show no movement of the stock price or only a slight decrease. We can understand that after stock prices continuously increase, the change in stock prices stops for a while and then decreases. In Table 34, the past increasing trend is significant. Nevertheless, as No.1 shows, there is no change in the stock price, and as No. 2 indicates, the change in the stock price can continuously decrease even if it was decreasing in the past.

Table 31 Rules of increase (Necessity Rules)

No.	Conditional Attribute(n times previous intraday trading)	Coverage (%)
1	(2 times = 0) (16 times = -10)	0.75%
2	(8 times = 0) (16 times = -8)	0.75%
3	(1 time = 1) (2 times = 0) (4 times = 0) (8 times = -4) (16 times = 0)	0.75%

Table 32 Rules of increase (Possibility Rules)

No.	Conditional Attribute(n times previous intraday trading)	Coverage (%)
1	(1 time = -1) (4 times = -2) (8 times = -2) (16 times = -4)	15.73%
2	(1 time = -1) (4 times = -2) (8 times = -4) (16 times = -4)	9.51%
3	(2 times= 0) (4 times = 2) (8 times = 2) (16 times = 4)	9.42%
4	(1 time = -1) (2 times = 0) (4 times = 0) (8 times = 2) (16 times = 4)	7.16%
5	(2 times = -2) (8 times = -4) (16 times = -6)	5.56%

Table 33 Rules of decrease (Necessity Rules)

No.	Conditional Attribute(n times previous intraday trading)	Coverage (%)
1	(2 times = 0) (4 times = -2) (16 times = 10)	0.78%
2	(4 times = -2) (16 times = 8)	0.78%

Table 34 Rules of decrease (Possibility Rules)

No.	Conditional Attribute(n times previous intraday trading)	Coverage (%)
1	(1 times = 1) (2 times = 0) (4 times = 0) (8 times = 0) (16 times = -2)	20.67%
2	(2 times = 0) (4 times = 0) (8 times = -2) (16 times = -4)	13.43%
3	(1 times = 1) (2 times = 0) (4 times = 0) (8 times = 2) (16 times = 4)	8.30%
4	(2 times = 0) (4 times = 2) (8 times = 2) (16 times = 0)	7.49%
5	(2 times = 2) (4 times = 2) (8 times = 2) (16 times = 4)	7.08%

6 Conclusions

In this research, we investigated whether we can forecast the future change frequency of intraday trading transactions based on the past change frequency of intraday trading transactions. We used the stock price of Fuji Heavy Industry Corporation for one month in June 2008. Using a rough sets analysis, we mine minimal decision rules on both future and past changes. The results showed that it can successfully forecast the future change of the intraday trading transaction based on the past. The rules obtained in the experiment illustrate that the future change has a decreasing trend when the past trend was increasing, and the future change has an increasing trend when the past trend was decreasing.

This chapter explained the rough set method to acquire the forecasting knowledge from time-series data. The emphasis should be placed on that the rough set method enables us to obtain IF-THEN type rules for forecasting stock prices. On the other hand, when we mind any knowledge by means of multivariate analysis and neural networks, only black-box type knowledge is obtained. It is not easy to understand the latent structure of the knowledge. In the rough set method, not only a system can forecast using decision rules, but also human employs the decision rules to forecast.

References

1. Azibi, R., Vanderpooten, D.: Construction of rule-based assignment models. European Journal of Operational Research 138(2), 274–293 (2002)
2. Beynon, M.J., Peel, M.J.: Variable precision rough set theory and data discrimination: an application to corporate failure prediction. Omega 29(6), 561–576 (2001)
3. Goh, C., Law, R.: Incorporation the rough sets theory. Chemometrics and Intelligent Laboratory Systems 47(1), 1–16 (2003)
4. Greco, S., Matarazzo, B., Slowinski, R.: Rough sets theory for multi-criteria decision analysis. European Journal of Operational Research 129(1), 1–47 (2001)
5. Gronhaug, K., Gilly, M.C.: A transaction cost approach to consumer dissatisfaction and complaint action. Journal of Economic Psychology 12(1), 165–183 (1991)
6. Harada, T., Tanaka, R.: Analysis of Specifications for Web Screen-Design Using Rough Sets. Journal of Advanced Computational Intelligence and Intelligent Informatics 10(5), 688–694 (2006)
7. Jhieh, Y., Tzeng, G., Wang, F.: Rough set Theory in Analyzing the Attributes of Combination Values for insurance market. Expert System with Applications 32(1) (2007)
8. Kim, D., Bang, S.Y.: IRIS Data Classification Using Tolerant Rough Sets. Journal of Advanced Computational Intelligence and Intelligent Informatics 4(5) (2000)
9. Li, R., Wang, Z.O.: Mining classification rules using rough set and neural networks. European Journal of Operational Research 157(2), 439–448 (2004)
10. Lin, C., Watada, J., Tzeng, G.: Rough sets theory and its application to management engineering. In: Proceedings, International Symposium of Management Engineering, Kitakyushu, Japan, pp. 170–176 (2008)
11. Matsumoto, Y., Watada, J.: Improvement of Chaotic Short-term Forecasting on Fuzzy Reasoning and Tuning on Genetic Algorithm. Japan Society Journal of Fuzzy Theory and Intelligent Informatics 16(1), 44–52 (2004) (in Japanese)
12. Mori, N., Tanaka, H., Inoue, K.: Rough sets and Kansei: Knowledge acquisition and reasoning from Kansei data (2004) (in Japanese)
13. Pawlak, Z.: Rough Sets. International Journal of Computer and Information Science 11(5), 341–356 (1982)
14. Pawlak, Z.: Rough Sets - Theoretical Aspects of Reasoning about Data. Kluwer Academic Publishers (1991)
15. Pawlak, Z.: Rough classification. International Journal of Human-Computer Studies 51(15), 369–383 (1999)
16. Prędki, B., Słowiński, R., Stefanowski, J., Susmaga, R., Wilk, S.: ROSE - Software Implementation of the Rough Set Theory. In: Polkowski, L., Skowron, A. (eds.) RSCTC 1998. LNCS (LNAI), vol. 1424, pp. 605–608. Springer, Heidelberg (1998)

17. Prędki, B., Wilk, S.: Rough set Based Data Exploration Using ROSE System. In: Raś, Z.W., Skowron, A. (eds.) ISMIS 1999. LNCS (LNAI), vol. 1609, pp. 172–180. Springer, Heidelberg (1999)
18. Quafafou, M.: α-RST: a generalization of rough set theory. Information Sciences 124(4), 301–316 (2000)
19. Tan, S., Cheng, X., Xu, H.: An efficient global optimization approach for rough set based dimensionality reduction. International Journal of Innovative Computing, Information and Control 3(3), 725–736 (2007)
20. Tanaka, H., Tsumoto, S.: Rough sets and Expert System. Mathematical Sciences 378, 76–83 (1994) (in Japanese)
21. Walczak, B., Massart, D.L.: Rough set theory. Chemometrics and Intelligent Laboratory 47(1), 1–16 (1999)

Chapter 15
A Best-Match Forecasting Model
for High-Order Fuzzy Time Series

Yi-Chung Cheng and Sheng-Tun Li

Abstract. An area of Fuzzy time series has attracted increasing interest in the past decade since Song and Chissom's pioneering work and Chen's milestone study. Various enhancements and generalizations have been subsequently proposed, including high-order fuzzy time series. One of the key steps in the Chen's framework is to derive fuzzy relationships existing in a fuzzy time series and to encode the relationships as IF-THEN production rules. A generic exact-match strategy is then applied to the forecasting process. However, the uncertainty and fuzziness characteristics inherent to the fuzzy relationships tend to be overlooked due to the nature of the matching strategies. This omission could lead to inferior forecasting outcomes, particularly in the case of high-order fuzzy time series. In this study, to overcome this shortcoming we propose a best-match strategy forecasting method based on the fuzzy similarity measure. The experiments concerning Taiwan Weighted Stock Index and Dow Jones Industrial Average are reported. We show the effectiveness of the model by running some comparative analysis using some models well-known in the literature.

Keywords: fuzzy time series, forecasting, high-order model, best-match, fuzzy similarity measure.

Yi-Chung Cheng
Department of International Business Management
Tainan University of Technology, Taiwan, R.O.C.
e-mail: t20042@mail.tut.edu.tw

Sheng-Tun Li
Department of Industrial and Information Management &
Institute of Information Management
National Cheng Kung University, Taiwan, R.O.C.
e-mail: stli@mail.ncku.edu.tw

W. Pedrycz & S.-M. Chen (Eds.): Time Series Analysis, Model. & Applications, ISRL 47, pp. 331–345.
DOI: 10.1007/ 978-3-642-33439-9_15   © Springer-Verlag Berlin Heidelberg 2013

1 Introduction

Song and Chissom (1993a, 1993b) were the first who proposed the definition of fuzzy time series and the framework of fuzzy time series forecasting which uses fuzzy relations to represent the temporal relationships in a fuzzy time series. The forecasting framework includes four steps: (1) determine and partition the universe of discourse into intervals, (2) define fuzzy sets from the universe of discourse and fuzzify the real time series, (3) derive fuzzy relationships existing in the fuzzified time series, (4) forecast and defuzzify the forecasting outputs. In order to alleviate the computational burden, Chen (1996) first constructed first-order fuzzy logic relationship as 'IF-THEN' rule group to modify the step (3). Various enhancing works based on IF-THEN rule relationship have been subsequently proposed. Chen (2002) extended his previous work to a high-order model. Chen and Hsu (2004) proposed first-order time-variant forecasting model. Own and Yu (2005) further modified Chen's model as a heuristic high-order fuzzy time series model, which depends strongly on the trend of fuzzy time series. Lee et al. (2006) extended Chen's (2002) model to allow two-variable forecasting. Lee et al. (2007) constructed two-factor high-order forecasting model based on fuzzy logical relationships and genetic algorithms. Li and Cheng (2007) proposed a backtracking scheme to construct a fuzzy logic rule base which solves the problem of determination of a suitable order. The work Huarng et al. (2007) developed a multivariate heuristic function to improve the forecasting accuracy, in which the heuristic is based on fuzzy logical relationships. Li and Cheng (2008) applied Fuzzy C-Means to enhance their proposed deterministic forecasting model whereas Li and Cheng (2009) modified the backtracking algorithm to improve the forecasting accuracy, and extended it to forecasting long-term forecasting (Li et. al, 2010). Chen and Wang (2010) presented a method of new fuzzy-trend logical relationship groups on fuzzy time series forecasting, and obtains higher average forecasting accuracy rate. Li et al. (2011) extended Li and Cheng's previous work from single fuzzy set forecasting to a fuzzy vector forecasting. Chen and Chen (2011) proposed models of rule base and variation scheme to enhance forecasting accuracy.

The aforementioned forecasting models constructed the relationships in a fuzzy time series built on the basis of IF-THEN fuzzy logical rule and then these high-order models all applied a basic exact-match strategy when forecasting. However, such an exact-match high-order forecasting model comes with the two major shortcomings. First it is a so-called rule redundancy in that the rules could be less useful when either the number of fuzzy sets or the order increases. The low hit ratio of the rules indicates that the fuzzy logical relationships are likely redundant (Li and Cheng, 2008). Secondly, the uncertainty and fuzziness characteristics inherent in the fuzzy relationships are ignored as using exact-match at forecasting (Pappis and Karacapilidis, 1993).

To tackle the aforementioned shortcomings, in this study, we propose a best-match forecasting model for high-order fuzzy time series which applies a similarity measure to the forecasting step. We demonstrate how the proposed forecasting model works by conducting experiments with the up and down values of Taiwan Weighted Stock Index (TWSI) and Dow Jones Industrial Average (INDU).

2   Fuzzy Time Series and Similarity Measurement

Song and Chissom (1993b) defined fuzzy time series as follows:

Definition 1. Let $Y(t)$ $(t = ..., 0, 1, 2, ...)$, a subset of R, be the universe of discourse on which fuzzy sets $\tilde{A}_i$ $(i = 1, 2, ...)$ are defined and let $F(t)$ be a collection of $\tilde{A}_i$. Then, $F(t)$ is called a fuzzy time series on $Y(t)$ $(t = ..., 0, 1, 2, ...)$.

Song and Chissom employed fuzzy first-order (Definition 2) and mth-order (Definition 3) relational equations to develop their forecasting model under the assumption that the observations at time t are dependent only upon the accumulated results of the observations at previous times, which is defined as follows:

Definition 2. Suppose $F(t)$ is caused by $F(t-1)$ or $F(t-2)$ or ... or $F(t-m)(m>0)$ only. The relation can be expressed as the following fuzzy relational equation:

$$F(t) = F(t-1) \circ R(t, t-1) \text{ or } F(t) = F(t-2) \circ R(t, t-2) \text{ or } ... \text{ or }$$
$$F(t) = F(t-m) \circ R(t, t-m)$$

or

$$F(t) = \left(F(t-1) \cup F(t-2) \cup \cdots \cup F(t-m)\right) \circ R(t, t-m)$$

where '$\cup$' is the union operator, and '$\circ$' is the composition. $R(t, t-m)$ is a relation matrix to describe the fuzzy relationship between $F(t-m)$ and $F(t)$.

Definition 3. Suppose $F(t)$ is caused by $F(t-1)$, $F(t-2)$, ..., and $F(t-m)(m>0)$. The relation can be expressed as the following fuzzy relational equation:

$$F(t) = \left(F(t-1) \times F(t-2) \times \cdots \times F(t-m)\right) \circ R_a(t, t-m).$$

$R_a(t, t-m)$ is a relation matrix to describe the fuzzy relationship between $F(t-1)$, $F(t-2), ..., F(t-m)$ and $F(t)$.

Let fuzzy time series $F(t) = \langle f_1 f_2 \dots f_t \dots f_n \rangle$, $f_t \in S$,
$S = \{\tilde{A}_1, \tilde{A}_2, \dots, \tilde{A}_i, \dots, \tilde{A}_m\}$. A fuzzy set $\tilde{A}_i$ in the universal of discourse U ,
$U = \{I_1, I_2, \dots, I_m\}$, $\bigcup_{i=1}^{m} I_i = U$, is presented as

$$\tilde{A}_i(x) = \begin{cases} \mu_{[l,\,c]}^i(x) & when\ l \le x \le c \\ \mu_{(c,\,u]}^i(x) & when\ c < x \le u ,\ i = 1, 2, \dots, m, \\ 0 & otherwise \end{cases} \tag{1}$$

$\tilde{A}_i(x)$ is the membership function of x belonging to fuzzy set $\tilde{A}_i$. $\mu_{[l,\,c]}^i(x)$
and $\mu_{(c,\,u]}^i(x)$ are the degree functions of fuzzy set $\tilde{A}_i$ within interval
$[l, u] = [l, c] \cup (c, u]$.

Hsu and Chen (1996) defined the similarity between fuzzy sets, denoting it
as $sim(\tilde{A}_i, \tilde{A}_j)$,

Definition 4. The fuzzy similarity measure between two fuzzy sets $\tilde{A}_i$ and $\tilde{A}_j$ is
defined as

$$sim(\tilde{A}_i, \tilde{A}_j) = \frac{\displaystyle\int_{x \in \tilde{A}_i \cup \tilde{A}_j} \min(\tilde{A}_i(x),\ \tilde{A}_j(x)) dx}{\displaystyle\int_{x \in \tilde{A}_i \cup \tilde{A}_j} \max(\tilde{A}_i(x),\ \tilde{A}_j(x)) dx} \tag{2}$$

The similarity measurement $sim(\tilde{A}_i, \tilde{A}_j)$ between fuzzy set $\tilde{A}_i$ and $\tilde{A}_j$ is
determined by the proportion of the overlapping area to the total area. If
$sim(\tilde{A}_i, \tilde{A}_j)$ =1, the two fuzzy sets are completely overlapping, that is,
$\tilde{A}_i = \tilde{A}_j$; If $0 < sim(\tilde{A}_i, \tilde{A}_j) < 1$, the two fuzzy sets are partially overlapping; If
$sim(\tilde{A}_i, \tilde{A}_j)$ =0, the overlapping area is zero. Therefore, $0 \le sim(\tilde{A}_i, \tilde{A}_j) \le 1$.

The similarity measure $sim(\tilde{A}_i, \tilde{A}_j)$ on a given domain
$S = \{\tilde{A}_1, \tilde{A}_2, \dots, \tilde{A}_i, \dots, \tilde{A}_m\}$ is a mapping $sim: S \times S \to [0,1]$, which satisfies the following properties:

Reflexivity: $sim\left(\tilde{A}_i, \tilde{A}_i\right) = 1$ for all $\tilde{A}_i \in S$.

Separability: $sim\left(\tilde{A}_i, \tilde{A}_j\right) = 1$ if and only if $\tilde{A}_i = \tilde{A}_j$

Bounded: $0 \le sim\left(\tilde{A}_i, \tilde{A}_j\right) \le 1$ for all $\tilde{A}_i, \tilde{A}_j \in S$.

Further, we define the similarity measure between two fuzzy sequences order in time.

Definition 5. Let $e_i^k = f_i f_{i+1} \cdots f_{i+k-1}$ and $e_j^k = f_j f_{j+1} \cdots f_{j+k-1}$ be subsequences of $F(t) = \langle f_1 f_2 \cdots f_t \cdots f_n \rangle$ with k successive elements ordered in time and starting from time i and j by suffix, respectively. $f_t \in S = \{\tilde{A}_1, \tilde{A}_2, \ldots, \tilde{A}_i, \ldots, \tilde{A}_m\}$. The similarity measure between sequence e_i^k and e_j^k is defined as

$$Sim\left(e_i^k, e_j^k\right) = \frac{1}{k} \sum_{p=0}^{k-1} sim\left(f_{i+p}, f_{j+p}\right) \tag{3}$$

$Sim\left(e_i^k, e_j^k\right)$ satisfies the following conditions:

(1) $0 \le Sim\left(e_i^k, e_j^k\right) \le 1$

(2) $Sim\left(e_i^k, e_i^k\right) = 1$ or $Sim\left(e_i^k, e_j^k\right) = 1$ if and only if $e_i^k = e_j^k$, for all fuzzy set e_i^k and e_j^k.

3   A Best-Match Model Based on Similarity Measure for Fuzzy Time Series

The novel forecasting model is developed based on the similarity measurement of Song and Chissom's (1993b) proposed framework. Its design comprises the following steps

Step 1: Partition the universe of discourse U into several intervals of equal length. In general, U is defined as $U = [D_{\min} - D_1, D_{\max} + D_2]$, where $D_{\min}$ and $D_{\max}$ are the minimal and maximal values of the historical data, and D_1 and D_2 are properly selected positive numbers Song and Chissom's (1993b). Then U is partitioned into m equal intervals, $I_1, I_2, \ldots, I_m$,

$I_i = [l_i, u_i] = [l_i, c_i] \cup (c_i, u_i]$, with length *length* defined as

$length = \dfrac{1}{m}[(D_{max} + D_2) - (D_{min} - D_1)].$

Step 2: Define fuzzy sets on the universe of discourse and fuzzify the time series. Given a traditional real time series, one needs a fuzzification procedure to obtain the corresponding fuzzy time series. For this, m fuzzy sets $\tilde{A}_1, \tilde{A}_2, \ldots, \tilde{A}_m$ can be defined as Eq (1).

$$\tilde{A}_i(x) = \begin{cases} \mu^i_{[l_i, c_i]}(x) & when\ l_i \leq x \leq c_i \\ \mu^i_{(c_i, u_i]}(x) & when\ c_i < x \leq u_i,\ i = 1, 2, \ldots, m. \\ 0 & otherwise \end{cases} \qquad (4)$$

Step 3: Fuzzify the precise time series data to a fuzzy time series. Pedrycz and Vasilakos (1999) proposed a linguistic approach to design a new category of fuzzy models. In order to fairly compare to other models, such as Lee's model, in this study, the precise data are fuzzified to fuzzy ones by the method proposed by Song and Chissom (1993a). If the real value $y_t \in I_i$, then y_t is fuzzified to fuzzy set $\tilde{A}_i$, and the membership function is defined by (4). Suppose that a time series $Y(t) = \langle y_1 y_2 \ldots y_t \ldots y_n \rangle$ is fuzzified to fuzzy time series $F(t) = \langle f_1 f_2 \ldots f_t \ldots f_n \rangle$, where $f_i \in \{\tilde{A}_1, \tilde{A}_2, \ldots, \tilde{A}_i, \ldots, \tilde{A}_m\}$.

Step 4: Derive k order fuzzy logical relationships from fuzzy time series $F(t) = \langle f_1 f_2 \ldots f_t \ldots f_n \rangle$, the form with 'IF-THEN' as

$$f_i f_{i+1} \cdots f_{i+k-1} \to f_{i+k},\ for\ i = 1, 2, \ldots, n-k$$

For an order k rule, the left hand side of rule is starting at time i and ending at time $i+k-1$, the right hand side of rule is next of the ending time $i+k$.

Step 5: Forecasting and defuzzifying. The similarity measurement between fuzzy time series, $Sim(e_i^k, e_j^k)$, is applied to the forecasting step. The firing rule as forecasting is determined by the value of $Sim(e_i^k, e_j^k)$ and a threshold α. If $Sim(e_i^k, e_j^k) \geq \alpha$, the rule is fired and the right hand side of firing rule is the forecasting result. The forecasting result $\tilde{A}_j$ is defuzzified with the average of the apex value of $\tilde{A}_j$.

For example, a time series of the enrollment data of the University of Alabama from 1971 to 1992 (Song and Chissom, 1993a), $Y(t) = <13055, 13563, 13867,$

14696, 15460, 15311, 15603, 15861, 16807, 16919, 16388, 15433, 15497, 15145, 15163, 15984, 16859, 18150, 18970, 19328, 19337, 18876>.

Step 1: The $D_{min} = 13055$ and $D_{max} = 19337$, $U = [13000, 20000]$ is partitioned into seven equal length intervals $I_1, I_2, I_3, I_4, I_5, I_6$, and I_7, where $I_1 = [13000, 14000]$, $I_2 = (14000, 15000]$, $I_3 = (15000, 16000]$, $I_4 = (16000, 17000]$, $I_5 = (17000, 18000]$, $I_6 = (18000, 19000]$, and $I_7 = (19000, 20000]$ with the interval length 1000.

Step 2: A fuzzy set $\tilde{A}_i$ of U is defined as

$$\tilde{A}_1(x) = \begin{cases} 1 & when\ 13000 \le x \le 14000 \\ -\dfrac{1}{1000}(x - 15000) & when\ 14000 < x \le 15000, \\ 0 & otherwise \end{cases}$$

$$\tilde{A}_i(x) = \begin{cases} \dfrac{1}{1500}(x - l_{i-1}) & when\ l_{i-1} \le x \le c_i \\ -\dfrac{1}{1500}(x - u_{i+1}) & when\ c_i < x \le u_{i+1},\ i = 2,3,...,6, \\ 0 & otherwise \end{cases}$$

$$\tilde{A}_7(x) = \begin{cases} \dfrac{1}{1000}(x - 18000) & when\ 18000 \le x \le 19000 \\ 1 & when\ 19000 < x \le 20000 \\ 0 & otherwise \end{cases}$$

where l_{i-1}, c_i, and u_{i+1} are lower bound, median value, and upper bound of intervals I_{i-1}, I_i, and I_{i+1}, respectively. Similar to Song and Chissom (1993a), the vocabulary variables are $\tilde{A}_1$ = (not many), $\tilde{A}_2$ = (not too many), $\tilde{A}_3$ = (many), $\tilde{A}_4$ = (many many), $\tilde{A}_5$ = (very many), $\tilde{A}_6$ = (too many), $\tilde{A}_7$ = (too many many).

Step 3: Fuzzify the time series $Y(t)$. For example, the enrollment 13055 is fuzzified as $\tilde{A}_1$ since the value is located at interval I_1 whereas the enrollment 14696 is fuzzified as $\tilde{A}_2$ due to located at interval I_2. In such a way, the numeric series is fuzzified as $F(t) = \langle \tilde{A}_1 \tilde{A}_1 \tilde{A}_1 \tilde{A}_2 \tilde{A}_3 \tilde{A}_3 \tilde{A}_3 \tilde{A}_3 \tilde{A}_4 \tilde{A}_4 \tilde{A}_4 \tilde{A}_3 \tilde{A}_3 \tilde{A}_3 \tilde{A}_3 \tilde{A}_3 \tilde{A}_4 \tilde{A}_6 \tilde{A}_6 \tilde{A}_7 \tilde{A}_7 \tilde{A}_6 \rangle$.

Step 4: Suppose that a three-order rule base is established from $F(t)$ which contains the following rules:

$$\tilde{A}_1\tilde{A}_1\tilde{A}_1 \to \tilde{A}_2, \ \tilde{A}_1\tilde{A}_1\tilde{A}_2 \to \tilde{A}_3, \ \tilde{A}_1\tilde{A}_2\tilde{A}_3 \to \tilde{A}_3, \ \tilde{A}_2\tilde{A}_3\tilde{A}_3 \to \tilde{A}_3,$$

$$\tilde{A}_2\tilde{A}_3\tilde{A}_3 \to \tilde{A}_3, \tilde{A}_4, \ \tilde{A}_3\tilde{A}_3\tilde{A}_4 \to \tilde{A}_4, \tilde{A}_6, \ \tilde{A}_3\tilde{A}_4\tilde{A}_4 \to \tilde{A}_4,$$

$$\tilde{A}_4\tilde{A}_4\tilde{A}_4 \to \tilde{A}_3, \ \tilde{A}_4\tilde{A}_4\tilde{A}_3 \to \tilde{A}_3, \ \tilde{A}_4\tilde{A}_3\tilde{A}_3 \to \tilde{A}_3, \ \tilde{A}_3\tilde{A}_4\tilde{A}_6 \to \tilde{A}_6,$$

$$\tilde{A}_4\tilde{A}_6\tilde{A}_6 \to \tilde{A}_7, \ \tilde{A}_6\tilde{A}_6\tilde{A}_7 \to \tilde{A}_7, \text{ and } \tilde{A}_6\tilde{A}_7\tilde{A}_7 \to \tilde{A}_6.$$

Step 5: The similarity measures, $sim(\tilde{A}_i, \tilde{A}_j)$, between fuzzy set $\tilde{A}_i$ and $\tilde{A}_j$ defined in Eq. (2) for $i, j = 1, 2, ..., 7$ are shown in Table 1.

Table 1 The similarity measure between fuzzy set $\tilde{A}_i$ and $\tilde{A}_j$

$sim(\tilde{A}_i, \tilde{A}_j)$	$\tilde{A}_1$	$\tilde{A}_2$	$\tilde{A}_3$	$\tilde{A}_4$	$\tilde{A}_5$	$\tilde{A}_6$	$\tilde{A}_7$
$\tilde{A}_1$	1	0.3636	0	0	0	0	0
$\tilde{A}_2$	0.3636	1	0.6667	0	0	0	0
$\tilde{A}_3$	0	0.6667	1	0.6667	0	0	0
$\tilde{A}_4$	0	0	0.6667	1	0.6667	0	0
$\tilde{A}_5$	0	0	0	0.6667	1	0.6667	0
$\tilde{A}_6$	0	0	0	0	0.6667	1	0.3636
$\tilde{A}_7$	0	0	0	0	0	0.3636	1

Assume that given an inquiry fuzzy time series $\tilde{A}_7\tilde{A}_7\tilde{A}_6$, one wants to forecast the next fuzzy set. From Eq.(3), we compute the similarity between fuzzy time series with order 3, the value of $Sim(e_i^3, e_j^3)$ between $\tilde{A}_7\tilde{A}_7\tilde{A}_6$ and other fuzzy time series shown in Table 2. Suppose the threshold under consideration $\alpha = 0.5$, then the two qualified firing rules are $\tilde{A}_4\tilde{A}_6\tilde{A}_6 \to \tilde{A}_7$ and $\tilde{A}_6\tilde{A}_6\tilde{A}_7 \to \tilde{A}_7$. Therefore, the forecasting result is $\tilde{A}_7$.

If the threshold α is set to 0.3, then firing rules are

$$\tilde{A}_3\tilde{A}_4\tilde{A}_6 \to \tilde{A}_6 \ , \quad \tilde{A}_4\tilde{A}_6\tilde{A}_6 \to \tilde{A}_7 \ , \quad \tilde{A}_6\tilde{A}_6\tilde{A}_7 \to \tilde{A}_7 \ , \quad \text{and}$$

$$\tilde{A}_6\tilde{A}_7\tilde{A}_7 \to \tilde{A}_6.$$

The forecasting results are possibly $\tilde{A}_6$ and $\tilde{A}_7$ meaning the next state of enrollments are "too many" or "too many many". If a numeric result of forecasting is necessary, then the defuzzified forecast is 19000, which is the average of the midpoint of I_6 and I_7 ($\frac{1}{2}(18500+19500)=19000$).

Table 2 The similarity measure between two fuzzy time series

Rule	e_i^3	e_j^3	$Sim\left(e_i^k, e_j^k\right)$
$\tilde{A}_1\tilde{A}_1\tilde{A}_1 \to \tilde{A}_2$	$\tilde{A}_1\tilde{A}_1\tilde{A}_1$		0
$\tilde{A}_1\tilde{A}_1\tilde{A}_2 \to \tilde{A}_3$	$\tilde{A}_1\tilde{A}_1\tilde{A}_2$		0
$\tilde{A}_1\tilde{A}_2\tilde{A}_3 \to \tilde{A}_3$	$\tilde{A}_1\tilde{A}_2\tilde{A}_3$		0
$\tilde{A}_2\tilde{A}_3\tilde{A}_3 \to \tilde{A}_3$	$\tilde{A}_2\tilde{A}_3\tilde{A}_3$		0
$\tilde{A}_2\tilde{A}_3\tilde{A}_3 \to \tilde{A}_3, \tilde{A}_4$	$\tilde{A}_2\tilde{A}_3\tilde{A}_3$		0
$\tilde{A}_3\tilde{A}_3\tilde{A}_4 \to \tilde{A}_4, \tilde{A}_6$	$\tilde{A}_3\tilde{A}_3\tilde{A}_4$		0
$\tilde{A}_3\tilde{A}_4\tilde{A}_4 \to \tilde{A}_4$	$\tilde{A}_3\tilde{A}_4\tilde{A}_4$	$\tilde{A}_7\tilde{A}_7\tilde{A}_6$	0
$\tilde{A}_4\tilde{A}_4\tilde{A}_4 \to \tilde{A}_3$	$\tilde{A}_4\tilde{A}_4\tilde{A}_4$		0
$\tilde{A}_4\tilde{A}_4\tilde{A}_3 \to \tilde{A}_3$	$\tilde{A}_4\tilde{A}_4\tilde{A}_3$		0
$\tilde{A}_4\tilde{A}_3\tilde{A}_3 \to \tilde{A}_3$	$\tilde{A}_4\tilde{A}_3\tilde{A}_3$		0
$\tilde{A}_3\tilde{A}_4\tilde{A}_6 \to \tilde{A}_6$	$\tilde{A}_3\tilde{A}_4\tilde{A}_6$		0.3333
$\tilde{A}_4\tilde{A}_6\tilde{A}_6 \to \tilde{A}_7$	$\tilde{A}_4\tilde{A}_6\tilde{A}_6$		0.6667
$\tilde{A}_6\tilde{A}_6\tilde{A}_7 \to \tilde{A}_7$	$\tilde{A}_6\tilde{A}_6\tilde{A}_7$		0.5758
$\tilde{A}_6\tilde{A}_7\tilde{A}_7 \to \tilde{A}_6$	$\tilde{A}_6\tilde{A}_7\tilde{A}_7$		0.3636

4   Experimental Results and Their Anaylsis

To demonstrate how the proposed forecasting model works, we conduct the experiments of forecasting up and down values of Taiwan Weighted Stock Index (TWSI) during the period from 2004 to 2006. The data set is divided into three parts, i.e., that collected in 2004 is used to construct the rule base, the collected data in 2005 is used as the training set to train the threshold α, and the collected data in 2006 is used as the testing set to evaluate forecasting performance. The forecasting performance is then evaluated by comparing to Lee's model (2006), considered as the high-order fuzzy time series benchmark in the literature. The forecasting accuracy is measured in terms of mean square error (MSE) and average forecasting errors ratio (AFER), which are defined as follows:

$$MSE = \frac{\sum_{i=1}^{N} (Forecasting\_Value_i - Actual\_Value_i)^2}{N}.$$

$$AFER = \frac{1}{N} \sum_{i=1}^{N} \frac{|Forecasting\_Value - Actual\_Value|}{Actual\_Value} \times 100\%$$

For the data set in 2004, the universe of discourse is defined as $U = [-500, 400]$, and is partitioned into 45 intervals of equal length, namely $u_1 = [-500, -480)$, $u_2 = [-480, -460)$, ..., $u_{44} = [360, 380)$ and $u_{45} = [380, 400]$. Six kinds of high-order models, i.e. order 3, 4, ..., 8, are considered and their respective rule bases are constructed. The set in 2005 is used to train the threshold α. Tables 3 and 4 show the results of MSE and AFER of different α ranging from 0.1 to 0.9 for order 3 to 8. For all high-order models except order 5, the lowest MSE and AFER are obtained at α=0.1, that illustrates the lower α obviously achieves better performance. Tables 5 and 6 show the comparison result between Lee's model and the proposed model at α=0.1 for the

Table 3 MSE of TWSI by the proposed model for the training set

α	Order 3	Order 4	Order 5	Order 6	Order 7	Order 8
0.9	4563.5	4613.1	4637.3	4643.1	4660.1	4613.1
0.8	4563.5	4616.1	4620.7	4643.1	4660.1	4613.1
0.7	4811.0	4818.0	4620.7	4643.1	4649.3	4613.1
0.6	5855.6	4818.0	4885.7	4583.7	4649.3	4596.3
0.5	5855.6	4663.4	4885.7	5119.8	4600.0	4679.0
0.4	6132.5	4663.4	4108.3	5119.8	5314.4	4768.9
0.3	2507.8	4120.3	4081.1	3513.2	3825.2	5538.9
0.2	2507.8	2475.2	**2475.8**	3022.8	3196.9	3026.9
0.1	**2501.5**	**2475.1**	2476.5	**2475.6**	**2430.0**	**2389.3**

testing data set in 2006. It indicates that the proposed model significantly outperforms the Lee's model both in terms of MSE and AFER for all orders. Furthermore, the superiority of the proposed model is also validated when comparing to the traditional time series forecasting model, autoregressive integrated moving average (ARIMA) model. The mean square errors of ARIMA(1, 1), ARMA(1, 2), and ARIMA(2, 2) are 6985.23, 6983.88, and 6976.58, respectively.

Table 4 AFER of TWSI by the proposed model for the training set

α	Order 3	Order 4	Order 5	Order 6	Order 7	Order 8
0.9	14.91%	14.94%	15.00%	15.03%	15.09%	15.15%
0.8	14.91%	14.94%	15.00%	15.03%	15.09%	15.15%
0.7	13.41%	14.99%	15.00%	15.03%	15.09%	15.15%
0.6	13.46%	14.99%	15.09%	14.98%	15.09%	15.14%
0.5	13.46%	7.14%	15.09%	15.66%	15.04%	15.13%
0.4	13.81%	7.14%	11.24%	15.66%	15.67%	15.54%
0.3	5.49%	7.06%	11.24%	10.81%	11.60%	17.50%
0.2	5.49%	**4.61%**	**4.07%**	8.14%	10.68%	9.50%
0.1	**5.48%**	**4.61%**	**4.07%**	**2.97%**	**2.93%**	**2.78%**

Table 5 MSE of TWSI for the testing set at $\alpha = 0$

	Order 3	Order 4	Order 5	Order 6	Order 7	Order 8
Lee's model	9245.0	9325.3	9277.2	9469.8	9479.2	9504.4
The proposed model	4904.8	4829.1	4777.0	4839.0	4862.9	4809.9

Table 6 AFER of TWSI for the testing set at α=0.1

	Order 3	Order 4	Order 5	Order 6	Order 7	Order 8
Lee's model	9.59%	9.67%	9.68%	9.71%	9.74%	9.77%
The proposed model	2.27%	2.20%	2.16%	1.62%	1.38%	1.35%

The second experiment is to forecast Dow Jones Industrial Average (INDU) from 2004 to 2006. Similarly to the previous experiment, the universe of discourse is partition into 45 intervals of equal length and the rule bases for order 3, 4, ..., 8 are constructed using the data set in 2004. The optimal value of α is determined empirically by the training set of the 2005 data. Tables 7 and 8 display MSE and AFER by the proposed model for the year of 2005 at α=0.1, 0.2... 0.9 and orders from 3 to 8. The MSE and AFER achieved the lowest ones when α=0.1 for all orders. The results shown in Tables 9 and 10 demonstrate that the proposed model at α=0.1 significantly outperforms than Lee's model (2006).

Table 7 MSE of INDU of the proposed model for the 2005 data set

α	Order 3	Order 4	Order 5	Order 6	Order 7	Order 8
0.9	0.7276	0.8819	0.9182	0.9237	0.9391	0.9322
0.8	0.7276	0.8819	0.6372	0.8303	0.8955	0.9249
0.7	0.7276	0.4789	0.6372	0.8303	0.7242	0.8172
0.6	0.4245	0.4789	0.4632	0.5451	0.7242	0.5999
0.5	0.4245	0.4312	0.4632	0.4454	0.4697	0.4416
0.4	0.4245	0.4312	0.4330	0.4454	0.4456	0.4416
0.3	**0.4193**	0.4312	0.4330	0.4316	0.4456	0.4360
0.2	**0.4193**	**0.4242**	**0.4236**	0.4316	0.4304	**0.4292**
0.1	**0.4193**	**0.4242**	**0.4236**	**0.4274**	**0.4282**	**0.4293**

Table 8 AFER of INDU for the proposed model for the 2005 data set

α	Order 3	Order 4	Order 5	Order 6	Order 7	Order 8
0.9	2.47%	2.81%	3.21%	3.19%	3.25%	3.24%
0.8	2.47%	2.81%	2.26%	2.84%	3.07%	3.23%
0.7	2.47%	1.45%	2.26%	2.84%	2.27%	2.92%
0.6	1.18%	1.45%	1.29%	1.72%	2.27%	1.96%
0.5	1.18%	1.06%	1.29%	1.23%	1.59%	1.25%
0.4	1.18%	1.06%	1.09%	1.23%	1.11%	1.25%
0.3	**1.05%**	1.06%	1.09%	1.04%	1.11%	1.08%
0.2	**1.05%**	**1.03%**	**1.01%**	1.04%	1.04%	1.01%
0.1	**1.05%**	**1.03%**	**1.01%**	**1.00%**	**1.01%**	**1.00%**

Table 9 The comparison of MSE of INDU for 2006 at $\alpha = 0.1$

	Order 3	Order 4	Order 5	Order 6	Order 7	Order 8
Lee's model	0.70	0.84	0.77	0.75	0.75	0.74
The proposed model	0.40	0.39	0.39	0.39	0.39	0.39

Table 10 The comparison of AFER of INDU for 2006 at $\alpha = 0.1$

	Order 3	Order 4	Order 5	Order 6	Order 7	Order 8
Lee's model	3.80%	4.52%	4.18%	4.09%	4.04%	3.92%
The proposed model	1.05%	1.04%	1.02%	1.01%	1.00%	1.00%

5 Conclusions and Future Works

In this study, we presented a novel forecasting model for high-order fuzzy time series which is based on the fuzzy similarity measurement. The proposed model is motivated by the shortcomings of the traditional 'exact-match' rule-based approaches and thus the uncertainty and fuzziness characteristics inherent in the fuzzy relationships are overlooked. In contrast to the exact-match, the best-match approach used in the proposed model outperforms its counterpart in terms of mean square errors and average forecasting errors ratio in the experiments of forecasting TWSI and INDU values. Future work may involve applying the proposed model to deal with more complicated applications and extending the model to handle the problem of multi-dimensional fuzzy time series.

Acknowledgments. This study was supported in part by NSC 99-2410-H-165-002-, Taiwan, R.O.C.

References

Chen, S.M.: Forecasting enrollments based on fuzzy time series. Fuzzy Sets and Systems 81, 311–319 (1996)

Chen, S.M.: Forecasting enrollments based on high-order fuzzy time series. Cybernetics and Systems: An International Journal 33, 1–16 (2002)

Chen, S.M., Chen, C.D.: TAIEX forecasting based on fuzzy time series and fuzzy variation groups. IEEE Transactions on Fuzzy Systems 19(1), 1–12 (2011)

Chen, S.M., Hsu, C.C.: A new method to forecast enrollments using fuzzy time series. International Journal of Applied Science and Engineering 2(3), 234–244 (2004)

Chen, S.M., Wang, N.Y.: Fuzzy forecasting based on fuzzy-trend logical relationship group. IEEE Transactions on Systems, Man, and Cybernetics-Part B: Cybernetics 40(5), 1343–1358 (2010)

Hsu, H.M., Chen, C.T.: Aggregation of fuzzy opinions under group decision making. Fuzzy Sets and Systems 79, 279–285 (1996)

Huarng, K.H., Yu, T.H.K., Hsu, Y.W.: A multivariate heuristic model for fuzzy time-series forecasting. IEEE Transactions on Systems, Man, and Cybernetics-Part B: Cybernetics 37(4), 836–846 (2007)

Lee, L.W., Wang, L.H., Chen, S.M.: Temperature prediction and TAIFEX forecasting based on fuzzy logical relationships and genetic algorithms. Expert Systems with Applications 33, 539–550 (2007)

Lee, L.W., Wang, L.H., Chen, S.M., Leu, Y.H.: Handling forecasting problems based on two-factors high-order fuzzy time series. IEEE Transaction on Fuzzy Systems 14(3), 468–477 (2006)

Li, S.-T., Cheng, Y.-C.: Deterministic fuzzy time series model for forecasting enrollments. Computers and Mathematics with Applications 53, 1904–1920 (2007)

Li, S.-T., Cheng, Y.-C.: A FCM-based deterministic forecasting model for fuzzy time series. Computers and Mathematics with Applications 56(12), 3052–3063 (2008)

Li, S.-T., Cheng, Y.-C.: An enhanced deterministic fuzzy time series forecasting model. Cybernetics and Systems: An International Journal 40(3), 211–235 (2009)

Li, S.-T., Kuo, S.-C., Cheng, Y.-C., Chen, C.-C.: Deterministic vector long-term forecasting for fuzzy time series. Fuzzy Sets and Systems 161(13), 1852–1870 (2010)

Li, S.-T., Kuo, S.-C., Cheng, Y.-C., Chen, C.-C.: A vector forecasting model for fuzzy time series. Applied Soft Computing 11, 3125–3134 (2011)

Own, C.M., Yu, P.T.: Forecasting fuzzy time series on a heuristic high-order model. Cybernetics and Systems: An International Journal 36, 705–717 (2005)

Pappis, C.P., Karacapilidis, N.I.: A comparative assessment of measures of similarity of fuzzy values. Fuzzy Sets and Systems 56(2), 171–174 (1993)

Pedrycz, Vasilakos, A.V.: Linguistic models and linguistic modeling. IEEE Transactions on Systems, Man, and Cybernetics-Part B: Cybernetics 29(6), 745–757 (1999)

Song, Q., Chissom, B.S.: Forecasting enrollments with fuzzy time series–Part I. Fuzzy Sets and Systems 54, 1–9 (1993a)

Song, Q., Chissom, B.S.: Fuzzy time series and its models. Fuzzy Sets and Systems 54, 269–277 (1993b)

Chapter 16
Building Fuzzy Autocorrelation Model and Its Application to the Analysis of Stock Price Time-Series Data

Yoshiyuki Yabuuchi and Junzo Watada

Abstract. The objective of economic analysis is to interpret the past, present or future economic state by analyzing economic data. Economic analyses are typically based on the time-series data or the cross-section data. Time-series analysis plays a pivotal role in analyzing time-series data. Nevertheless, economic systems are complex ones because they involve human behaviors and are affected by many factors. When a system includes substantial uncertainty, such as those concerning human behaviors, it is advantageous to employ a fuzzy system approach to such analysis. In this paper, we compare two fuzzy time-series models, namely a fuzzy autoregressive model proposed by Ozawa *et al.* and a fuzzy autocorrelation model proposed by Yabuuchi and Watada. Both models are built based on the concepts of fuzzy systems. In an analysis of the Nikkei Stock Average, we compare the effectiveness of the two models. Finally, we analyze tick-by-tick data of stock dealing by applying fuzzy autocorrelation model.

Keywords: fuzzy AR model, fuzzy autocorrelation, possibility, economic analysis.

1 Introduction

Many econometric models including the time-series model have been proposed for evaluating economic systems. In this paper, we propose the fuzzy time-series model, which is employed in the analysis of an economic system.

Yoshiyuki Yabuuchi
Shimonoseki City University, Faculty of Economics; 2-1-1 Daigaku-cho, Shimonoseki, Yamaguchi 751-8510 Japan
e-mail: yabuuchi@shimonoseki-cu.ac.jp

Junzo Watada
Waseda University, Graduate School of Information, Production and Systems;
2-4 Hibikino, Wakamatsu, Kitakyushu, Fukuoka 808-0196 Japan
e-mail: junzow@osb.att.ne.jp

W. Pedrycz & S.-M. Chen (Eds.): Time Series Analysis, Model. & Applications, ISRL 47, pp. 347–367.
DOI: 10.1007/978-3-642-33439-9_16   © Springer-Verlag Berlin Heidelberg 2013

The objective of economic analysis is to precisely understand the past and present states of an economic system based on the statistical data [1, 21]. However, in an economic system, the state is closely related to many factors that are triggered by the aggregation of numerous human behaviors [24].

Therefore, it becomes insufficient to interpret such an economic system with the use of conventional statistical methods. Instead, it is desirable to apply the concept of the fuzzy system theory which can handle the ambiguity of a structure when we analyze an economic system with many vague factors.

In addition, illustrating a time-series system and making a prediction by a time-series model, estimates are far from observed values. It is a natural interpretation of one reason that a possibility of a time-series system makes data fluctuating. Under the consideration that a Box-Jenkins model can describe a time-series with high accuracy, we propose a fuzzy autocorrelation model based on a Box-Jenkins model to describe a time-series system.

Ozawa et al. have proposed a fuzzy autoregressive model [11]. Their model describe the system in terms of error term instead of the coefficient mentioned above. The Ozawa's model, the autoregressive parameter is constructed by including time-series data, minimizing the vagueness of the model, and has real values. In addition, their model express the vagueness of the time-series system by the error term written by fuzzy numbers. The characteristics of these two models are compared by using numerical examples.

We analyze the Nikkei stock average by employing both the fuzzy autoregressive model which was first proposed by Ozawa et al. [11], and the fuzzy autocorrelation model, which is being proposed in this paper. Furthermore, the fuzzy autocorrelation model is applied to an economic analysis by analyzing the tick-by-tick data of stock prices. This enables us to forecast a future trend by a sequential prediction that fits the present condition.

The structure of this chapter is organized as follows: In Section 2, a fuzzy time-series analysis based on the fuzzy auto-correlation model will be built by expanding the Box-Jenkins model. In Section 3, we compare the characteristics of the fuzzy autoregressive model with those of the fuzzy autocorrelation model by analyzing the Nikkei stock average. In Section 4, we analyze the tick-by-tick data of stocks by applying the fuzzy autocorrelation model.

2 Fuzzy Time-Series Model

This section elaborates on various fuzzy time-series analysis models. Then, the fuzzy autoregressive model proposed by Ozawa et al. and our fuzzy autocorrelation model are described.

2.1 Various Fuzzy Time-Series Analysis Models

Let us review several fuzzy time-series models before discussing our fuzzy autocorrelation model and the conventional fuzzy autoregressive model.

L.A. Zadeh proposed fuzzy set theory [29] in 1965. It helps deal with qualitative data such as linguistic ones by using membership functions.

Based on these concepts, a fuzzy time-series analysis can be built. The models are classified into three groups: (1) a regression model-based analysis, (2) a Box-Jenkins model-based analysis and (3) a fuzzy reasoning(If-Then rule)-based analysis.

These three groups and others are reviewed below.

2.1.1 Regressive Model-Based Analysis of Fuzzy Time-Series Data

In general, a fuzzy regression-based time-series analysis employs the fuzzy regression model proposed by Tanaka *et al.* [16], in which the vagueness included in the target system is addressed with the use of the fuzzy regression coefficients

$$Y_i = (a_1, c_1)x_{i1} + (a_2, c_2)x_{i2} + \cdots + (a_p, c_p)x_{ip} = (\mathbf{a}, \mathbf{c})\mathbf{x}_i \qquad (1)$$

where, $\mathbf{x}_i$ are explanatory variables, $\mathbf{a} = [a_1, a_2, \cdots, a_p]$ denotes the center position of the fuzzy coefficients in the vector and $\mathbf{c} = [c_1, c_2, \cdots, c_p]$ are the widths of the fuzzy coefficients. The fuzzy regression model basically deals with the vagueness included in the system and treated with intervals that can express all of the possibilities by including all of the samples. Therefore, the estimated values, Y_i, of the observed values y_i are expressed as fuzzy numbers, and the coefficients are viewed as fuzzy coefficients. In this formulation, the fuzzy coefficients $(\mathbf{a}, \mathbf{c})$ can be obtained by solving a certain Linear Programming (LP) problem.

Watada *et al.* employ the fuzzy regression model to time-series data, and using fuzzy coefficients, they build a fuzzy regression model that includes all the vagueness of time-series system [22, 23]. However, they change the formulation of the fuzzy regression model to account the smoothing of time-series data.

2.1.2 Box-Jenkins Model-Based Analysis of Fuzzy Time-Series Data

Similarly to Box-Jenkins model, fuzzy time-series analysis includes two models based on fuzzy numbers and one model dealing with numeric data. Models dealing with fuzzy numbers are proposed by Yabuuchi *et al.* and Ozawa *et al..* Yabuuchi *et al.* interpreted time-series data from a possibilistic point of view, define fuzzy autocorrelation coefficients based on fuzzified data, and build a fuzzy autocorrelation model using these values [24].

The fuzzy autocorrelation model aims to capture the present state or the future state of the time-series process by using past fuzzy time-series data similar to the Box-Jenkins model. Therefore, the fuzzy autocorrelation model is obtained from a fuzzy autocorrelation coefficient that expresses the relationship between fuzzy time-series data by a fuzzy operation. The proposed fuzzy autocorrelation model is explained below.

In contrast to the above model, Ozawa *et al.* proposed a fuzzy autoregressive model [11] that expresses the possibilities of fuzzified difference sequences. The fuzzy autoregressive model is written down as follows:

$$\left.\begin{array}{l} \tilde{Z}_t = \phi_1 Z_{t-1} + \cdots + \phi_p Z_{t-p} + \mathbf{u} \\ Z_t \subseteq \tilde{Z}_t, \ \tilde{Z}_t = (\tilde{\alpha}_t, \tilde{\beta}_t, \tilde{\delta}_t) \end{array}\right\} \tag{2}$$

A time-series data is fuzzified to be used by the fuzzy autoregressive model:

$$\mathbf{Y}_t = \left(z_t \min_{i=1}^{3} z_{t+2-i} / \max_{i=1}^{3} z_{t+2-i}, \ z_t, \ z_t \max_{i=1}^{3} z_{t+2-i} / \min_{i=1}^{3} z_{t+2-i} \right) \tag{3}$$

Using the differenced series Z_t, which transformed fuzzy time-series data $\mathbf{Y}_t$ to detrending series, a fuzzy autoregressive model is formed. The fuzzy autoregressive model illustrate the relationship between fuzzy time-series data by a real-valued autoregressive parameter ϕ, and the vagueness of the system by introducing a triangular fuzzy number $\mathbf{u} = (u_\alpha, u_\beta, u_\gamma)$.

F.M. Tseng et al. proposed a fuzzy ARIMA model [18]. The fuzzy ARIMA model expresses the possibilities of the time-series system with fuzzy coefficients of the model, which is similar to other fuzzy time-series models.

The fuzzy ARIMA model is expressed as follows:

$$\begin{aligned} \tilde{z}_t = &(a_1, c_1) z_{t-1} + \cdots + (a_p, c_p) z_{t-p} \\ &+ \varepsilon_t - (a_{p+1}, c_{p+1}) \varepsilon_{t-1} - \cdots - (a_{p+q}, c_{p+q}) \varepsilon_{t-q} \end{aligned} \tag{4}$$

where $z_t = \nabla_d(Z_t - \mu)$, α_i denotes the center position of the fuzzy number and c_i stands for the vagueness of the fuzzy number. In addition, ε_t are independent and identically distributed normal random variables with a zero mean and variance of σ^2.

Although the fuzzy ARIMA model does not deal with seasonal variations, the fuzzy ARIMA model is developed as a fuzzy regression model and has fuzzy parameters [19]. Therefore, F.M. Tseng et al. proposed the fuzzy seasonal ARIMA model to deal with the seasonal variations. The fuzzy ARIMA model combines the fuzzy regression model and the seasonal ARIMA model [19].

This Box-Jenkins model based fuzzy time-series analysis models have a fuzzy numbers output to illustrate the possibilities of the system. Additionally, because these models are formulated by the LP problem, the required number of time-series data is less than that required to construct a statistical model.

2.1.3 Fuzzy Reasoning(If-Then Rule)-Based Analysis

Song et al. proposed a fuzzy reasoning (IF-Then rule)-based model, which is based on a fuzzy time-series model [13, 14, 15].

Song and Chisson describe the relationships between fuzzy time-series data using IF-Then rules and build a fuzzy time-series model to express these rules [13, 14, 15]. Various models have been proposed based on Song et al.'s fuzzy time-series model. Yu proposed the method to employ Song et al.'s fuzzy time-series model after assigning weighs to emphasize data near to the time point in the fuzzy time-series data [28].

Teoh *et al.* improved a model to accept intuitive and subjective opinions using fuzzy logical rules proposed by Song *et al.*'s model. Cheng *et al.* indicated that Song *et al.* did not consider any logical relations among rules and proposed a novel forecasting model that considers the similarities among fuzzy logical relations [3].

Teoh *et al.* created fuzzy logical relations based on rough sets theory and built a model with the rules generated by Song *et al.*'s model [17].

Additional fuzzy time-series models that improve upon Song *et al.*'s model have been proposed as well [2, 6]

2.1.4 Some Other Fuzzy Time-Series Models

Many other useful fuzzy time-series models have been proposed. R. Dong *et al.* proposed a granular time series approach that employs a fuzzy clustering technique to construct an original series [4]. This model uses long-term forecasting and trend forecasting.

M. Khashei *et al.* proposed a hybrid model, in which the ARIMA models are integrated with artificial neural networks and fuzzy logic to move beyond the linearity of a model [8]. However, O. Valenzuela *et al.*'s model is an integrated ARMA model, using both neural networks and fuzzy logic [20]. K. Lukoseviciute *et al.*'s model is employed using an evolutionary algorithm [10].

T. Partal *et al.* proposed a method that combines a discrete wavelet transform and a neuro-fuzzy method [12].

C.H.L. Lee *et al.*'s approach employs the Japanese candlestick theory [9]. The theory assumes that the candlestick patterns reflect the psychology of the market, and the investors can make their investment decision based on the identified candlestick patterns. With this approach, a vague candlestick patterns is transformed by fuzzy linguistic variables, and the financial time series data is transformed by fuzzy candlestick patterns. The objective of this approach is to understand the vagueness of investors and markets, which is expressed with fuzzy linguistic variables.

J.T Yao *et al.* proposed the rough set model [27]. The rough set model captures the vagueness and uncertainty of time-series data.

2.2 Fuzzy Autoregressive Model

Let us assume that all the fuzzy time-series data $\mathbf{Z}_t$ be defined by triangular fuzzy numbers. Triangular fuzzy numbers are defined by three parameters, and are denoted as $\mathbf{Z}_t = (\alpha_t, \beta_t, \delta_t), (\alpha_t \leq \beta_t \leq \delta_t)$. The inclusion relation of triangular fuzzy numbers is defined through the following inequalities:

$$\mathbf{Z}_t \subseteq \mathbf{Z}_s \Leftrightarrow \{\alpha_t \geq \alpha_s, \ \delta_t \leq \delta_s\} \tag{5}$$

In the same way, the arithmetic operations are defined as follows:

$$\mathbf{Z}_t + \mathbf{Z}_s = (\alpha_t + \alpha_s, \beta_t + \beta_s, \delta_t + \delta_s)$$
$$\mathbf{Z}_t - \mathbf{Z}_s = (\alpha_t - \alpha_s, \beta_t - \beta_s, \delta_t - \delta_s)$$
$$p \cdot \mathbf{Z}_t = \begin{cases} (p \times \alpha_t, p \times \beta_t, p \times \delta_t), & p \geq 0 \\ (p \times \delta_t, p \times \beta_t, p \times \alpha_t), & p < 0, \end{cases}$$

where p is a real number.

A fuzzy autoregressive model is specified as follows:

$$\tilde{\mathbf{Z}}_t = \phi_1 \mathbf{Z}_{t-1} + \cdots + \phi_p \mathbf{Z}_{t-p} + \mathbf{u}$$
$$\mathbf{Z}_t \subseteq \tilde{\mathbf{Z}}_t, \ \tilde{\mathbf{Z}}_t = (\tilde{\alpha}_t, \tilde{\beta}_t, \tilde{\delta}_t) \tag{6}$$

Based on (5) and (6), it is clear that the following relations hold.

$$\alpha_t \geq \tilde{\alpha}_t, \delta_t \leq \tilde{\delta}_t$$

Namely, the fuzzy time-series model includes all of the fuzzy data. The autoregressive parameters $\phi_1, \phi_2, \cdots, \phi_p$ have real values, and show the degree the fuzzy time-series data depend on the past. An error term is the constant specific to the model and refers to the part of the fuzzy data that do not depend on the past data. This term is defined as a triangular fuzzy number:

$$\mathbf{u} = (u_\alpha, u_\beta, u_\delta)$$

A fuzzy autoregressive model results through solving a problem of linear programming that minimizes that ambiguity of the model according to the inclusion condition (6) as follows:

$$\begin{aligned} \text{minimize} \quad & \sum_{t=p+1}^{n} (\tilde{\delta}_t - \tilde{\alpha}_t) \\ \text{subject to} \quad & \alpha_t \geq \tilde{\alpha}_t, \\ & \delta_t \leq \tilde{\delta}_t \\ & (t = p+1, p+2, \cdots, n) \\ & u_\alpha \leq u_\delta \end{aligned} \tag{7}$$

2.3 Fuzzy Autocorrelation Model

Even if we had described the behavior of time-series system by using a time-series model, the estimated values should have a near value from the observed data. The understanding of the behavior is shaped so as that the possibility of the time-series system is making it natural. Our model describes the possibility of the time-series system by the coefficients. Let us employ a triangular fuzzy number here, since it is manageable.

When different sequences are employed, trend and noise can be easily removed. The autocorrelation makes the model describing the behaviors easily. Therefore, we fuzzify the Box-Jenkins model.

In the fuzzy autocorrelation model, the time-series data z_t are transformed into a fuzzy number to express the possibilities of the data. The following fuzzy equation shows the case in which only one time point before and after the time point t is taken into consideration in building a fuzzy number [23].

$$\mathbf{Y}_t = (Y_t^L, Y_t^C, Y_t^U) = (\min(z_{t-1}, z_t, z_{t+1}), z_t, \max(z_{t-1}, z_t, z_{t+1})) \tag{8}$$

Next, we employ a calculus of finite differences to filter out the time-series trend data, which enables us to use the first-order difference-equation to write the following:

$$\mathbf{Z}_t = (Z_t^L, Z_t^C, Z_t^U) = (\min(\mathbf{Y}_t - \mathbf{Y}_{t-1}), Y_t^C - Y_{t-1}^C, \max(\mathbf{Y}_t - \mathbf{Y}_{t-1})) \tag{9}$$

Generally, if we take the finite differences then we reduce the trend variation, and only an irregular pattern is included in the difference series. However, when we use the fuzzy operation, the ambiguity may increase, and the value of an autocorrelation coefficient may take values not lower than 1 or not greater than -1. To solve this problem in the case of the fuzzy operation, we adjust the width of a fuzzy number using α-cut when determing the difference series. An α-cut level h is determined from the value of the autocorrelation. When we calculate the fuzzy autocorrelation, we employ the usual fuzzy operation under the condition that the fuzzy autocorrelation of lag 0 is set to $\rho_0 = \lambda_0/\lambda_0 = (1,1,1)$, which results in the following linear programming to decide the value at the α-cut level. When we set the α-cut level to 1, the ambiguity of the fuzzy autocorrelation is the smallest, but we cannot obtain the fuzzy autocorrelation that reflects the possibility of the system. Therefore, we maximize the width of the autocorrelation. However, the size of the width is decided automatically as the value of autocorrelation should be included in [-1,1].

$$\begin{aligned}
&\underset{h}{\text{maximize}} \quad \sum_{i=1}^{p}(\rho_i^U - \rho_i^L) \\
&\text{subject to} \quad \rho_i^U \leq 1 \\
&\qquad\qquad\quad \rho_i^L \geq -1 \\
&\qquad\qquad\quad \rho_i^L \leq \rho_i^C \leq \rho_i^U \\
&\qquad\qquad\quad (i = 1, 2, \cdots, p)
\end{aligned} \tag{10}$$

We can define the fuzzy covariance and the fuzzy autocorrelation as follows:

$$\Lambda_k \equiv Cov[\mathbf{Z}_t \mathbf{Z}_{t-k}] = E[\mathbf{Z}_t \mathbf{Z}_{t-k}] = [\lambda_k^L, \lambda_k^C, \lambda_k^U]$$
$$\mathbf{r}_k = \Lambda_k/\Lambda_0 = [\rho_k^L, \rho_k^C, \rho_k^U]$$

We adjust the ambiguity of the difference series by employing the α-cut level h, which is obtained by solving the above linear programming. Using the fuzzy autocorrelation coefficient which is calculated by employing α-cut level h, we redefine the Yule-Walker equations as in linear programming and calculate the partial autocorrelation.

We form the following autoregressive process.

$$\mathbf{Z}_t = \Phi_1 \mathbf{Z}_{t-1} + \Phi_2 \mathbf{Z}_{t-2} + \cdots + \Phi_p \mathbf{Z}_{t-p}$$

where $\Phi = [\phi^L, \phi^C, \phi^U]$ is a fuzzy partial autoregressive coefficient.

As mentioned above, the next observation value either exceeds the observed value at present by the size of the value of autocorrelation or it is less than the observed value. For this reason, autocorrelation is important to the time-series analysis. Therefore, we build a model that illustrates the ambiguity of the system captured by the fuzzy autocorrelation. The reason for the autocorrelation is also fuzzy autocorrelation, Yule-Walker equations can also be viewed as the fuzzy equation in the same way.

$$\mathbf{R}_t = \Phi_1 \mathbf{r}_{t-1} + \Phi_2 \mathbf{r}_{t-2} + \cdots + \Phi_p \mathbf{r}_{t-p} \tag{11}$$

Φ shown in (11) is an unknown coefficient. We are building the model in terms of fuzzy autocorrelation which can describe the ambiguity of the system. However, when the ambiguity of a model is large, the relationship between a model and a system becomes ambiguous. Therefore, the possibility of the system cannot be properly described. Therefore, to obtain the fuzzy partial autocorrelation coefficient, for which the ambiguity of a time-series model should be minimized, we have the following linear programming:

$$
\begin{aligned}
\text{minimize} \quad & \sum_{i=1}^{p} (\rho_t^U - \rho_t^L) \\
\text{subject to} \quad & R_t^U \geq \rho_t^U \\
& R_t^C = \rho_t^C \\
& R_t^L \leq \rho_t^L \\
& \rho_t^L \leq \rho_t^C \leq \rho_t^U \\
& (t = 1, 2, \cdots, p)
\end{aligned}
\tag{12}
$$

As mentioned above $\mathbf{R}$ is obtained by the fuzzy operation employing the fuzzy autocorrelation $\mathbf{r}$ and fuzzy partial autocorrelation Φ. R^L, R^C and R^U represent the lower limit, the center, and the upper limit of $\mathbf{R}$, respectively.

A fuzzy autocorrelation model expresses the possibility that the change of the system is realized in the data, which is different from the conventional statistical method. We are building a model that can show an ambiguous portion called a possibility that has not been clearly expressed through conventional statistics techniques.

The time-series data was fuzzified by using Equation (8), and a fuzzy operation was employed to calculate any coefficients. Then, the center of all coefficients is coincident to the non-fuzzy coefficient. In addition that, the center of our model is coincident to the non-fuzzy model, the autoregressive model by the constraint of LP problem (12).

3 A Numerical Example

In this section, we employ the Nikkei stock average which indicates the trend of the whole stock market as an index of the Japanese stock market. We use the monthly data from 1970 to 1998.

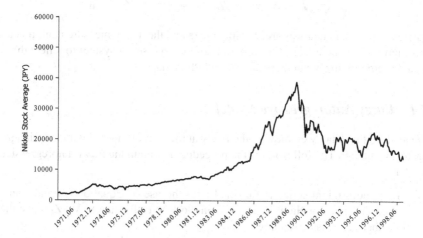

Fig. 1 Nikkei Stock Average in 1970 to 1998

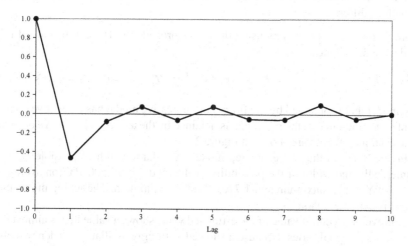

Fig. 2 Autocorrelation of Nikkei Stock Average

We show the sample autocorrelation coefficient at each time lag (Figure 2) to determine the order. Figure 2 shows the negative correlation in lags 1 and 2 where the sign of the autocorrelation coefficient changes from minus to plus. Because of this result, we analyze the Nikkei stock average by employing the AR(2) model of the second-order.

Furthermore, because of the existing seasonal variation in this data, we employ the calculation $\nabla^2 \nabla_{12} z_t$ which take the first-order seasonal difference of every 12-month period after taking the second-order difference.

$$\nabla^2 \nabla_{12} \tilde{Z}_t = \phi_1 \nabla^2 \nabla_{12} Z_{t-1} + \phi_2 \nabla^2 \nabla_{12} Z_{t-2} + \mathbf{u}$$

where the data z_t in analysis are statistical data and the actual measurement. $\mathbf{u}$ is an error term of the model. For the ambiguity of the time-series system to reflect these data, we employ fuzzy numbers to deal with these data.

3.1 Fuzzy Autoregressive Model

We analyze the data by employing the fuzzy autoregressive model with the triangular fuzzy number. The following is the procedure to obtain the fuzzy autoregressive model.

Step 1.   The original series is fuzzified with the use of Equation (3) and transformed difference without trends. We employ the difference $\nabla^2 \nabla_{12} Z_t$ since the differebce series is detrending.

Step 2.   In order to minimize the vagueness of the model, the autoregressive parameters ϕ_1, ϕ_2 and an error term $(u_\alpha, u_\beta, u_\delta)$ was determined by LP problem (7). Here, the constraint of LP problem (7) is expressed that the time-series model can include difference series.

Ozawa's model is obtained by using the above procedures. The coefficients of this model are determined as follows:

$$\nabla^2 \nabla_{12} \tilde{Z}_t = -0.749 \nabla^2 \nabla_{12} Z_{t-1} - 0.348 \nabla^2 \nabla_{12} Z_{t-2} + (-0.143, -0.013, 0.117)$$

The model that is obtained by the fuzzy autoregressive model has a negative coefficient that is the same as the result that is obtained by the autocorrelation. An original series and its estimate are shown in Figure 3.

Figure 3 shows that the estimated model has a large width of possible values. Numerically, the width of the possibility of the model is 12,500JPY (on average), 41,340JPY at the maximum and 1,770JPY at the minimum. The ambiguity of this model is extremely large.

However, the central value of the estimated value shows a value that is almost the same as the original series. Because it showed a strongly oscillating tendency in the past, these results are with the large width of the model estimation. This can also be understood from the section of the error term of the model.

A result of the sequential prediction of this model is shown in Figure 4.

Fig. 3 The result of the Fuzzy Autoregressive Model

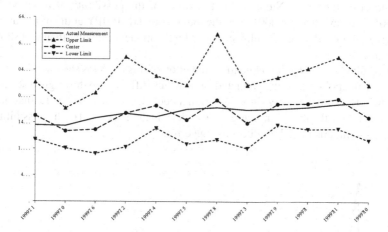

Fig. 4 The prediction result obtained by the Fuzzy Autoregressive Model (1999)

3.2 Fuzzy Autocorrelation Model

Next, we analyze the Nikkei stock average by employing the fuzzy autocorrelation model which is proposed in this paper. The following is the procedure to obtain our model.

Step 1.   The original series is fuzzified by Equation (8) and transformed difference without trends. We employ the difference $\nabla^2 \nabla_{12} \mathbf{Z}_t$ since the differebce series is detrending.

Step 2.   The fuzzy autocorrelation coefficients was obtained by fuzzy operation. The fuzzy autocorrelation coefficient has a large vagueness because fuzzy operation. Therefore, the vagueness is managed using an α-cut method without removing characteristics of a fuzzy autocorrelation coefficients. The α−cut value h is determined by LP problem (10). In this model, we set the α−cut level 0.978. The fuzzy autocorrelation showed minus in the correlation of lag 2 is similar to the case of the autocorrelation.

Step 3.   The fuzzy autocorrelation coefficients is determined by LP problem (12).

In the estimated fuzzy autocorrelation model, the coefficient was determined as follows:

$$\nabla^2 \nabla_{12} \tilde{\mathbf{Z}}_t = (-1, -0.642, -0.642) \nabla^2 \nabla_{12}\mathbf{Z}_{t-1} \\ + (-0.607, -0.380, -0.321) \nabla^2 \nabla_{12}\mathbf{Z}_{t-2}$$

The model that is obtained by the fuzzy autocorrelation model has a negative coefficient that is the same as the result that is obtained by the fuzzy autoregressive model.

The original series and estimated series are shown in Figure 6.

As shown in Figure 6, the estimated model has a small width and results in the low level of fuzziness. Numerically, the width of the possibility of the model is 2,500JPY on average, 18,000JPY at the maximum and 100JPY at the minimum. In Figure 6, the width of the results produced by the model is smaller than that shown in Figure 3.

A result obtained by the sequential prediction of this model is shown in Figure 7.

There is a point that is the predicted value which differs from the original series in Figure 7. Because the fuzzy autocorrelation model places stress on the fluctuation of a system unlike the fuzzy autoregressive model which includes all of the possibilities of the system, this model produces a large error.

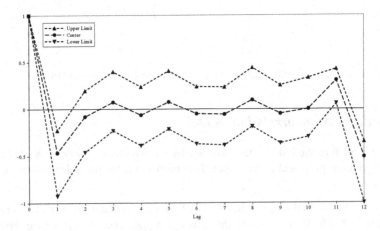

Fig. 5 Fuzzy Autocorrelation

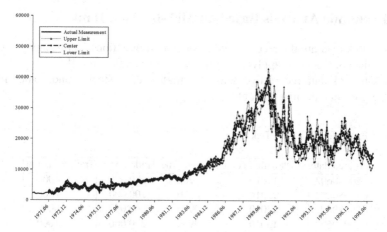

Fig. 6 The result produced by the Fuzzy Autocorrelation Model

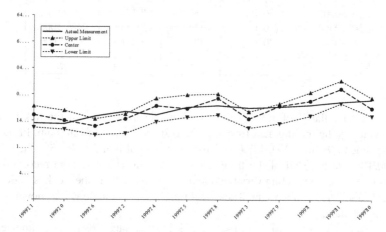

Fig. 7 The predicted result - Fuzzy Autocorrelation Model (1999)

Here, the difference between the fuzzy autoregressive model and our model are summarized as followings:

The fuzzy autoregressive model   The fuzzy autoregressive model is determined by including relation between this and difference series. At that time, non-fuzzy coefficients ϕ and fuzzy error term **u** are determined by LP in order to minimize the vagueness of model.

Our fuzzy autocorrelation model   First, the fuzzy autocorrelation coefficient is obtained from fuzzy difference series by LP. Next, the fuzzy autoregressive coefficient is obtained from a fuzzy autocorrelation coefficient in order to describe the behavior of a fuzzy difference series with high dimensional accuracy by LP.

The difference of two models is illustrated in Figures 3 and 6.

4   Economic Analysis Based on Tick-by-Tick Data

Tick-by-tick data are the record of stock dealing transactions. Let us analyze tick-by-tick data of stocks by applying the fuzzy autocorrelation model.

Tick-by-tick data record five items: the trading date, stock brand, traded time, dealing price and dealing amount.

Table 1 Example of Tick-by-Tick Data

Traded date	Stock name record	Traded time	Traded price	Traded amount
20080602	13010	0900	200.000	6000
20080602	13010	0900	200.000	4000
20080602	13010	0900	200.000	10000
20080602	13010	0900	200.000	1000
20080602	13010	0900	200.000	4000
$\vdots$	$\vdots$	$\vdots$	$\vdots$	$\vdots$
20080602	13010	0900	201.000	1000
20080602	13010	0900	201.000	1000
20080602	13010	0901	202.000	1000
$\vdots$	$\vdots$	$\vdots$	$\vdots$	$\vdots$

We will use the stock trading record. Table 1 shows some of the tick-by-tick data, which is the trading data from 2nd June 2008 recorded from 9 am. From the stating time at 9 am, 6,000, 4,000 and 10,000 shares of brand code 13010 is traded at 200JPY. We have all of the trading records. At a later time, the records have different values. The trading variables change in real time and the stock is traded by the real time price and amount. These real time trading situations are shown in the tick-by-tick data.

However, it is not possible to use the tick-by-tick data directly in a time-series analysis, and the upper, lower and average prices of the same stock at the same time are employed to create fuzzy numbers of time-series data.

In this chapter, we employ one week tick-by-tick data from 7th July 2008 (Monday) to 11th July 2008 (Friday). When nothing was traded for a certain time period, we used the previous price.

First, we apply AR model to analyze the tick-by-tick data. When we took a 5 minutes difference of the central value, we could remove its trend. Figure 8 shows the correlogram time series figure of the tick-by-tick data. As it shows one-time previous value ($\Delta_5 Z_{t-1}$), two-time previous value ($\Delta_5 Z_{t-2}$) and three-time previous value ($\Delta_5 Z_{t-3}$) have a large correlation, the following AR(3) model was employed.

$$\Delta_5 \tilde{Z}_t = \phi_1 \Delta_5 Z_{t-1} + \phi_2 \Delta_5 Z_{t-2} + \phi_3 \Delta_5 Z_{t-3}$$

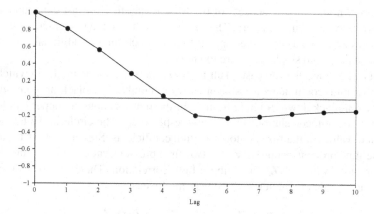

Fig. 8 Correlogram of AR model

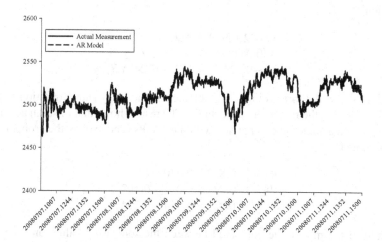

Fig. 9 Tick-by-tick data and the AR(3) model

Solving the Yule-Walker equation we form the following AR(3) model.

$$\Delta_5 \tilde{Z}_t = 0.970 \Delta_5 Z_{t-1} - 0.041 \Delta_5 Z_{t-2} - 0.224 \Delta_5 Z_{t-3}$$

$$\phi_1 = 0.970, \ \phi_2 = -0.041, \ \phi_3 = -0.224$$

Figure 9 shows the forecasted result based on the AR(3) model. Tick-by-tick data are given as fuzzy numbers but when using interval values, it is not possible to distinguish between the observed and forecasted values. Therefore, Figure 9 shows only the center value of the traded values and the forecasted values of the AR(3) model.

Figure 9 highlights the success of the forecast but the forecast delay and some error are recognized in the figure. Therefore, in the figure the movement of lines of real prices and forecasted values appear to have thick lines. Additionally, in some places the predicted stock prices are too high.

Figure 10 shows the forecast result by fuzzy autocorrelation model in which the autocorrelation coefficients are shown as fuzzy numbers. A black up-pointing triangle and a black down-pointing triangle in the figure denote the upper and lower values of the autocorrelation coefficients, respectively. The circle in the boxes are the center values of the fuzzy autocorrelation coefficients. Similar to the AR model, the one-time previous value ($\Delta_5 Z_{t-1}$), two-time previous value ($\Delta_5 Z_{t-2}$) and three-time previous value ($\Delta_5 Z_{t-3}$) exhibit a high correlation. Therefore, the following fuzzy autocorrelation model, FAR(3) was employed.

$$\tilde{Z}_t = \Phi_1 \Delta_5 Z_{t-1} + \Phi_2 \Delta_5 Z_{t-2} + \Phi_3 \Delta_5 Z_{t-3}$$

The fuzzy autocorrelation model was obtained by solving the fuzzy Yule-Walker equation.

$$\begin{aligned}\tilde{Z}_t = &[0.892, 0.970, 1]\Delta_5 Z_{t-1} \\ &+ [0.041, 0.041, 0.103]\Delta_5 Z_{t-2} \\ &+ [-0.224, -0.224, -0.224]\Delta_5 Z_{t-3}\end{aligned}$$

Figure 11 shows the forecasting result obtained by the fuzzy autocorrelation model. The figure shows the results are acceptable. The result shows that the high accuracy of the predicted values in comparison with the results produced by the probabilistic AR(3) model.

$$\begin{aligned}\Phi_1 &= [\quad 0.892, \quad 0.970, \quad 1 \quad] \\ \Phi_2 &= [-0.041, -0.041, \quad 0.103] \\ \Phi_3 &= [-0.224, -0.224, -0.224]\end{aligned}$$

To validate the model, we compared between the forecasted values and the real traded result for the last 30 minutes from 14:30 to 15:00 on 11th July 2008 and the next 30 minutes from 9:00 to 9:30 on 14th July 2008. Figures 12 and 13 show the results of the AR(3) model and the fuzzy autocorrelation model. In these figures, the values forecasted by the model based on the tick-by-tick data from 7th July to 11th July are applied to 14th July. Shown are the latter 30 minutes. In each figure, the observed values are shown by a solid line and the forecasted values by a dotted line.

First, the result using AR(3) looks one timing proceeding in Figure 12. At some points, when the real stock price came down, the forecasted value increased. At another points when the real stock price does not increase, the model forecasted the increasing of the price. Additionally when the stock price fell, the forecasted stock price decreased too rapidly.

However, Figure 13 shows the interval values because the fuzzy autocorrelation model is an interval model. Figure 13 illustrates looks one timing proceeding similar

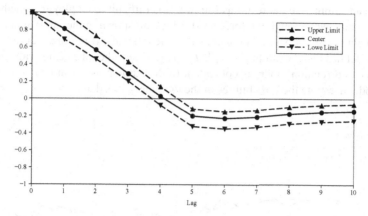

Fig. 10 Correlogram of Fuzzy Autocorrelation Model

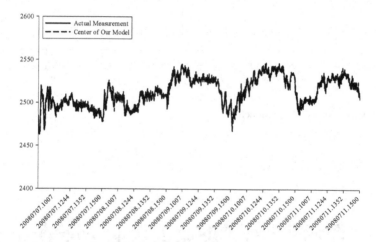

Fig. 11 Center Values of Fuzzy Autocorrelation Model of Tick-by-Tick Data

to the AR(3). Furthermore, for some points, the change of stock prices is forecasted to be too large.

For example, at the closing time on 11th July, the forecasting error looks small at the center of the model but the vagueness of the forecasted value is large. We interpret this as the last minute push of dealings that disturbed the forecasting of the time series system. That is, even though the stock price stopped moving, the forecasted value increased or decreased with the real value. This tendency of the forecasted result explicitly shows the forecasted value of the stocks starting on 14th July.

Overall it was found that the fuzzy autocorrelation model could describe the movement of the time series system well.

We can summarize the result as follows: It is difficult to employ this method to stock trading even though the forecasted values are shown in the interval and even though almost all the values are included in the intervals. If the fuzzy autocorrelation model has the constraint $R_t^C = \rho_t^C (t = 1, 2, \cdots, p)$, then the center value of the fuzzy autocorrelation model is coincident to the autoregressive model. Therefore, our model illustrate the possibilities of the the time-series data.

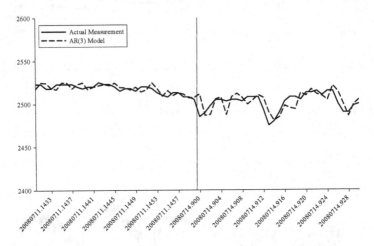

Fig. 12 Validation of AR(3) Model

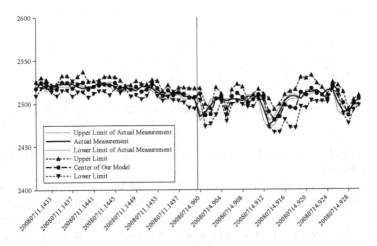

Fig. 13 Validation of the Fuzzy Autocorrelation Model

5 Model Interpretation

In this section, we interpret two fuzzy time-series models, the fuzzy autoregressive model proposed by Ozawa *et al.* and the fuzzy autocorrelation model proposed by Yabuuchi and Watada.

Let us compare the results between a fuzzy autoregressive model and a fuzzy autocorrelation model employing the triangular fuzzy number, which are illustrated in Figures 3 and 6. These figures show that the fuzzy autoregressive model is estimated as the model with a large width of possibility and the fuzzy autocorrelation is estimated as the model with a small width. The ambiguity of the fuzzy autocorrelation model is little, and this model estimated the original series more correctly. However, the fuzzy autoregressive model includes the entire fluctuation of an original series by estimating its fluctuation with a large value.

Let us compare the results of sequential prediction which are illustrated in Figures 4 and 7. Though the width of the possibility of the fuzzy autoregressive model was larger, it should be predicted that the estimated central value corresponds to the actual measurement. In the case of the fuzzy autocorrelation model, the estimated value is different from the original series at three points. Because the fuzzy autocorrelation describes the fluctuation of a system, this is considered to be the cause that an error produces.

The return and the amount of funds for an investor, a company, etc. are greatly influenced by the Nikkei stock average. Therefore, if a prediction is greatly different from a reality, it can cause great damage; in particular, in the case of a company, it may cause a business downsizing or the laying off of company employees. Sometimes, the Japanese economy may be inactive because of the influence of a wrong prediction. Therefore, the lower such a risk is, the more favorable it becomes for the company and the investor when they consider the future return, future prospects, and so on. However, the monthly fluctuation of the real Nikkei stock average is almost less than 10,000JPY. Therefore, for these reasons, we could regard a fuzzy autocorrelation model as being suitable for estimation and prediction of the Nikkei stock average.

In the case of the fuzzy autocorrelation model, we should not include all of the economic data within the model, but we should construct the model by employing the fuzzy autocorrelation to include the possibility of the fluctuation of data. We could show the effectiveness of the economic analysis using the Nikkei stock average by the fuzzy autocorrelation model, because the proposed model could illustrate the fluctuation in the system.

Next, the forecast of stock prices based on the tick-by-tick data of stocks are provided using both the autoregressive model and probabilistic autocorrelation model.

In the analysis of the tick-by-tick data of stocks, past traded data from three times, $\Delta_5 Z_{t-1}$, $\Delta_5 Z_{t-2}$, $\Delta_5 Z_{t-3}$ were used to describe successfully the movement of stock prices.

The purpose of employing tick-by-tick data for stock price forecasting is to validate the forecasting precision of the fuzzy autocorrelation model in stock trading. Figure 13 showed the stock prices of both the model building data and the last 30 minutes for validation. In this validation the figures showed that the fuzzy autocorrelation model could describe successfully the price movement of the time series system.

6 Conclusions

In this paper, we proposed a fuzzy autocorrelation model, which is based on a Box-Jenkins model and describes a time-series system by using fuzzy autocorrelation coefficients. The proposed model has compared with a fuzzy autoregressive model proposed by Ozawa *et al.* through numerical examples, the Nikkei stock average and the tick-by-tick data of stocks.

We showed the width of a fuzzy autocorrelation model is smaller than the one of a fuzzy autoregressive model in the analysis of the Nikkei stock average. But, the center of a fuzzy autoregressive model could estimate the original series.

The fuzzy autocorrelation model had the higher accuracy of the predicted values in the analysis of the tick-by-tick data of stock.

Finally, we can conclude that the fuzzy autocorrelation model can describe the movement of the time-series system well.

References

1. Box, G.P., Jenkins, G.M.: Time Series Analysis: Forcasting and Control. Holden-day Inc., San Francisco (1976)
2. Chen, S.M.: Forecasting enrollments based on fuzzy time series. Fuzzy Sets and Systems 81(3), 311–319 (1996)
3. Cheng, Y.C., Li, S.T., Li, H.Y., Chen, P.C.: A Novel Forecasting Model Based on Similarity Measure for Fuzzy Time Series. Proceedings of Business and Information 8 (2011)
4. Dong, R., Pedrycz, W.: A granular time series approach to long-term forecasting and trend forecasting. Physica A: Statistical Mechanics and its Applications 387(13), 3253–3270 (2008)
5. Dubois, D., Prade, H.: Fuzzy Sets and Systems, Theory and Application. Academic Press, New York (1980)
6. Ismail, Z., Efendi, R.: Enrollment Forecasting based on Modified Weight Fuzzy Time Series. Journal of Artificial Intelligence 4(1), 110–118 (2011)
7. Kariya, T., Liu, R.Y.: Asset Pricing -Discrete Time Approach. Kluwer Academic Publishers (2002)
8. Khashei, M., Bijari, M., Ardali, G.A.R.: Improvement of Auto-Regressive Integrated Moving Average models using Fuzzy logic and Artificial Neural Networks (ANNs). Neurocomputing 72(4-6), 956–967 (2009)
9. Lee, C.H.L., Liu, A., Chen, W.S.: Pattern discovery of fuzzy time series for financial prediction. IEEE Transactions on Knowledge and Data Engineering 18(5), 613–625 (2006)
10. Lukoseviciute, K., Ragulskis, M.: Evolutionary algorithms for the selection of time lags for time series forecasting by fuzzy inference systems. Neurocomputing 12(10-12), 2077–2088 (2010)
11. Ozawa, K., Watanabe, T., Kanke, M.: Forecasting Fuzzy Time Series with Fuzzy AR Model. In: Proceedings, the 20th International Conference on Computers & Industrial Engineering, pp. 105–108 (1996)

12. Partal, T., Kişi, Ö.: Wavelet and neuro-fuzzy conjunction model for precipitation forecasting. Journal of Hydrology 342(1-2), 199–212 (2007)
13. Song, Q., Chissom, B.S.: Forecasting enrollments with fuzzy time series – Part I. Fuzzy Sets and Systems 54(1), 1–9 (1993)
14. Song, Q., Chissom, B.S.: Fuzzy times series and its models. Fuzzy Sets and Systems 54(3), 269–277 (1993)
15. Song, Q., Chissom, B.S.: Forecasting enrollments with fuzzy time series – Part II. Fuzzy Sets and Systems 62(1), 1–8 (1994)
16. Tanaka, H., Watada, J.: Possibilistic Linear Systems and Their Application to The Linear Regression Model. Fuzzy Sets and Systems 27, 275–289 (1988)
17. Teoh, H.J., Cheng, C.H., Chu, H.H., Chen, J.S.: Fuzzy time series model based on probabilistic approach and rough set rule induction for empirical research in stock markets. Data & Knowledge Engineering 67, 103–117 (2008)
18. Tseng, F.M., Tzeng, G.H., Yu, H.C., Yuan, B.J.C.: Fuzzy ARIMA Model for forecasting the foreign exchange market. Fuzzy Sets and Systems 118(1), 9–19 (2001)
19. Tseng, F.M., Tzeng, G.H.: A fuzzy seasonal ARIMA model for forecasting. Fuzzy Sets and Systems 126(3), 367–376 (2002)
20. Valenzuela, O., Rojas, I., Rojas, F., Pomares, H., Herrera, L.J., Guillen, A., Marquez, L., Pasadas, M.: Hybridization of intelligent techniques and ARIMA models for time series prediction. Fuzzy Sets and Systems 159(7), 821–845 (2008)
21. Vandaele, W.: Applied Time Series and Box-Jenkins Model. Academic Press (1983)
22. Watada, J.: Fuzzy time series analysis and forecasting of sales volume. In: Kacprzyk, J., Fedrizzi, M. (eds.) Fuzzy Regression Analysis, pp. 211–227. Omnitec Press, Physica-Verlag, Warsaw, Heidelberg (1992)
23. Watada, J.: Possibilistic Time-series Analysis and Its Analysis of Consumption. In: Dubois, D., Prade, H., Yager, R.R. (eds.), pp. 187–217. John Wiley & Sons, INC. (1996)
24. Yabuuchi, Y., Watada, J., Toyoura, Y.: Fuzzy AR model of stock price. Scientiae Mathematicae Japonicae 60(2), 303–310 (2004)
25. Yabuuchi, Y., Watada, J.: Possibilistic Forecasting Model and its Application to Analyze the Economy in Japan. In: Proceedings of Eighth International Conference on Knowledge-Based Intelligent Information & Engineering Systems, Part.III, pp. 151–158 (2004)
26. Yabuuchi, Y., Watada, J.: Fuzzy Robust Regression Model by Possibility Maximization. Journal of Advanced Computational Intelligence and Intelligent Informatics 15(4), 479–484 (2011)
27. Yao, J.T., Herbert, J.P.: Financial time-series analysis with rough sets. Applied Soft Computing 9(3), 1000–1007 (2009)
28. Yu, H.K.: Weighted fuzzy time series models for TAIEX forecasting. Phisica A: Statistical Mechanics and its Applications 349(3-4), 609–624 (2005)
29. Zadeh, L.A.: Fuzzy sets. Information and Control 8(3), 338–353 (1965)

Chapter 17
Predicting Hourly Ozone Concentration Time Series in Dali Area of Taichung City Based on Seven Types of GM (1, 1) Model

Tzu-Yi Pai[*], Su-Hwa Lin, Pei-Yu Yang, Dyi-Huey Chang, and Jui-Ling Kuo

Abstract. In this study, seven types of first-order and one-variable grey differential equation model (abbreviated as GM (1, 1) model) were used to predict hourly ozone concentrations in Dali area of Taichung City, Taiwan. The results indicated that the minimum mean absolute percentage error (MAPE), mean squared error (MSE), root mean squared error (RMSE), and maximum correlation coefficient (R) were 19.00 %, 45.27, 6.73, and 0.91, respectively. All statistical values revealed that the prediction performance of GM $(1, 1, x^{(0)})$, GM $(1, 1, a)$, and GM $(1, 1, b)$ is better than the performance of other GM (1, 1) models. The GM (1, 1) model required a very small sample size, as low as four samples, but the modeling could result in very high prediction accuracy. It is also revealed that GM (1, 1) GM (1, 1) was an efficiently early warning tool to provide ozone information to inhabitants.

Keywords: grey system theory, GM (1, 1), hourly ozone, air quality.

Tzu-Yi Pai · Su-Hwa Lin
Department of Science Application and Dissemination,
National Taichung University of Education, Taichung, 40306, Taiwan, ROC
e-mail: bai@ms6.hinet.net

Tzu-Yi Pai · Pei-Yu Yang · Dyi-Huey Chang · Jui-Ling Kuo
Department of Environmental Engineering and Management,
Chaoyang University of Technology, Wufeng, Taichung, 41349, Taiwan, ROC

[*] Corresponding authors.

W. Pedrycz & S.-M. Chen (Eds.): Time Series Analysis, Model. & Applications, ISRL 47, pp. 369–383.
DOI: 10.1007/ 978-3-642-33439-9_17   © Springer-Verlag Berlin Heidelberg 2013

1 Introduction

In the past two decades, air pollution has improved in most cities in Western Europe, North American as well as Japan. Air pollution reductions have resulted mainly from greater efficiency and pollution-control technologies in factories, power plants, and other facilities (Cunningham and Cunningham, 2006). Although improvements are also achieved in transportation, the regulation efficiencies of O_3 pollution sources are not as significant as those of other pollution sources because of their emitted and reactive characteristics (Faiz et al., 1995; Fischer et al., 2000; Kingham et al., 2000; Lipfert et al., 2006; Pai et al., 2007).

Among all air pollutants, the elevated O_3 concentrations at ground level are of particular concern, because of the harm to human health and vegetation. Gao and Niemeier (2008) indicated that ozone pollution was caused by photochemical reactions of precursor volatile organic compounds (often called non-methane hydrocarbons, NMHC) and nitrogen oxides, of which transportation emissions are the single major source. Several references showed that the mobile sources had a significant influence on ozone formation (Gao, 2007; Gao and Niemeier, 2007; Wang et al., 2009). In addition, the emissions of NMHC are one of the main contributors to ozone formation (Delucchi et al., 1994).

The relationship between ozone and its precursors is complicated due to the fact that meteorological and chemical reaction rates range from very fast to very slow. Such relationships between meteorological condition and ozone concentrations have been explored in several studies which have utilized statistical regression, graphical analysis, fuzzy theory, and cluster analysis.

Typically, environmental data are very complex for modeling because interrelations between various components result in a complicated combination of relations. Models providing reasonable accuracy have to consider physical and chemical relations among O_3 and other pollutants under various meteorological conditions simultaneously. However, the uncertainty problem will occur when above modeling approaches were adopted. One of the most important problems is the uncertainty of input data, including source identification, meteorological conditions, and relevant reaction mechanisms. No matter how good the inventory investigation was carried out in a large-scale modeling analysis, the uncertainties of input data in the mechanistic modeling process cannot be completely eliminated.

Many other attempts to model the interrelations have also been carried out. Linear regression methods, for instance, have been widely employed for decades (Abdul-Wahab et al., 2005). Additionally, to adequately model complex, non-linear phenomena and chemical procedures, artificial neural networks (ANN) and fuzzy logic approach have been widely applied because of their ability to model nonlinear data well (Gautam et al. 2008; Cai et al., 2009).

Although ANNs could predict air pollutant concentrations successfully, they require a large amount of training data. In order to simplify statistical complexity and gain consistent results from the investigation data for predicting air pollutant, the grey system theory (GST) offers a suite of methods.

The GST can resolve the problem of incomplete data and has been applied in our previous studies (Deng, 2002, 2005; Pai et al., 2007 a, b; Pai et al., 2008 a, b, c; Pai et al., 2010; Pai et al., 2010 a, b). GST focuses on the relational analysis, model construction, and prediction of the incomplete information. It requires only a small amount of data and the better prediction results can be obtained.

There are many methods of analysis in GST including grey model (GM). GM can be used to establish the relationship between many sequences of data. Among all air pollutants, the O_3 concentrations at ground level are of particular concern because of the serious harm to human health, especially in a short-time period. If an efficient method could be developed to predict the short-time O_3 concentrations, a better control strategy could be sought. Since the hourly data of particulate matter (PM) were predicted successfully using GM presented in our previous work (Pai et al., 2011), GM could be used to predict the hourly O_3 concentrations.

The objectives of this study are as follows: (1) Construct seven types of first-order and one-variable grey differential equation model (abbreviated as GM (1, 1) model) for predicting hourly O_3 concentrations in Dali area of Taichung City in Taiwan, (2) Compare the prediction performance of seven types of GM (1, 1) model.

2   Materials and Methods

2.1   Data Set

The monitoring data from air quality monitoring station locating in Dali area of Taichung City was selected in this study (Figure 1). The concentrations of O_3 from 29th of July to 16th of August 2008 were investigated. They were sampled and investigated every hour. The total number of data was 456. Among the data, 384 data points were used to estimate the coefficients of the models and 72 data points were used as the observed values when evaluating the performance of the model. The maximum, minimum, mean value and standard deviation of O_3 series were 100.2, 1.1, 25.0, and 21.2 ppb, respectively. The meteorological condition was ignored in this study.

2.2   Grey Modeling Process

In a situation where information is lacking, using fewer (at least 4) system information, one can create a GM to describe the behavior of the few outputs. By means of accumulated generating operation (AGO), the disorderly and the unsystematic data may become exponentially behaved such that a first-order differential equation can be used to characterize the system behavior. Solving the differential equation will yield a time response solution for prediction. Through inverse

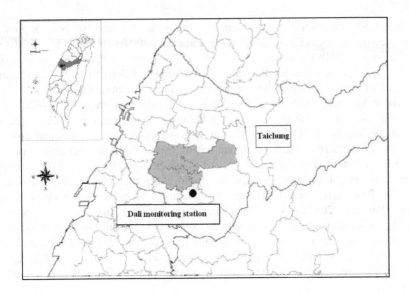

Fig. 1 Dali area

accumulated generating operation (IAGO), the forecast can be transformed back to the sequence of original series. A grey modeling process is described as follows.

Assume that a series of data with n samples is expressed as:

$$X^{\{0\}} = (x^{(0)}(1), x^{(0)}(2), \cdots, x^{(0)}(n)), \tag{1}$$

where the superscript (0) of $X^{(0)}$ represents the original series. Let $X^{(1)}$ be the first-order AGO of $X^{(0)}$, whose elements are generated from $X^{(0)}$:

$$X^{(1)} = (x^{(1)}(1), x^{(1)}(2), \cdots, x^{(1)}(n)), \tag{2}$$

where $x^{(1)}(k) = \sum_{i=1}^{k} x^{(0)}(i)$, for $k = 1, 2, \cdots, n$. Further operation of AGO can be conducted to develop the r-order AGO series, $X^{(r)}$:

$$X^{\{r\}} = (x^{(r)}(1), x^{(r)}(2), \cdots, x^{(r)}(n)), \tag{3}$$

where $x^{(r)}(k) = \sum_{i=1}^{k} x^{(r-1)}(i)$, for $k = 1, 2, \cdots, n$. The IAGO is the inverse operation of AGO. It transforms the AGO-operational series back to the one of a lower order. The operation of IAGO for the first-order series is defined as follows: $x^{(0)}(1) = x^{(1)}(1)$ and $x^{(0)}(k) = x^{(1)}(k) - x^{(1)}(k-1)$ for $k = 2, 3, \cdots, n$. After extending this representation to the IAGO of r-order series, we have

$x^{(r-1)}(k) = x^r(k) - x^r(k-1)$ for $k = 2,3,\cdots,n$. The tendency of AGO can be approximated by an exponential function. Its dynamic behavior resembles differential equation. The grey model GM (1, 1) thus adopts a first order differential equation to fit the AGO series,

$$\frac{dx^{(1)}}{dt} + ax^{(1)} = b \tag{4}$$

where the parameter a is the developing coefficient and b is the grey input. According to the definition, GM (1, 1) is that the order in grey differential equation is equal to 1 and defined as follows:

$$x^{(0)}(k) + az^{(1)}(k) = b \tag{5}$$

where $z^{(1)}(k) = 0.5x^{(1)}(k-1) + 0.5x^{(1)}(k)$ $k = 2, 3, 4, \ldots, n$. Expanding (5), we have

$$\begin{aligned}
x^{(0)}(2) + az^{(1)}(2) &= b \\
x^{(0)}(3) + az^{(1)}(3) &= b \\
&\vdots \quad \vdots \\
x^{(0)}(n) + az^{(1)}(n) &= b
\end{aligned} \tag{6}$$

Transforming (6) into a matrix form, we have

$$\begin{bmatrix} x^{(0)}(2) \\ x^{(0)}(3) \\ \vdots \\ x^{(0)}(n) \end{bmatrix} = \begin{bmatrix} -z^{(1)}(2) & 1 \\ -z^{(1)}(3) & 1 \\ \vdots & \vdots \\ -z^{(1)}(n) & 1 \end{bmatrix} \begin{bmatrix} a \\ b \end{bmatrix} \tag{7}$$

Then the coefficients can be estimated by solving the matrix relationship,

$$p = \begin{bmatrix} a \\ b \end{bmatrix} = (B^T B)^{-1} B^T Y,$$

where $p = \begin{bmatrix} a \\ b \end{bmatrix}$, $B = \begin{bmatrix} -z^{(1)}(2) & 1 \\ -z^{(1)}(3) & 1 \\ \vdots & \vdots \\ -z^{(1)}(n) & 1 \end{bmatrix}$, and $Y = \begin{bmatrix} x^{(0)}(2) \\ x^{(0)}(3) \\ \vdots \\ x^{(0)}(n) \end{bmatrix}$.

Sometimes, singularity would be encountered when treating the increasingly accumulated data. Then the inverse matrix could not be determined. Once this situation occurs, Computational Intelligence techniques could be applied. In this study, the increasingly accumulated data would not result in singularity due to their values and numbers were not too high. Additionally, the whitening type of GM (1, 1) model (or in terms of GM (1, 1, W)) that can be used for prediction is described as:

$$\hat{x}_1^{(1)}(k+1) = (x^{(0)}(1) - \frac{b}{a}) \cdot e^{-ak} + \frac{b}{a} \tag{8}$$

$$\hat{x}^{(0)}(k+1) = \hat{x}^{(1)}(k+1) - \hat{x}^{(1)}(k) \tag{9}$$

Additionally, there are several types of GM (1, 1) model which are derived from (4) as follows.

Connotation type of GM (1, 1): GM (1, 1, C)

$$x^{(0)}(k) = \left(\frac{1-0.5a}{1+0.5a}\right)^{k-2} \frac{b-ax^{(0)}(1)}{1+0.5a} \tag{10}$$

Grey difference type of GM (1, 1): GM (1, 1, $x^{(1)}$)

$$x^{(0)}(k) = \beta - \alpha x^{(1)}(k-1) \tag{11}$$

where $\beta = \dfrac{b}{1+0.5a}$ and $\alpha = \dfrac{a}{1+0.5a}$.

IAGO type of GM (1, 1): GM (1, 1, $x^{(0)}$)

$$x^{(0)}(k) = (1-\alpha)x^{(0)}(k-1) \tag{12}$$

Parameter-a type of GM (1, 1): GM (1, 1, a)

$$x^{(0)}(k) = \frac{1-0.5a}{1+0.5a} x^{(0)}(k-1) \tag{13}$$

Parameter-b type of GM (1, 1): GM (1, 1, b)

$$x^{(0)}(k) = \frac{x^{(1)}(k)-0.5b}{x^{(1)}(k-1)+0.5b} x^{(0)}(k-1) \tag{14}$$

Exponent type of GM (1, 1): GM (1, 1, e)

$$x^{(0)}(k) = x^{(0)}(3)e^{(k-3)\ln(1-\alpha)} \tag{15}$$

When adopting GM (1, 1, $x^{(0)}$), GM (1, 1, a), GM (1, 1, b), and GM (1, 1, e), $x^{(0)}(2)$ has to be calculated as follows:

$$x^{(0)}(2) = \beta - \alpha x^{(0)}(1) \tag{16}$$

All seven types of the GM (1, 1) model and their denotation are summarized in Table 1. The detailed derivation of these GM (1, 1) models can be found in Deng (2002, 2005).

2.3   Error Analysis

In order to evaluate the prediction accuracy of GM (1, 1), the mean absolute percentage error (MAPE), mean square error (MSE), root mean square error (RMSE), and correlation coefficient (R) were employed,

$$MAPE = \frac{1}{n}\sum_{i=1}^{n}\left|\frac{obs_i - pre_i}{obs_i}\right| \times 100\% \tag{17}$$

$$MSE = \frac{1}{n}\sum_{i=1}^{n}(obs_i - pre_i)^2 \tag{18}$$

$$RMSE = \sqrt{\frac{1}{n}\sum_{i=1}^{n}(obs_i - pre_i)^2} \tag{19}$$

Table 1 Seven types of GM (1, 1) model

Type	Denotation	Prediction equation
Whitening type	GM (1, 1, W)	$$\hat{x}_1^{(1)}(k+1) = (x^{(0)}(1) - \frac{b}{a}) \cdot e^{-ak} + \frac{b}{a}$$ $$\hat{x}^{(0)}(k+1) = \hat{x}^{(1)}(k+1) - \hat{x}^{(1)}(k)$$
Connotation type	GM (1, 1, C)	$$x^{(0)}(k) = \left(\frac{1-0.5a}{1+0.5a}\right)^{k-2} \frac{b - ax^{(0)}(1)}{1+0.5a}$$
Grey difference type	GM (1, 1, $x^{(1)}$)	$$x^{(0)}(k) = \beta - \alpha x^{(1)}(k-1)$$ $$\beta = \frac{b}{1+0.5a}, \quad \alpha = \frac{a}{1+0.5a}$$
IAGO type	GM (1, 1, $x^{(0)}$)	$$x^{(0)}(k) = (1-\alpha)x^{(0)}(k-1)$$ $$x^{(0)}(2) = \beta - \alpha x^{(0)}(1)$$
Parameter-a type	GM (1, 1, a)	$$x^{(0)}(k) = \frac{1-0.5a}{1+0.5a}x^{(0)}(k-1)$$ $$x^{(0)}(2) = \beta - \alpha x^{(0)}(1)$$
Parameter-b type	GM (1, 1, b)	$$x^{(0)}(k) = \frac{x^{(1)}(k) - 0.5b}{x^{(1)}(k-1) + 0.5b}x^{(0)}(k-1)$$ $$x^{(0)}(2) = \beta - \alpha x^{(0)}(1)$$
Exponent type	GM (1, 1, e)	$$x^{(0)}(k) = x^{(0)}(3)e^{(k-3)\ln(1-\alpha)}$$ $$x^{(0)}(2) = \beta - \alpha x^{(0)}(1)$$

$$R = \frac{\sum_{i=1}^{n}(obs_i - \overline{obs})(pre_i - \overline{pre})}{\sqrt{\sum_{i=1}^{n}(obs_i - \overline{obs})^2 \sum_{i=1}^{n}(pre_i - \overline{pre})^2}} \qquad (20)$$

where obs_i is the observed value, pre_i is the result of prediction, $\overline{obs}$ and $\overline{pre}$ are the average values of observed values and prediction values, respectively.

3   Results and Discussion

3.1   Determination of Grey Parameters

For determining the parameters of GM (1, 1), the observed O_3 data were plugged into (6) and the grey parameters were calculated by solving (7). When predicting, the values of the parameters a and b were equal to -0.00090492 and 23.404, respectively. According to (4), the parameter a (developing coefficient) will determine the predicting trend meanwhile parameter b (grey input) will determine the interception of (4).

3.2   Simulation of O_3

Table 2 shows all the values of MAPE, MSE, RMSE and R using seven types of GM (1, 1) model. The 1st to 384th data were used for constructing model, 385th to 456th data were used to evaluate the fitness. All values of the performance indexes revealed that the predicting performance of GM (1, 1, $x^{(0)}$), GM (1, 1, a), and GM (1, 1, b) prevailed. Figure 2 (a), (b), and (c) depict the prediction results of O_3 using seven types of GM (1, 1) model.

As shown in Table 2, when constructing, MAPEs between the predicted and observed values of O_3 were between 29.03 % and 29.30 % using GM (1, 1, $x^{(0)}$), GM (1, 1, a), and GM (1, 1, b), but they were 153.60 % - 220.96 % using other GM (1, 1) models. When predicting, the MAPEs were 19.00 % - 19.06 % when adopting GM (1, 1, $x^{(0)}$), GM (1, 1, a), and GM (1, 1, b), but they were between 94.66 % and 147.43 % when using other GM (1, 1) models.

The MSE values of 78.85 - 79.48 using GM (1, 1, $x^{(0)}$), GM (1, 1, a), and GM (1, 1, b) were lower than those of 440.64 – 541.01 using other GM (1, 1) models when model constructing. When predicting, the values of 45.27 - 45.41 using GM (1, 1, $x^{(0)}$), GM (1, 1, a), and GM (1, 1, b) were also lower than those of 300.11 – 586.04 using other GM (1, 1) models. When constructing, the RMSE values of 8.88 – 8.92 using GM (1, 1, $x^{(0)}$), GM (1, 1, a), and GM (1, 1, b) were lower than

Table 2 The prediction performance for O_3 using seven types of GM (1, 1) model

	MAPE		MSE		RMSE		R	
	Constructing	Predicting	Constructing	Predicting	Constructing	Predicting	Constructing	Predicting
GM (1, 1, W)	154.46	95.40	440.64	302.28	20.99	17.39	0.14	-0.19
GM (1, 1, C)	154.30	95.28	440.64	301.93	20.99	17.38	0.14	-0.19
GM (1, 1, x(1))	153.60	94.66	441.07	300.11	21.00	17.32	0.13	-0.17
GM (1, 1, x(0))	29.30	19.06	79.48	45.41	8.92	6.74	0.91	0.91
GM (1, 1, a)	29.30	19.06	79.48	45.41	8.92	6.74	0.91	0.91
GM (1, 1, b)	29.03	19.00	78.85	45.27	8.88	6.73	0.91	0.91
GM (1, 1, e)	220.96	147.43	541.01	586.04	23.26	24.21	0.14	-0.19

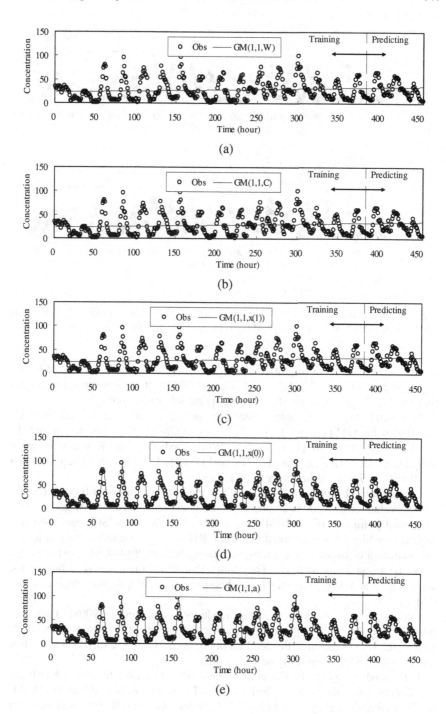

(a)

(b)

(c)

(d)

(e)

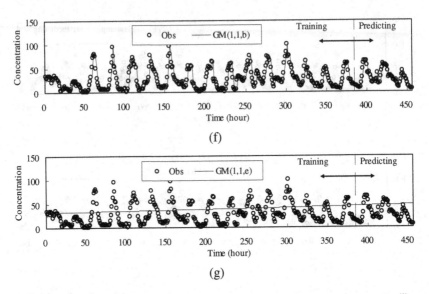

Fig. 2 Prediction results of O$_3$. (a) GM (1, 1, W), (b) GM (1, 1, C), (c) GM (1, 1, x$^{(1)}$), (d) GM (1, 1, x$^{(0)}$), (e) GM (1, 1, a), (f) GM (1, 1, b), (g) GM (1, 1, e)

those of 20.99 – 23.26 using other GM (1, 1) models. The RMSE value of 6.73 – 6.74 using GM (1, 1, x$^{(0)}$), GM (1, 1, a), and GM (1, 1, b) were also lower than those of 17.32 – 24.21 using other GM (1, 1) models when predicting.

When constructing, R value between the predicted and observed values of O$_3$ was 0.91 using GM (1, 1, x$^{(0)}$), GM (1, 1, a), and GM (1, 1, b), but they were 0.13 – 0.14 using other GM (1, 1) models. When predicting, the R was 0.91 when adopting GM (1, 1, x$^{(0)}$), GM (1, 1, a), and GM (1, 1, b), but they were between -0.19 and -0.17 when using other GM (1, 1) models.

Comparable observations were similarly made by Abdul-Wahab et al. (2005). Abdul-Wahab et al. (2005) employed data on the concentrations of seven environmental pollutants (CH$_4$, NMHC, CO, CO$_2$, NO, NO$_2$ and SO$_2$) and meteorological variables (WS and direction, Temp, RH and solar radiation) to predict the concentration of ozone in the atmosphere using both multiple linear and principal component regression methods. They found that R^2 for the day and night periods, were of 0.82 and 0.76, respectively. In this study, the R of 0.84 was obtained using GM.

Comparable observations were also made by Gautam et al. (2008). They proposed a new algorithm to predict the chaotic time series of O$_3$ based on the ANN technique. They found that the MAPEs lay between 12.26 – 24.01 % using ANN and 9.46 – 13.55 % even using new algorithm.

In the study proposed by Cai et al. (2009), ANN was used to predict hourly air pollutant concentrations near urban arterials. The results indicated that the MAPE for predicting O$_3$ fell in the range of 32.93 % and 45.15 %, RMSE were between 9.5 and 10.3, and R lay between 0.941 and 0.951, respectively.

In our previous study, seven types of GM (1, 1) models were used to predict hourly PM including PM_{10} and $PM_{2.5}$ concentrations in Banciao City of Taiwan (Pai et al., 2011). The results indicated that the minimum MAPE, MSE, RMSE, and maximum R was 14.10 %, 25.62, 5.06, and 0.96, respectively when predicting PM_{10}. When predicting $PM_{2.5}$, the minimum MAPE, MSE, RMSE, and maximum R value of 15.24 %, 11.57, 3.40, and 0.93, respectively could be achieved. In this study, the minimum MAPE, MSE, RMSE, and maximum R was 19.00 %, 45.27, 6.73, and 0.91, respectively.

According to both results, the GM (1, 1) model required a very small sample size, as little as four sample points, however the modeling could result in very high prediction accuracy. Furthermore, the parameter estimation in GM (1, 1) model was only a procedure to fit a simple regression. Therefore, GM could be applied successfully in predicting O_3 when the information was not sufficient.

In addition, the source identification, meteorological conditions, and relevant reaction mechanisms were taken as the input variables when using fuzzy or neural network models. But the source identification, meteorological conditions, and relevant reaction mechanisms did not be taken into account when using GM (1, 1). Although the mechanisms were unclear, the whitening part of the GM (1, 1) model could serve as useful reference to help observer realize more O_3 variation.

4   Conclusions

Seven types of GM (1, 1) model were used to predict hourly O_3 concentrations in Dali area of Taiwan. Their prediction performance was also compared. The conclusions can be drawn as follows. All statistical values revealed that the predicting performance of GM (1, 1, $x^{(0)}$), GM (1, 1, a), and GM (1, 1, b) outperformed other models. When predicting O_3, the minimum MAPE, MSE, RMSE, and maximum R was 19.00 %, 45.27, 6.73, and 0.91, respectively. According to the results, it is shown that GM (1, 1) could predict the hourly O_3 variation. Additionally, GM (1, 1) was an efficiently early warning tool for providing timely O_3 information.

References

Abdul-Wahab, S.A., Bakheit, C.S., Al-Alawi, S.M.: Principal component and multiple regression analysis in modelling of ground-level ozone and factors affecting its concentrations. Environmental Modelling and Software 20(10), 1263–1271 (2005)

Cai, M., Yin, Y., Xie, M.: Prediction of hourly air pollutant concentra-tions near urban arterials using artificial neural network approach. Transportation Research Part D: Transport and Environment 14, 32 (2009)

Cunningham, W.P., Cunningham, M.A.: Principles of Environmental Science. McGraw-Hill Company, New York (2006)

Deng, J.: The Foundation of Grey Theory. Huazhang University of Science and Technology Press, Wuhan (2002)

Deng, J.: The Primary Methods of Grey System Theory. Huazhang University of Science and Technology Press, Wuhan (2005)

Delucchi, M.A., Greene, D.L., Wang, M.Q.: Motor-vehicle fuel economy: The forgotten hydrocarbon control strategy. Transportation Research Part A: Policy and Practice 28(3), 223–244 (1994)

Faiz, A., Gautam, S., Burki, E.: Air pollution from motor vehicles: issues and options for Latin American countries. The Science of the Total Environment 169(1-3), 303–310 (1995)

Fischer, P.H., Hoek, G., van Reeuwijk, H., Briggs, D.J., Lebret, E., van Wijnen, J.H., Kingham, S.: Traffic-related differences in outdoor and indoor concentrations of particles and volatile organic compounds in Amsterdam. Atmospheric Environment 34(22), 3713–3722 (2000)

Gao, H.O.: Day of week effects on diurnal ozone/NOx cycles and transportation emissions in Southern California. Transportation Research Part D: Transport and Environment 12(4), 292–305 (2007)

Gao, H.O., Niemeier, D.A.: The impact of rush hour traffic and mix on the ozone weekend effect in southern California. Transportation Research Part D: Transport and Environment 12(2), 83–98 (2007)

Gautam, A.K., Chelani, A.B., Jain, V.K., Devotta, S.: A new scheme to predict chaotic time series of air pollutant concentrations using artificial neural network and nearest neighbor searching. Atmospheric Environment 42, 4409 (2008)

Kingham, S., Briggs, D., Elliott, P., Fischer, P., Erik, L.: Spatial variations in the concentrations of traffic-related pollutants in indoor and outdoor air in Huddersfield, England. Atmospheric Environment 34(6), 905–916 (2000)

Lipfert, F.W., Wyzga, R.E., Baty, J.D., Miller, J.P.: Traffic density as a surrogate measure of environmental exposures in studies of air pollution health effects: long-term mortality in a cohort of US veterans. Atmospheric Environment 40(1), 154–169 (2006)

Pai, T.Y., Hanaki, K., Ho, H.H., Hsieh, C.M.: Using grey system theory to evaluate transportation on air quality trends in Japan. Transportation Research Part D: Transport and Environ-ment 12(3), 158–166 (2007a)

Pai, T.Y., Tsai, Y.P., Lo, H.M., Tsai, C.H., Lin, C.Y.: Grey and neural network prediction of suspended solids and chemical oxygen demand in hospital wastewater treatment plant effluent. Computers & Chemical Engineering 31(10), 1272–1281 (2007b)

Pai, T.Y., Chiou, R.J., Wen, H.H.: Evaluating impact level of different factors in environmental impact assessment for incinerator plants using GM (1, N) model. Waste Management 28(10), 1915–1922 (2008a)

Pai, T.Y., Chuang, S.H., Ho, H.H., Yu, L.F., Su, H.C., Hu, H.C.: Predicting performance of grey and neural network in industrial effluent using online monitoring parameters. Process Biochemistry 43(2), 199–205 (2008b)

Pai, T.Y., Chuang, S.H., Wan, T.J., Lo, H.M., Tsai, Y.P., Su, H.C., Yu, L.F., Hu, H.C., Sung, P.J.: Comparisons of grey and neural network prediction of industrial park wastewater effluent using influent quality and online monitoring parameters. Environmental Monitoring and Assessment 146(1-3), 51–66 (2008c)

Pai, T.Y., Chang, T.C., Chen, H.H., Ouyang, C.F.: Using grey relation analysis to evaluate the reuse potential of municipal wastewater treatment plant effluent based on quality and quantity. Journal of Environmental Engineering and Management 20(2), 85–90 (2010)

Pai, T.Y., Lin, K.L., Shie, J.L., Chang, T.C., Chen, B.Y.: Predicting the co-melting temperatures of municipal solid waste incinerator fly ash and sewage sludge ash using grey model and neural network. Waste Management & Research (2011a) (in press)

Pai, T.Y., Ho, C.L., Chen, S.W., Lo, H.M., Sung, P.J., Lin, S.W., Lai, W.J., Tseng, S.C., Ciou, S.P., Kuo, J.L., Kao, J.T.: Using seven types of GM (1, 1) model to forecast hourly particulate matter concentration in Banciao City of Taiwan. Water, Air, and Soil Pollution 217(1-4), 25–33 (2011b)

Wang, G., Bai, S., Ogden, J.M.: Identifying contributions of on-road motor vehicles to urban air pollution using travel demand model data. Transportation Research Part D: Transport and Environment 14(3), 168–179 (2009)

Chapter 18
Nonlinear Time Series Prediction of Atmospheric Visibility in Shanghai

Jian Yao and Wei Liu

Abstract. Atmospheric visibility has recently become more essential to both the aviation safety and environmental pollution studies. Due to the characteristic of nonlinear time series, the visibility is difficult to predict by traditional statistical method. In this study, fuzzy time series models are used to predict the atmospheric visibility in Shanghai. The irregular dynamic of visibility was firstly investigated by the histogram as well as the autocorrelation analysis to identify the long-term memory of its behavior. Observed single-variable time series data were used to construct the fuzzy forecasting model. Parameters needed to construct the model were chosen to extract the rule of visibility variation. The results revealed that fuzzy time series could well predict the variation of visibility. The relative error between model outputs and observations was within the practically acceptable limits, which points out that atmospheric visibility could be explained and well predicted by the fuzzy time series.

1   Introduction

1.1   Application of Time Series Analysis to Environmental Research

Prediction of air quality takes place in the environmental management. Among the various statistical prediction method, time series analysis is one or the most useful tool for simulation of pollutant concentration. Time series analysis applies the previous data to find the regularities and estimate future values in consecutive time moments. It has been widely used in the field of economics, finance, and

Jian Yao · Wei Liu
Key Laboratory of Nuclear Analysis Techniques, Shanghai Institute of Applied Physics,
Chinese Academy of Sciences, Shanghai 201800, China
e-mail: yaojian@sinap.ac.cn

W. Pedrycz & S.-M. Chen (Eds.): Time Series Analysis, Model. & Applications, ISRL 47, pp. 385–399.
DOI: 10.1007/ 978-3-642-33439-9_18   © Springer-Verlag Berlin Heidelberg 2013

market forecasting, especially for the phenomenon with periodical cycle. However, the way to predict the non-linear data or those without distinct relationship remains unclear.

A time series is a sequence of statistical data arranged over time. It is obtained at determined time moments from environmental system of interest. This analysis is fundamental to scientific, engineering, and business research. A time series can be divided into two categories, single variable and multiple variables. The major model of single variable time series is auto-regressions model (ARs), and the model for multivariable time series is vector auto-regressions model (VARs). In mathematical expression, the stochastic process is defined as the behavior of a random variable { z, t $\in$ T}, where T is the range of the variable of t. If T = (-∞, ∞), then the stochastic process can be represented by {Z_t, -∞ < t < ∞}.

A time series can be represented as follows:

$$X = \{x_t, t = 1,\dots, N\} \qquad (1)$$

where t is the time index and N is the total number of observations.

The stationary time series is an important random series with the following properties:

$$E(Y_t) = E(T_{t-s}) = \mu_y \qquad (2)$$

$$V(Y_t) = Var(Y_{t-s}) = \sigma_y^2 \qquad (3)$$

$$Cov(Y_t, Y_{t-s}) = Cov(Y_{t-j}, Y_{t-j-s}) = \gamma_s, \text{ for all t, t-s, t-j-s} \qquad (4)$$

where μ_y, σ_y^2, and γ_s are some constants. If time series is not stationary, then the process exhibits changes over time.

The disadvantage present in traditional time series analysis is that the time series should be converted to stationary and periodic series prior to its analysis (Kantz and Schreiber 1997). To alleviate this problem, we explored the fuzzy time series and established its inference engine. The prediction results of fuzzy time series were compared with those formed by the traditional time series method to determine the feasibility of this fuzzy time series. The visibility concentration was analyzed by the fuzzy time series with data mining technique, which was useful in extracting the hidden knowledge and characteristic pattern from existing database.

Mining Fuzzy time series forms the focus of this paper. The main purpose of time series data mining is to abstract the phase space of time delay and represent it by mathematical equation. Fuzzy theory is broadly used to forecast the phenomenon with uncertainty. This study presents the inference engine of fuzzy time series deduced from previous information or other important predictors. The fuzzy time series use the fuzzy interval to cope with the random perturbation of observed data. An important information extraction method, data mining, is the technique to analyze data with hidden patterns. The trend of visibility can be captured, as being revealed by the experimental analysis.

1.2 Definition of Atmospheric Visibility and Its Importance

Visibility is defined as the greatest distance in a given direction at which an object can be visually identified with unaided eyes. The object could be a dark object positioned prominently against the sky on the horizon in the daytime, or a known, preferably unfocused moderately intense light source at nighttime (Wark et al. 1998). Visibility impairment is a basic form of air pollution that people can see and recognize without special instruments.

As we all known, impairment of atmospheric visibility constitutes many common and vexing problems for different public authorities in multiple countries throughout the world. First, low visibility is obviously a problem for traffic safety. Secondly, reduced visibility is a cause of delays and disruption in air, sea and ground transportation for passengers and freight. Of cause, impaired visibility is also a symptom of environmental problems because it is evidence of air pollution (Hyslop 2009); in addition, it has been shown that impaired visibility in urban environment and mortality are correlated (Thach et al. 2010). Therefore, visibility degradation is a major problem in atmospheric pollution in many mega cities around the world. Impairment of visibility is not just an aesthetic problem, but could also be used as a visual indicator of ambient air quality in urban areas (Watson 2002). Improvement of visibility requires an understanding of what constituents in the atmosphere impair visibility as well as the origins of those constituents.

From 1973 to 2007, visibility had decreased substantially over the globe except for Europe (Wang et al. 2009). In the Asian region, dozens of studies have reported a severe decline in visibility (Vingarzan and Li 2006; Chang et al. 2009; Tsai et al. 2003). Many analysts have been conducted worldwide (Dzubay et al. 1982; Larson et al. 1988; Johnson et al. 1990; Wilson and Suh 1997; Kim et al. 2001; Clancy et al. 2002). Furthermore, visibility impairment due to urban aerosol has been the subject of numerous air pollution studies around the world over the past several decades (Chan et al. 1999; Lee and Sequeira 2002; Zhang et al. 2004). It is known that the impairment of visibility is attributed primarily to the scattering and absorption of visible light by suspended particles, as well as by gaseous pollutants (e.g. NO_2) in the atmosphere (Appel et al. 1985; Hodkinson 1966; Groblicki et al. 1981; Latha and Badarinath 2003). Among them, fine particulates, which include sulfates, nitrates, organic and elemental carbon, and soil, effectively scatter or absorb visible light and thus reduce visibility (Malm et al. 1994, 1996; Sisler and Malm1994; Latha and Badarinath 2003; Kim et al. 2006; Tan et al. 2009a, 2009b).

Previous studies revealed that the size, chemical composition, and mass concentration of airborne particles substantially affect visibility (Conner et al. 1991; Malm and Pitchford 1997). Although the extinction of visible light from gaseous species can also impair visibility, such species have a much weaker influence (Chan et al. 1999; Dzubay et al. 1982). Also the PM Science Assessment Report (North American Research Strategy for Tropospheric Ozone (NARSTO) 2004) published recently by the NARSTO suggested that the chemical and physical properties of PM with an aerodynamic diameter less than 2.5 μ m have to be better characterized, as they are responsible for adverse health effects linked to chronic respiratory diseases (Dockery and Pope 1994) and visibility impairment (Malm 1999).

In addition to air pollutants, many meteorological elements such as relative humidity (RH), pressure, wind, and temperature may directly or indirectly contribute to the degradation of visibility (Lee 1990; Green et al. 1992; Malm et al. 1994; Raunemaa et al. 1994; Tsai and Cheng 1998, 1999). Relative humidity in and of itself does not reduce visibility, but as RH increases, hygroscopic particles progressively absorb more water, thus increasing their scattering cross section and proportionately reducing visibility. Therefore RH directly affects the particles that contribute to visibility reduction.

However, other meteorological variables, such as wind speed, temperature, and barometric pressure, have little to no direct effect on visibility but may have an effect on the concentration of atmospheric particles because of atmospheric dispersive characteristics. According to studies by Chang (1999) and Tsai and Cheng (1999), lower wind speeds cause particulates to gather and subsequently prevent them from spreading, which in turn indirectly affects air quality.

Nevertheless, either fine particulates, or meteorological parameters are hardly to be controlled and forecasted. It becomes necessary to explore a new method to predict atmospheric visibility.

1.3 Stochastic Property for Environmental Phenomenon and Visibility

The analysis of visibility time series starts from the following two definition.

Definition 1: Visibility is a chaotic occurrence in the environmental system.

Definition 2: Since we get the measured visibility data in the monitoring station, we call the observational results in time domain as a "visibility time series".

The chaotic nature exists in many fields such as air pollution concentration, stock price index, rainfall, and earthquake. The nonlinear of visibility comes from many reasons such as the scale-invariant and clustering characteristics. Because the times of visibility are a scale-dependent process, it is not easy to extract the information by the traditional way. The irregular dynamic behavior, or chaos, could be explained by the influence of some non-linear interdependent parameters in the system.

The chaotic behavior could be investigated by many tools such as histograms or spectral analysis. The chaotic indicator, the correlation dimension was the tool for the evaluation of pollution concentration to know the possible chaotic characteristic. The autocorrelation could identify the long-term memory and the possibility of scale invariance.

2   Model Developments

2.1   *Theory of Fuzzy Time Series Analysis*

A visibility time series can be defined as the following.

Definition 3: Visibility variation is a sequence of numerical data represented as follows:

$$X_{vis} = \{ x_{vis\ t},\ t = 1,\ldots\ldots,N \} \tag{5}$$

Definition 4: A fuzzy time series of visibility can be estimated by the event characterization function, ECF.

The event of time series is defined by the event characterization function, ECF, as follows

$$ECF = g\ (t) = g\ (X_i, X_{i-1}, \ldots\ldots X_l) \tag{6}$$

The event characterization function is defined in such a way that its value at t time index correlates highly with the occurrence of an event at some specified time in the future (Povinelli 1999).

In analyzing the event in time series data mining, g (t) $=X_{i+1}$, the ECF can be chosen as:

$$g(t) = \frac{X_{i+1} - X_i}{X_i} \tag{7}$$

(7) offers the clear relationships of X_{i+1} and X_i in predicting the visibility.

Event characterization functions can be defined by different ways for event predictions at different time series. The event characterization function varies according to the objective of prediction. For example, if x_t represents today's monitoring results and the target is to predict the change of tomorrow's visibility, then the event characterization function can be defined in the form given above.

The membership function of the fuzzy time series was specified in the form $f(x,\ a,\ c)$ with two parameters a and c, and it is a mapping on a vector x. Depending on the sign of the parameter a, it is appropriate for representing concepts such as "very large" or "very small".

$$f(x,a,c) = \frac{1}{1 + e^{-a(x-c)}} \tag{8}$$

The value of X_{vis} at time step (n+1) is determined by the membership function $f(x,\ a,c)$ and the value of its previous step, as shown below.

$$X_{vis}(t_{n+1}) = X_{vis}(t_n) + f(x,a,c) \cdot [\ X_{vis}(t_{n+1}) - X_{vis}(t_n)\ ] \tag{9}$$

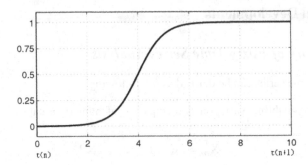

Fig. 1 The membership function of fuzzy time series in the interval $[t_n, t_{n+1}]$

2.2 Data Mining Technique of Visibility Time Series

How to extract the useful knowledge from time series data? The data mining could be a good answer. Data mining technique can extract hidden and useful information using various effective ways using pattern recognition, machine learning, artificial intelligence, and statistical methods. (Han and Kambe 2005).

Data mining is the analysis of data with the goal of uncovering hidden patterns. It is defined as extracting useful and meaningful information using statistic, machine learning, artificial intelligence and pattern recognition techniques from large data sets. (Han and Kamber 2005) Povinelli defines it as "combining of data mining, time series analysis and genetic (Povinelli 1999). Weiss and Indurkhya defined it as "the search for valuable information in large volumes of data". (Weiss and Indurkhya 1998).

Data mining is the process of discovering hidden and useful information from huge data. The data mining technique and nonlinear time series analysis to analyze a time series were combined in time series data mining. The event is considered as an interesting pattern when data mining is applied to time series data. The prediction algorithm based on data mining of fuzzy time series is shown in figure 2.

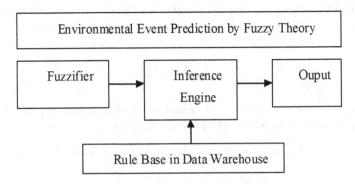

Fig. 2 The prediction algorithm of fuzzy time series with data mining technique

2.3   Prediction of Visibility by ARIMA Model

The model proposed by Box-Jenkins, with the method of Autoregressive Integrated Moving Average (ARIMA), is the most frequent used traditional time series analysis methods. A seasonal univariate $\text{ARIMA}(p, d, q)(P, D, Q)s$ model is given by

$$\Phi(B)[\Delta y_t - \mu] = \Theta(B) a_t, \qquad t = 1, \cdots\cdots, N \tag{10}$$

Where

$$\Phi(B) = \varphi p(B) \Phi P(B) \tag{11}$$

$$\Theta(B) = \theta q(B) \Theta Q(B) \tag{12}$$

and μ is an optional model constant. It is also called the stationary series mean, assuming that, after differencing, the series is stationary. An optional log scale transformation can be applied to y_t before the model is fitted. In this section, the same symbol, y_t, is used to denote the series either before or after log scale transformation.

Independent variables $x_1, x_2, ..., x_m$ can also be included in the model. The model with independent variables is given by

$$\Phi(B) \left[\Delta \left(y_t - \sum_{i=1}^{m} c_i x_{it} \right) - \mu \right] = \Theta(B) a_t \tag{13}$$

where $c_i, i = 1, 2, ..., m$, are the regression coefficients for the independent variables.

Basically, two different estimation algorithms are used to compute maximum likelihood (ML) estimates for the parameters in an ARIMA model. Melard's algorithm is used for the estimation when there is no missing data in the time series. The algorithm computes the maximum likelihood estimates of the model parameters. The details of the algorithm are described in (Melard 1984), (Pearlman 1980), and (Morf et al. 1974). A Kalman filtering algorithm is used for the estimation when some observations in the time series are missing. The algorithm efficiently computes the marginal likelihood of an ARIMA model with missing observations. The details of the algorithm are described in the following literature: (Kohn and Ansley 1986) and (Kohn and Ansley 1985).

The Conditional least square, CLS, IS used as the forecasting method in the ARIMA model. Define $\hat{y}_t(l)$, the l-step-ahead forecast of y_{t+l} at the time t, can be described as:

$$\hat{y}_t(l) = D(B) \hat{y}_{t+l} + \Phi(B) \mu + \Theta(B) \hat{a}_{t+l} + \sum_{i=1}^{m} c_i \Phi(B) \Delta x_{i,t+1} \tag{14}$$

Note that

$$\hat{y}_{t+l-i} = \begin{cases} y_{t+l-i} & \text{if } l \leq i \\ \hat{y}_t(l-i) & \text{if } l > i \end{cases} \tag{15}$$

$$\hat{a}_{t+l-j} = \begin{cases} y_{t+l-i} - \hat{y}_{t+l-i-1}(1) & \text{if } l \leq i \\ 0 & \text{if } l > i \end{cases}$$

3   Results and Discussion of Numerical Experiment

3.1   *Experimentals*

Three experiments were performed in this study. The first is the statistical analysis. The second was tradition forecasting method by ARIMA model. And the third was the fuzzy forecasting. The statistical analyses calculate the basic attributes of this time series which helps us to clarify its feature. The fuzzy forecasting applied the method in section 2.1 and 2.2 to find the rules in the data mine. The results were compared with the results from ARIMA model. Meanwhile, the trend of the time series was represents by the dimensionless time series together with the autocorrelation analysis for the long term memory in the data mining.

Model evaluation performance was compared by the observed data with the forecasted data. The scattering diagram was shown and the statistical value such as the root mean square error (RMSE); mean absolute percentage error (MAPE); maximum absolute percentage error (MaxAPE); mean absolute error (MAE); and maximum absolute error (MaxAE) were calculated as well.

3.2   *Study Area and Data Collection*

A time series of hourly average visibility observations, which was obtained from the monitoring station at Shanghai, was analyzed by descriptive statistics and statistical methods and fuzzy model to examine the temporal structures of visibility. The length of time for analysis was one year, which is enough to discriminate the most important feature of this time series.

3.3   *The Correlation between Model and Observed Values*

The model performance evaluation was accomplished by the comparison of forecasted value with the observed value. The correlation coefficient, which represents the relationship between the two quantities, was calculated as follows:

$$r = \frac{\sum_{i=1}^{n}\left(X_i - \overline{X}\right)\left(Y_i - \overline{Y}\right)}{\sqrt{\sum_{i=1}^{n}\left(X_i - \overline{X}\right)^2}\sqrt{\sum_{i=1}^{n}\left(Y_i - \overline{Y}\right)^2}} \tag{16}$$

Where $\overline{X}$ is the sample mean.

The values of the correlation coefficient were shown in Table 1.

Table 1 The values of the correlation coefficient for different variables

Parameter	Description	R value
T	Temperature	0.486[a]
V	Wind velocity	0.214[a]
RH	Relative Humidity	0.162[a]
PM$_{2.5}$	Particulate matter with diameter less than 2.5 μ m	-0.125[a]

Note: Confidence level: a: 0.01

3.4 The Trend of Time Series Data

The non-dimensional time series were used to compare the trend of time series data. The dimensionless time series were represented by

$$X = \left[\hat{x}_i, i = t, \cdots, n\right], \hat{x}_i = \frac{x_i}{\overline{x}}. \tag{17}$$

Where, X is the domain of time series, x_i is the individual value, $\overline{x}$ is the average value of the time series.

Figure 3 shows the trend of visibility and its related variables, the occurrence sequence of the hourly averaged visibility data in this study. This figure reveals that the characteristic of stochastic perturbation is obvious, and this is also the reason why a simple linear regression cannot be used in the prediction of atmospheric visibility.

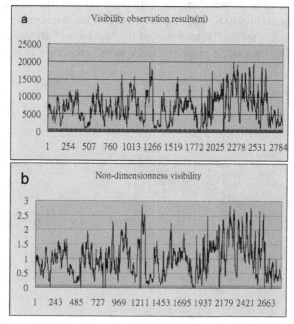

Fig. 3 The trend of visibility (a) is the visibility observation results and (b) is the non-dimensionless plot of the results

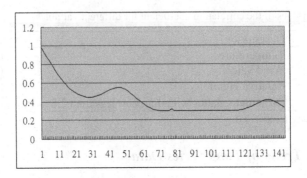

Fig. 4 Autocorrelation coefficients of visibility time series

The autocorrelation of this study is shown in figure 4. As shown here, the value of autocorrelation in decreasing. The results reveal that the long term memory effect is not obvious. There are probably many factors which will influence the variation of atmospheric visibility. The possible influence factors were listed in Table 1.

3.5 Analysis of Data by Fuzzy Time Series

The comparison of observed value and prediction results were shown in figure 5. The observed value is shown along the horizontal axis and the forecasted value is presented on the vertical axis. The results reveal that the relationship was fine. The fuzzy model could explain the tendency of the variation of atmospheric visibility time series.   The forecasting results of fuzzy model are shown in Figure 6.

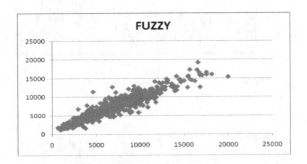

Fig. 5 Observed value and predicted value by fuzzy time series

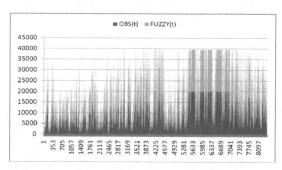

Fig. 6 The forecasting results of fuzzy model

3.6   Comparison of ARIMA Model

Table 2 is the summary statistics of the results by ARIMA (1, 1, 1) model. Figure 7 is the prediction results of ARIMA model. This figure also shows that the ARIMA model could also predict the atmospheric visibility time series.

In order to have a more clear comparison of the forecasting, the errors of these two methods were shown in figure 8. The values of error for these two models were converting into a 100 percents scale. The results produced by the fuzzy forecasting model are shown in grey and the ARIMA results were shown in black. It is seen that these two models are similar. In some place the results of fuzzy is better while in some place the ARIMA model is better. The results reveal that both models are capable to predict the non-linear characteristics of the atmospheric visibilities.

Table 2 Summary statistics of the results by ARIMA model

Summary Statistics	Aver	Min	Max	Percentile				
				5	10	50	90	95
Stationary R square	.253	.253	.253	.253	.253	.253	.253	.253
R square	.959	.959	.959	.959	.959	.959	.959	.959
RMSE	719 .380	719 .380	719 .380	719 .380	719 .380	719 .380	719 .380	719 .380
MAPE	9 .079	9 .079	9 .079	9 .079	9 .079	9 .079	9 .079	9 .079
MaxAPE	81 .433	81 .433	81 .433	81 .433	81 .433	81 .433	81 .433	81 .433
MAE	517 .445	517 .445	517 .445	517 .445	517 .445	517 .445	517 .445	517 .445
MaxAE	3096 .390	3096 .390	3096 .390	3096 .390	3096 .390	3096 .390	3096 .390	3096 .390
Normalized BIC	13 .237	13 .237	13 .237	13 .237	13 .237	13 .237	13 .237	13 .237

Note: RMSE: root mean square error; MAPE: mean absolute percentage error; MaxAPE: maximum absolute percentage error; MAE: mean absolute error; MaxAE: maximum absolute error

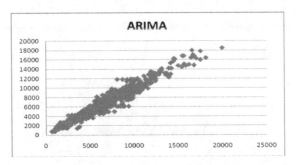

Fig. 7 Observed value and predicted value by ARIMA

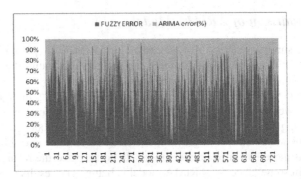

Fig. 8 The comparison between fuzzy time series and ARIMA

4   Conclusions

The study presents the results of atmospheric visibility forecasting by fuzzy time series. The proposed method, time series data mining, is based on the fuzzy logic to find the hidden rule in previous data. The fuzzy inference engine was obtained by the event characterization function, ECF, to abstract the previous information of variation in this study. In order to know the performance of this method, the forecasting was compared with the ARIMA model. The results reveal that fuzzy time series can handle the non-linear characteristics as effectively as ARIMA. However, the prediction of environmental event by fuzzy time series is simpler and does not include complex relationships. Comparing to the traditional forecasting method, the method can describe complicated nature of the environmental phenomenon, especially for non-periodical, non-cyclical data.

References

Appel, B.R., Tokiwa, Y., Hsu, J., Kothny, E.I., Hahn, E.: Visibility as related to atmospheric aerosol constituents. Atmos. Environ. 19, 1525–1534 (1985)

Ansley, C.F., Kohn, R.: Estimation, filtering, and smoothing in state space models with incompletely specified initial conditions. Ann. Stat. 13(4), 1286–1316 (1985)

Chan, Y.C., Simpson, R.W., Mctainsh, G.H., Vowles, P.D., Cohen, D.D., Bailey, G.M.: Source apportionment of visibility degradation problems in Brisbane (Australia)—using the multiple linear regression techniques. Atmos. Environ. 33, 3237–3250 (1999)

Chang, D., Song, Y., Liu, B.: Visibility trends in six megacities in China 1973–2007. Atmos. Res. 94(2), 161–167 (2009)

Chang, J.: The relationship of visibility with physical and chemical characteristics of suspended particles in Kaohsiung City. Master Thesis, National Sun Yat-Sen University, Kaohsiung, Taiwan (1999)

Clancy, L., Goodman, P., Sinclair, H., Dockery, D.W.: Effect of air-pollution control on death rates in Dublin, Ireland: intervention study. Lancet 360, 1210 (2002)

Conner, W.D., Bennett, R.L., Weathers, W.S., Wilson, W.E.: Particulate characteristics and visual effects of the atmosphere at Research Triangle Park. J. Air Waste Manage. Assoc. 41, 154–160 (1991)

Dockery, D.W., Pope, C.A.: Acute respiratory effects of particulate air pollution. Annu. Rev. Publ. Health 15, 107–132 (1994)

Dzubay, T.G., Stevens, R.K., Lewis, C.W., Hern, D.H., Courtney, W.J., Tesch, J.W., et al.: Visibility and aerosol composition in Houston, Texas. Environ. Sci. Technol. 16, 514–525 (1982)

Green, M.C., Flocchini, R.G., Myrup, L.O.: The relationship of the extinction coefficient distribution to wind field patterns in southern California. Atmos. Environ. 26, 827–840 (1992)

Groblicki, P.J., Wolff, G.T., Countess, R.J.: Visibility reducing species in the Denver Brown Cloud—1. Relationships Between Extinction and Chemical Composition. Atmos. Environ. 15, 2473–2484 (1981)

Han, J., Kamber, M.: Data Mining: Concepts and Techniques, p. 800. Academic Press, San Francisco (2005)

Hodkinson, J.R.: Calculations of color and visibility in urban atmospheres polluted by gaseous NO_2. Int. J. Air Water Pollut. 10, 137–144 (1966)

Hyslop, N.P.: Impaired visibility: the air pollution people see. Atmos. Environ. 43(1), 182–195 (2009)

Johnson, K.G., Gideon, R.A., Luftsgaarden, D.O.: Montana air pollution study: children's health effects. J. Off. Stat. 5, 391–408 (1990)

Kantz, H., Schreiber, T.: Nonlinear Time Series Analysis, p. 388. Cambridge University Press, Cambridge (1997)

Kim, K.W., Kim, Y.J., Oh, S.J.: Visibility impairment during yellow sand periods in the urban atmosphere of Kwangju, Korea. Atmos. Environ. 35, 5157–5167 (2001)

Kim, Y.J., Kim, K.W., Kim, S.D., Lee, B.K., Han, J.S.: Fine particulate matter characteristics and its impact on visibility impairment at two urban sites in Korea: Seoul and Incheon. Atmos. Environ. 40, 593–605 (2006)

Kohn, R., Ansley, C.F.: Estimation, prediction, and interpolation for ARIMA models with missing data. J. Am. Stat. Assoc. 81(395), 751–761 (1986)

Larson, S.M., Cass, G.R., Hussey, K.J., Luce, F.: Verification of image processing based visibility models. Environ. Sci. Technol. 22, 629–637 (1988)

Latha, K.M., Badarinath, K.V.S.: Black carbon aerosols over tropical urban environment—a case study. Atmos. Res. 69, 125–133 (2003)

Lee, D.O.: The influence of wind direction, circulation type and air pollution emissions on summer visibility trends in southern England. Atmos. Environ. 24A, 195–201 (1990)

Lee, Y.L., Sequeira, R.: Water-soluble aerosol and visibility degradation in Hong Kong during autumn and early winter, 1998. Environ. Pollut. 116, 225–233 (2002)

Malm, W.C.: Introduction to visibility, cooperative institute for research in the atmosphere (1999), http://vista.cira.colostate.edu/improve/Education/ intro_to_visibility.pdf (accessed December 15, 2011)

Malm, W.C., Pitchford, M.L.: Comparison of calculated sulfate scattering efficiencies as estimated from size-resolved particle measurements at three national locations. Atmos. Environ. 31, 1315–1325 (1997)

Malm, W.C., Sisler, J.F., Huffman, D., Eldred, R.A., Cahill, T.A.: Spatial and seasonal trends in particle concentration and optical extinction in the United States. J. Geophys. Res. 99 (D1), 1347–1370 (1994)

Melard, G.: Algorithm AS 197: A fast algorithm for the exact likelihood of autoregressive-moving average models. Appl. Stat. 33(1), 104–114 (1984)

Morf, M., Sidhu, G.S., Kailath, T.: Some new algorithms for recursive estimation on constant, linear, discrete-time systems. IEEE Trans. Auto. Control AC-19, 315–323 (1974)

NARSTO.: Particulate Matter Assessment for Policy Makers: A NARSTO Assessment. In: McMurry, P., Shepherd, M., Vickery, J. (eds.). Cambridge University Press, Cambridge (2004)

Raunemaa, T., Kikas, U., Bernotas, T.: Observation of submicron aerosol, black carbon and visibility degradation in remote area at temperature range from −24 to 20 °C. Atmos. Environ. 28, 865–871 (1994)

Pearlman, J.G.: An algorithm for the exact likelihood of a high-order autoregressive-moving average process. Viometrika 67, 232–233 (1980)

Povinelli, R.J.: Time Series Data Mining: Identifying Temporal Patterns for Characterization and Prediction of Time Series Events. Ph.D. Dissertation, Marquette University, p.180 (1999)

Sisler, J.F., Malm, W.C.: The relative importance of soluble aerosols to spatial and seasonal trends of impaired visibility in the United States. Atmos. Environ. 28, 851–862 (1994)

Tan, J.H., Duan, J.C., Chen, D.H., Wang, X.H., Guo, S.J., Bi, X.H., et al.: Chemical characteristics of haze during summer and winter in Guangzhou. Atmos. Res. 94, 238–245 (2009a)

Tan, J.H., Duan, J.C., He, K.B., Ma, Y.L., Duan, F.K., Chen, Y., et al.: Chemical characteristics of $PM_{2.5}$ during a typical haze episode in Guangzhou. J. Environ. Sci. 21, 774–781 (2009b)

Thach, T.Q., Wong, C.M., Chan, K.P., Chau, Y.K., Chung, Y.N., Ou, C.Q., et al.: Daily visibility and mortality: assessment of health benefits from improved visibility in Hong Kong. Environ. Res. 110(6), 617–623 (2010)

Tsai, Y.I., Cheng, M.T.: Effects of sulfate and humidity on visibility in the Taichuang harbor area (Taiwan). J. Aero. Sci. 29, 1213–1214 (1998)

Tsai, Y.I., Cheng, M.T.: Visibility and aerosol chemical compositions near the coastal area in central Taiwan. Sci. Total Environ. 231, 37–51 (1999)

Tsai, Y.I., Lin, Y.H., Lee, S.Z.: Visibility variation with air qualities in the metropolitan area in southern Taiwan. Water Air Soil Poll. 144, 22 (2003)

Vingarzan, R., Li, S.M.: The Pacific 2001 Air Quality Study–synthesis of findings and policy implications. Atmos. Environ. 40(15), 2637–2649 (2006)

Wang, K., Dickinson, R.E., Liang, S.: Clear sky visibility has decreased over land globally from 1973 to 2007. Science 323(5920), 1468–1470 (2009)

Wark, K., Warner, C.F., Davis, W.T.: Air Pollution—Its Origin and Control, 3rd edn. Addison-Wesley Longman, Reading (1998)

Watson, J.G.: Visibility: science and regulation. J. Air Waste Manage. Assoc. 52, 628–713 (2002)

Weiss, S.M., Indurkhya, N.: Predictive Data Mining: A practical Guide, p. 228. Morgan Kaufmann, San Fransisco (1998)

Wilson, W.E., Suh, H.H.: Fine particles and coarse particles: concentration relationships relevant to epidemiologic studies. J. Air Waste Manage. Assoc. 47, 1238–1249 (1997)

Zhang, R., Wang, M., Sheng, L., Kanai, Y., Ohta, A.: Seasonal characterization of dust days, mass concentration and dry deposition of atmospheric aerosols over Qingdao, China. China Particuology 2(5), 196–199 (2004)

Abbreviations

ARs	auto-regressions model
VARs	vector auto-regressions model
NARSTO	North American Research Strategy for Tropospheric Ozone
RH	relative humidity
ECF	event characterization function
ARIMA	autoregressive integrated moving average
ML	maximum likelihood
CLS	Conditional least squares
FTS	fuzzy time series
RMSE	root mean square error
MAPE	mean absolute percentage error
MaxAPE	maximum absolute percentage error
MAE	mean absolute error
MaxAE	maximum absolute error

Author Index

Subject Index